Lecture Notes in Electrical Engineering

Volume 958

The book series *Lecture Notes in Electrical Engineering* (LNEE) publishes the latest developments in Electrical Engineering—quickly, informally and in high quality. While original research reported in proceedings and monographs has traditionally formed the core of LNEE, we also encourage authors to submit books devoted to supporting student education and professional training in the various fields and applications areas of electrical engineering. The series cover classical and emerging topics concerning:

- Communication Engineering, Information Theory and Networks
- Electronics Engineering and Microelectronics
- Signal, Image and Speech Processing
- Wireless and Mobile Communication
- Circuits and Systems
- Energy Systems, Power Electronics and Electrical Machines
- Electro-optical Engineering
- Instrumentation Engineering
- Avionics Engineering
- Control Systems
- Internet-of-Things and Cybersecurity
- Biomedical Devices, MEMS and NEMS

For general information about this book series, comments or suggestions, please contact leontina.dicecco@springer.com.

To submit a proposal or request further information, please contact the Publishing Editor in your country:

China

Jasmine Dou, Editor (jasmine.dou@springer.com)

India, Japan, Rest of Asia

Swati Meherishi, Editorial Director (Swati.Meherishi@springer.com)

Southeast Asia, Australia, New Zealand

Ramesh Nath Premnath, Editor (ramesh.premnath@springernature.com)

USA, Canada

Michael Luby, Senior Editor (michael.luby@springer.com)

All other Countries

Leontina Di Cecco, Senior Editor (leontina.dicecco@springer.com)

**** This series is indexed by EI Compendex and Scopus databases. ****

Vančo B. Litovski

Lecture Notes in Analog Electronics

Discrete and Integrated Large Signal Amplifiers

Springer

Vančo B. Litovski
University of Niš
Niš, Serbia

ISSN 1876-1100 ISSN 1876-1119 (electronic)
Lecture Notes in Electrical Engineering
ISBN 978-981-19-6530-2 ISBN 978-981-19-6528-9 (eBook)
https://doi.org/10.1007/978-981-19-6528-9

This Springer imprint is published by the registered company Springer Nature Singapore Pte Ltd.
The registered company address is: 152 Beach Road, #21-01/04 Gateway East, Singapore 189721, Singapore

Preface

Large signal electronic amplification, alike other modern electronic technologies, is conservative in that within modern circuits one may find even the oldest technologies such as electronic valves. On the other side, the most advanced components such as SiCMOS or HEMT are taking their own place in the variety of circuit architectures depending on issues such as amount of delivered power, high-frequency response, efficiency, and distortions. To all power electronic components (except the electronic valves) will be devoted Chap. 4.2 of the book.

A wide variety of topologies and principle of operation are implemented in circuit synthesis of large signal and power amplifiers. We try in Chap. 4.3 to cover as much as possible in some cases going into details while in others giving fundamental information only. We really hope that this chapter will be a good start for deeper specialization in power and large signal amplifier design. We start with Class A amplifiers and go through Classes B, C, D, G, E, F, H to reach Class S.

To exemplify in Chap. 4.4, we are addressing discrete and integrated audio power amplifier circuits. The goal here is to merge the low-power (voltage amplifying) part of the system (described in LNAE_Book 2) with the ones given here in Chap. 4.3. The schematic given is demonstrational only and by no means is recommended for final practical implementation.

We proceed with exemplification of large signal amplification in Chap. 4.5 where mostly integrated operational and transconductance amplifiers are described. As it was with Chap. 4.4, here again a variety of technologies is visited, and solutions are offered with bipolar and MOS components. The operational and the operational transconductance amplifiers are modeled, and circuit and model parameters are defined as well as numerically exemplified. Again, the schematics given are demonstrational only and by no means are directly related to any available circuit on the market.

Implementation of operational amplifiers and operational transconductance amplifiers is discussed in Chap. 4.6, named: "Analogue Computation". The terminology here comes from the time of analogue computers which were intensively used in the fifties and sixties of the last century to solve differential equations and to model control systems. Limited set of linear and nonlinear analogue operators is considered so making a basis for further study only.

After the literature (due to the target audience of the book here the literature items are not referred to in the text) four appendices are given.

In Appendix 4.A a very short vocabulary is given related to the technology of electronic components production. Appendix 4.B contains a set of worked problems helping better understanding the main subject given in Chap. 4.3. Appendix 4.C is similar in the sense that it further helps the understanding of the content of Chap. 4.3 but here more complex situations are considered so that SPICE analysis is used to tackle the problems. Time domain waveforms and frequency domain characteristics are extracted as well as spectral analyses enabling calculation of harmonic distortion coefficients. Finally, Appendix 4.D is related to a specific procedure of handmade (or calculator made) harmonic distortion analysis of nonlinear circuits.

Niš, Serbia Vančo B. Litovski

Introduction to the Lecture Notes of Analogue Electronics (LNAE) Series

The electron is a phenomenon exhibiting divine properties. We all know it exists but no-one can see it. Obviously, we consider Electronics inherits his divinity.

Electronics is, nowadays, ubiquitous. There is no aspect of human life being not supported by electronics. None would deny this claim. Even so, when comes to studying electronics, and especially analogue electronics, it is mostly avoided by many talented students since it is considered difficult to study. Thus, it is our intention to produce a series of books that will be student oriented and probably not understandable at first glance, electronics oriented. In that way we expect to help attracting more young and talented people to study electronics and to further contribute to the human society. In this pamphlet we will consider two issues. We will first try to answer the question as to why electronics is difficult to study, and then we will give the rationale for preparing a series of books related to analogue electronics only.

Let us start with the difficulty. Ever since Nikola Tesla introduced the alternating current (AC) circuits, the primary circuit elements became the coil, the transformer, the AC motor, and the transmission line. All these were described by voltage equations which originate from the discovery of the Russian physicist Heinrich Lenz of the directional relationships between induced magnetic fields, voltage, and current when a conductor is passed within the lines of force of a magnetic field. Lenz's law states: *"An induced electromotive force generates a current that induces a counter magnetic field that opposes the magnetic field generating the current."* As a consequence, voltage equations were introduced everywhere including the exclusive use of the Kirchhoff's voltage law. That we call electrical engineering approach to the circuit.

The things went so far that even the Ohm's law which should read $i = G \cdot v$ (currents are on both sides) was transformed into his voltage version $v = R \cdot i$. In that, two serious problems were introduced which concerns the modern student of electronics. First, in $v = R \cdot i$ the independent variable is on the left-hand side of the equation while the function (the consequence) is on the right. So, one is supposed first to reorder the equation to come to the solution. That means that, for evaluation of the current, division is supposed to be performed as opposed to the much more natural arithmetic action of multiplication. Of course, to make things still more difficult, instead of the natural quantity which is conductance, resistance was introduced. For example, if a voltage source $v = 1$ V is connected to a resistor whose conductance is $G = 25$ S one would immediately pronounce the value of the current to be $i = 25$ A. The same resistor has a resistance of $R = 40$ mΩ, and to come to the current becomes a serious computational task. The transition was so merciless that no schematic symbol for the conductance exists.

The conflict becomes still more sever when comes to electronic circuits. Namely, all (no exception) electronic components are modeled by current equation (voltage controlled current sources) and, for electronic circuit analysis, the Kirchhoff's current low is the most natural means for equation formulation which are known as nodal equations. A student who is trained to think of voltage equations, however, is immediately frustrated since he/she does not know what to do with the resistor now. Even a well-trained former student, as is the author of this text, is forced to use reciprocals of the resistances all the time and, at the end of the analysis, is forced again to rearrange the obtained expressions to make them tractable.

Of course, modern electronic circuit simulation programs are based on nodal analysis (or the so-called Modified Nodal Analysis) which we would call electronic approach to the circuits. Nevertheless, due to the voltage background, understanding of their functionality is not easy, obstructing their efficient use and, again, is frustrating to many.

Next, the students trying to enter into the subject of electronics are facing the duality in electronics which, as we see it, is twofold. To start with, we will mention the synergy (or the conflict) between the DC and AC signals in electronic circuits. In some analyses these are independent and in other mutually coupled. The fact that the DC supply is a source of energy which is transferred to the useful signal is part of the explanation on what is going on in the circuit. One should understand, however, why even small AC signals (bearing almost no power) are depending on the value of the voltage of the power supply which, in most cases of analysis of incremental circuits, is absent from the schematic. On the other side, one is supposed to understand that, in some electronic circuits, voltages larger than the power supply voltage are produced and limit the applicability of some devices.

The other aspect of duality is related to the mutual synergy and antagonism of the time and frequency domain. There is no linear electronic component. That means that the component's current is a nonlinear function of a single or several voltages or currents. So, it is normal, when starting from basic electrotechnics, for nonlinear differential equations to be created to enable the analysis of an electronic circuit. Solving nonlinear differential equation, however, is a task which extremely rarely may be performed in closed form. Practically, there is no time domain electronic circuit analysis without a computer. That is especially true since one has no complete information about the boundary conditions necessary for solving the nonlinear differential equations. That imposes a need for development of incremental models of the electronic devices to allow analysis for small signals. The problems is that, usually, the two version of the same circuit (the large signal nonlinear and the incremental linear) are not taught to students as a single complete. To learn the analysis of incremental circuits the students are kept in the so-called frequency domain where, all of a sudden complex arithmetic is necessary. In that, a complex variable is introduced so that the real frequency of the signal is represented as the imaginary part of the complex frequency. Even the interpretation of the last sentence is confusing if not read several times.

This is not the end, however. The students are not rarely facing textbooks written by teachers that want to promote themselves as the one having strong practical experience. In that a slang is used as if they are talking to some repair person. Avoiding universal professional terminology and especially verbs representing phenomena within the circuit invalidates the knowledge delivered and makes the further progress of the student difficult.

Having all these in mind it becomes understandable why not only the student but even the teachers are facing wild windmills as if they are the Spanish fiction hero Don Quixote.

Now, what we want to do by this series of books. We are not in a position to change the unchangeable. The resistors will stay in the schematics as are everywhere and the imaginary part of the complex frequency will still represent the real frequency of the signal. What we can do is to cover the analogue electronics as a whole using the same vocabulary which we consider universal. Also, we would like to enable a start from the very beginning in the sense that only very basic knowledge of the student to be necessary to follow the texts. Finally, we want to give a complete knowledge starting with fundamental physics and ending with simulation and optimization, and, especially, with the sustainability aspects of electronic engineering.

The series will contain the following issues with the probability for some of them to become merged or split in two, according to the running inspiration of the author:

Book 1. Semiconductor Components
Book 2. Electronic Signal Amplification and Linear Oscillators
Book 3. Semiconductor Technology and Specific Electronic Components
Book 4. Discrete and Integrated Large Signal Amplifiers
Book 5. Noise in Electronic Circuits
Book 6. Power Electronic Circuits and Diagnosis
Book 7. Analog Electronic Testing
Book 8. Analog Electronic Modelling, Simulation, and Optimization
Book 9. Sustainable Electronic Design

The titles listed here are subject to refinements, too.

We really hope that these contents will fit in the contemporary views on the subject and will inspire authors to build on or to make improvements in a way they find better than what we did.

Contents

About the Author

Prof. Vančo B. Litovski was born in 1947 in Rakita, South Macedonia, Greece. He graduated from the Faculty of Electronic Engineering in Niš in 1970, and obtained his M.Sc. in 1974, and his Ph.D. in 1977. He was appointed as a teaching assistant at the Faculty of Electronic Engineering in 1970 and became a Full Professor at the same faculty in 1987. He was elected as Visiting Professor (honoris causa) at the University of Southampton in 1999. From 1987 until 1990, he was a consultant to the CEO of Ei, and was Head of the Chair of Electronics at the Faculty of Electronic Engineering in Niš for 12 years. From 2015 to 2017, he was a researcher at the University of Bath. He has taught courses related to analogue electronics, electronic circuit design, and artificial intelligence at the electro-technical faculties in Priština, Skopje, Sarajevo, Banja Luka, and Novi Sad. He received several awards including from the Faculty of Electronic Engineering (Charter in 1980, Charter in 1985, and a Special Recognition in 1995) and the University of Niš (Plaque 1985). Professor Litovski has published six monographs, over 400 articles in international and national journals and at conferences, 25 textbooks, and more than 40 professional reports and studies.

His research interests include electronic and electrical design and design for sustainability, and he led the design of the first custom commercial digital and research-oriented analogue CMOS circuit in Serbia. He has also headed eight strategic projects financed by the Serbian and Yugoslav governments and the JNA and has participated in several European projects funded by the governments of Germany, Austria, UK, and Spain, and the EC as well as the Black See Organization of Economic Cooperation (BSEC).

4.1 Introduction

Large signal amplifiers are dealing with signals whose magnitude is such that the operation of the active element can no longer be considered linear. They are usually designed to get as much power gain and efficiency as possible. That is why they are often called power amplifiers. To make the text shorter in the sequel we will use the abbreviation PA to denote *large signal amplifiers*. This amplification is usually done in the last stage of signal amplification applications just before accessing the (usually non-electronic) load.

In this book we will consider two implementations of PA. First, it is of interest to obtain large signals (current or voltage) at the output of a cascade of direct-coupled amplifiers. In this case linearity, frequency response, and speed are the most important requirements. Second are real power amplifiers where the power delivered to the load is of primary interest. Of course, efficiency, linearity, and high-frequency response are of interest, too.

Obtaining the highest possible output power of the alternating signal, which is called the useful power, requires the use of the entire active area of operation of the active element. Therefore, large nonlinear distortions may be expected in the output signal. In the case of a multistage amplifier whose last amplifier stage is a PA, practically all nonlinear distortions of the entire system are concentrated in the last stage.

The active region of the transistor characteristic is limited by several factors, the most important of which is maximum dissipation. It is obvious that working conditions should be chosen in such a way that they would not be exceeded even in the worst case. As the useful power increases, so does the importance of the efficiency. The lower the efficiency, the greater the part of the battery power that would be dissipated on the active element. This would mean that if we need to get large useful power, we will need to spend large power on the active element, which makes the amplifier uneconomical.

The load resistance in power amplifiers is connected in two ways. A transformer may be used at both the input and output of the amplifier to ensure power matching. Modern microelectronic circuits, however, do not use transformers and matching is

V. B. Litovski, *Lecture Notes in Analog Electronics*, Lecture Notes in Electrical Engineering 958, https://doi.org/10.1007/978-981-19-6528-9_1

achieved in other ways. It should be noted that the transformer usually has a worse frequency response than the rest of the circuit, and its dimensions are large.

The following is mainly the case when choosing the position of the quiescent operating point of the active element of the PA.

When the quiescent operating point of the active element is chosen so that current flows at all times during the full cycle of the input signal, the power amplifier is known as Class A power amplifier. When the collector current flows only during the positive half-cycle of the input signal, the power amplifier is known as Class B power amplifier. When the collector current flows for less than half cycle of the input signal, the power amplifier is known as Class C power amplifier. A transitional Class AB is frequently in use. The resulting circuits differ in their capabilities to deliver monochromatic output, large power, small distortions, and large efficiency. All these will be discussed below.

Modern electronics recognizes power amplifiers in classes beyond Class C. A common properties of these circuits is that the active element is operated as a switch while the output signal is obtained after filtering. The switching scheme is defining the class be D, G, E, F, H to reach Class S.

As for the power amplifiers as such one may distinguish audio power amplifiers and radio frequency power amplifiers, the names being self-explanatory. Different circuit structures and design strategies are implemented, however, some of which will be discussed here.

The book is organized in four parts. First, a completion of the story of electronic components will be given. Namely, in LNAE Book 1, low-power components and sensors were considered. Here, we will go beyond that and discuss the components designed so that they can sustain large power. In addition, in order not to split the lecture on power components in two, we will consider some devices which are exclusively used as power switches as is the thyristor. Of course, the switching properties of the amplifying devices will be considered, too.

Then the theory of analysis, design, and implementation of large signal amplifiers will be given. That will enable entering the story of large signal amplifiers implemented in integrated operational and operational transconductance amplifiers which will be discussed next. The audio and RF power amplifiers will be discussed last.

The book is accompanied with a set of appendices which cover: solved problems, SPICE simulation result for selected set of circuits, ad a short repetitorium on microelectronic technology.

4.2 Power Electronic Devices

The high-voltage and high-current prospective.

Abstract Fundamentally a power electronic component is the one which can sustain high power in a given time interval. Increased dissipation on the components, as already mentioned, results in, above all, increase in temperature. The development of components that have a very low temperature coefficient, i.e., whose temperature remains within acceptable limits even for large dissipations, is a serious problem. We call such components power components. Modern power transistors allow dissipations of several tens or even hundreds of watts with continuous excitation. There are specific situations, however, when the component is extremely stressed while the dissipation is not so large. The first of these cases is a situation where the voltage on the transistor is relatively low but its current is high. The transistor in this situation acts as a closed switch. The second situation arises when we require the voltages on the transistor to be extremely high at reasonably high currents. In both cases, the dissipation is not by itself a limitation of the operation of the transistor, but still components that meet such requirements are called power components. In this chapter we will consider all three cases: high-power, high-current, and high-voltage components. An effort will be made for covering the most frequently encountered components of three categories: diodes, transistors (active devices), and thyristors. Power components introduced most recently especially interesting for automotive and power switching applications will be introduced together with "classical" components encountered in power integrated circuits. Having that in mind here we will consider their properties for both linear amplification and switching. The latter will be exploited in the book LNAE_Book 7: Power electronics, where the main implementation of thyristors will be found, too. To make the story complete we will start with a short repetitorium of the device physics which will be necessary to follow the explanations of the functionality and limitations encountered. We will try to give enough information for the reader to become capable to start a "specialist" study of the subject.

V. B. Litovski, *Lecture Notes in Analog Electronics*, Lecture Notes in Electrical Engineering 958, https://doi.org/10.1007/978-981-19-6528-9_2

4.2.1 Basic Semiconductor Device Physics

While in a single atom electrons can occupy prescribed discrete energy levels, in a solid body due to the mutual influence of the atoms, the electrons occupy energies which belong to regions called bands. For study of the electrical properties of solid body of prime importance are two energy bands: the valence band and the conductance band. We denote the highest energy in the valence band as E_v and the lowest energy of the conduction band as E_c as depicted in Fig. 4.2.1. Electrons having energies belonging to the valence band are tied to the atom while the ones having energies belonging to the conduction band are free to move in the solid body and to interchange atoms. At $T = 0$ K all electrons have low energies that belong to the valence band. The largest energy one electron may occupy at $T = 0$ K is called the Fermi energy or the Fermi level (E_F). Accordingly, at $T = 0$ K all electrons have energy lower than the Fermi level. The energy intervals between the bands are called gaps. In fact we are interested in only one of them, the gap between the valence and the conductance band, and usually it is referred to as if it is the only one ($E_g = E_c - E_v$). With the rise of the temperature the electrons are thermally excited and probability exists that some of them will get energy larger than the maximal energy of the valence band (E_v) plus the energy of the gap (E_g). In other words, there are electrons which may have energy which belongs to the conduction band (above E_c). The process of transferring an electron from the valence band into the conduction band is called generation (Fig. 4.2.1).

These electrons, referred to as free electrons, are available for conduction of electrical current and are of prime importance for the electrical properties of the material. Their locations in the valence band (positive ionized atoms), after generation, are referred to as holes. Holes are attributed by a positive charge. The excited electrons eventualy (after a short time named life time) emit their excessive energy and "collapse" back into the valence band to "fill" the hole. This process is referred to as recombination.

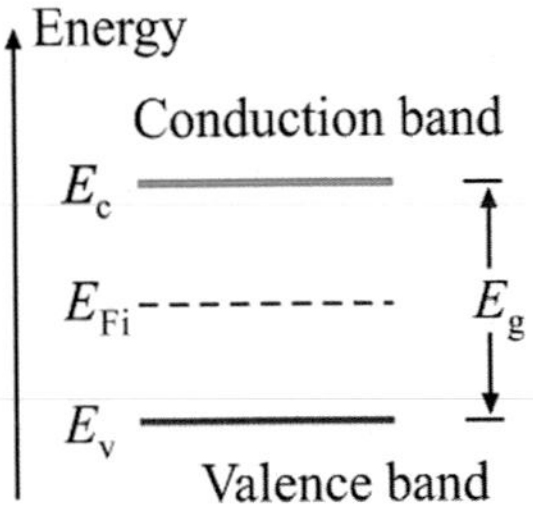

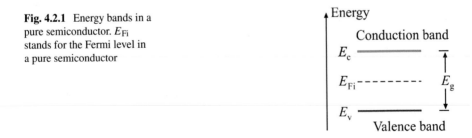

Fig. 4.2.1 Energy bands in a pure semiconductor. E_{Fi} stands for the Fermi level in a pure semiconductor

4.2.1.1 Intrinsic Concentration and the Energy Gap

The number of free electrons in a pure material, which is called intrinsic, is usually denoted as n_i. The number of holes is denoted by p_i and $p_i = n_i$. Depending on the value of the energy gap (which, in turn, is decreasing with temperature) we have different situations from the electrical conductance point of view of different materials. When E_g is very small or zero we have a situation where the valence and conductance bands are merged and all electrons from the valence band are available for conductance at very low temperatures. Such materials are called conductors.

At the opposite end are materials having very large E_g and they are called insulators since, at room temperature (usually approximated as $T_0 = 300$ K), there are practically no electrons available for conduction. In between are the semiconductors which may have a small or large E_g. The larger E_g the smaller the number of available electrons for conduction or, in other words, the larger the resistivity of the material. Silicon is the mostly used semiconductor in microelectronics nowadays and for it $n_i(300\text{ K}) = 1.5 \times 10^{10}$ electrons/cm^3 as depicted in Table 4.2.1. This means that (approximately) only 10^{-10}% of the atoms are ionized at room temperature, the density of the solid body being approximately of 5×10^{22} atoms/cm^{-3}.

Semiconductors belong to the middle of the periodic table of elements part of which is depicted in Figs. 4.2.2 and Fig. 4.2.3. If they are of single element, they are of valence four (Si and Ge). There are compounds behaving as semiconductors, however. Some of the type III–V (e.g., GaP, InSb) others of the type II-VI (e.g., ZnS, CdS, CdTe). There are other combinations which are of no importance for the subject of this study. Recently, semiconductor materials such as silicon carbide (4H-SiC, 4H- stands for the variant of this material which is most convenient for use in power electronic devices) and gallium nitride (GaN) re-emerged and became among the most important ones for production of high-power electronic devices. Their basic properties are depicted in Table 4.2.1. K_c stands for the breakdown electric field.

It is of fundamental importance to have in mind that the electrical properties of a pure material (i.e., n_i) are strongly dependent on temperature. It was shown that it may be expressed as

$$n_i^2 = A \cdot T^3 \cdot e^{-E_g/(kT)} \qquad (4.2.1)$$

where A is constant and $kT_{|T=300\text{ K}} = 0.026$ eV. In this expression there are three origins of the rise of n_i with rise of the temperature: rise of T^3, rise of T in the denominator of the negative exponent, and decrease of E_g in the nominator of the negative exponent. Namely, for silicon the value of the energy gap may be approximated by the following

$$E_g(T) = 1.21 - 3.6 \times 10^{-4} \cdot T \quad \text{(eV)} \qquad (4.2.2)$$

while for GaN and 4H-SiC one may use

Table 4.2.1 Basic properties of selected semiconductor materials

Material	$E_{g\,\vert T=300\,K}$ (eV)	$n_{i\,\vert T=300\,K}$ ($10^{10}\bullet cm^{-3}$)	K_c $\left(\frac{MV}{m}\right)$	T_{max} (°C)	Θ $\left(\frac{W}{cm\cdot K}\right)$	μ_n $\left(\frac{cm^2}{V\cdot s}\right)$	v_{sat} (10^7 m/s)
Si	1.12	1.45	300	200	1.5	1450	1
4H-SiC	3.26	10^{-4}	3500	600	4.9	900	2
GaN	3.5	10^{-10}	3300	400–900	1.3	2000	2.5
Ge	0.67	2500	0.1	937	0.64	3900	0.8

	5 **B**	6 **C**	7 **N**	
	13 **Al**	14 **Si**	15 **P**	16 **S**
39 **Zn**	31 **Ga**	32 **Ge**	33 **As**	34 **Se**
48 **Cd**	49 **In**	50 **Sn**	51 **SB**	52 **Te**

Fig. 4.2.2 Part of the periodic table of elements

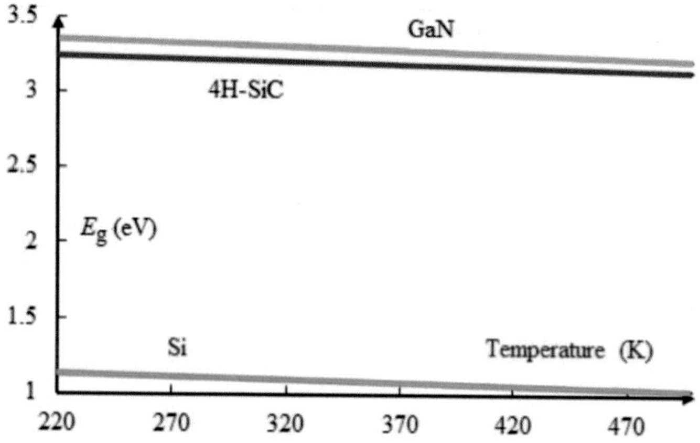

Fig. 4.2.3 Bandgap as a function of temperature for Si, SiC, and GaN

$$E_g(\text{eV}) = E_g(0) - \frac{\gamma \cdot T^2}{T + \beta}. \tag{4.2.3}$$

Figure 4.2.4 depicts the dependences of the energy gap of Si, GaN, and 4H-SiC according to (4.2.2) and (4.2.3) with parameter values from Table 4.2.2. The corresponding dependence of the intrinsic concentration for these two materials is given in Fig. 4.2.5.

To make the conductivity of the semiconductors independent of temperature (in a reasonable temperature range around the room temperature) the materials are purposely contaminated with impurities. The impurity materials are called dopants. The concentration of the impurities usually is 10^6 times smaller than the density of the basic material (one milligram per kilogram) so that the dopant atoms are largely surrounded by the atoms of the basic material. For silicon, usually, dopants are B, P, and As, while for SiC may be N, P, Al, B, Ga, or Be. For GaN one use Si, Ge, Se, O, Mg, Be, and Zn.

When the valence of a dopant is larger than the valence of the basic material it has a "redundant" valence bond (since all the surrounding atoms are "well supplied" with their own valence bonds) and at relatively low temperature creates one free

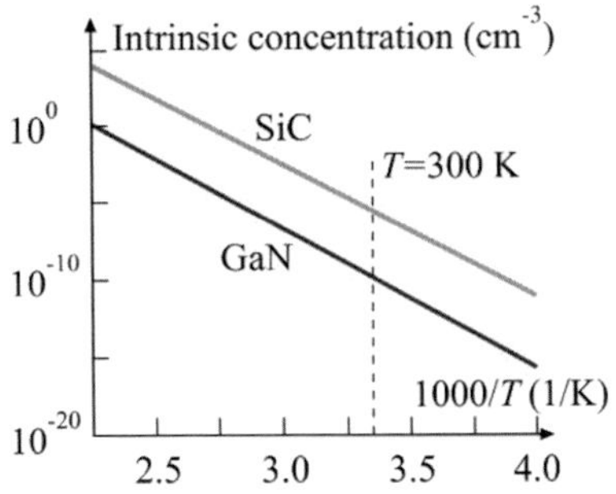

Fig. 4.2.4 Intrinsic carrier concentration in SiC and GaN as a function of temperature

Table 4.2.2 Temperature dependence of E_g for GaN and 4H-SiC

Material	$E_g(0)$ (eV)	γ	β (K)
GaN	3.396	9.39×10^{-4}	772
4H-SiC	3.285	3.5×10^{-4}	1100

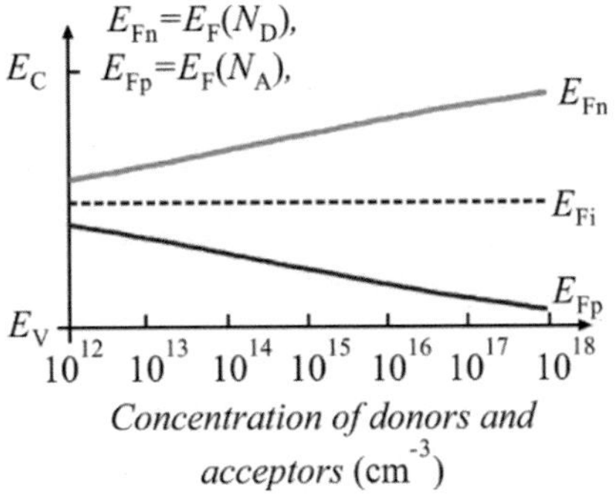

Fig. 4.2.5 Variation of the Fermi Level in n-type (E_{Fn}) and p-type (E_{Fp}) semiconductors as a function of impurity concentration at $T = 300$ K

electron. The number of these electrons is now equal to the concentration of the dopants called donors and denoted by N_D and is temperature independent in a very large temperature interval. It is obvious that $N_D \gg n_i$ meaning that the overall concentration of free electrons is approximately equal to N_D and is temperature independent. The semiconductor materials in which electrons are the main carriers are called n-type semiconductors.

In the case when the valence of the impurity is lower than the valence of the basic material, one valence bond is left "not filled" by an electron so that a *hole* is created. The number of holes at room temperature is equal to the number of impurity atoms and is denoted as N_A the impurity atoms being named acceptors. Again, the overall number of holes at room temperature is approximately equal to N_A and is not temperature dependent. The semiconductors in which holes are the main carriers are called p-type semiconductors.

For all, intrinsic, n-type, and p-type semiconductors the following is valid

$$n_i^2 = n \cdot p. \tag{4.2.4}$$

In n-type semiconductors this leads to $n_n = N_D$ and $p_n = n_i^2/N_D$ while for the p-type we have $p_p = N_A$ and $n_p = n_i^2/N_A$. This means that p_n and n_p are drastically smaller than n_i. This is why the electrons in n-type and the holes in p-type semicond uctor are referred to as majority caries. In the opposite, the holes in n-type and the electrons in p-type semiconductor are called minority carriers (of electrical charge).

Equation (4.2.5) may be used to define the temperature at which the intrinsic concentration becomes equal to the dopant concentrations:

$$T_m = \frac{E_g}{2k} \cdot \frac{1}{\mathbf{ln}(N_x/N_D)} \tag{4.2.5}$$

where N_x is a constant (for a given material) and $k = 8.61 \times 10^{-5}$ eV/K (the Boltzmann constant). Above this temperature the semiconductor is again intrinsic and of no use for its primary function. T_m is used as a base to define the maximum temperature of use of a semiconductor or part of a semiconductor component. Usually, the maximum temperature is lower than T_m. For example, for silicon with $N_D = 10^{15}$ cm^{-3} one gets $T_m = 277$ °C while usually one claims that the maximum temperature of use of a silicon slab is $T_{max} = 200$ °C. For 4H-SiC the intrinsic concentration is about 10^{-9} cm^{-3} which leads to much larger maximum temperature as shown in Table 4.2.1. It is obvious, however, that the maximum temperature to use of a semiconductor chip is set by its part having the lowest impurity level.

It is obvious that by adding impurities the resistivity of a semiconductor is (dramatically) changed. In silicon, for example, the intrinsic resistivity is 2.3×10^5 Ω cm while for $N_D = 10^{16}$ cm^{-3}, one gets a resistivity of only 0.53 Ω cm. For 4H-SiC for $N_D = 10^{18}$ cm^{-3} one gets approximately 1 Ω cm while its intrinsic value is higher than 10^9 Ω cm. Similar figures are obtained for GaN.

The dependence of the mobility of the carriers on the electric field on the surface (under the oxide) of the semiconductor will be considered next. Namely, due to additional scattering mechanisms, μ_n and μ_p on the surface changes significantly. The mobility on the surface decreases with increasing the voltage at the gate and the electric field in the oxide layer. In an N-channel transistor, a large electric field attracts electrons closer to the Si–SiO_2 surface and the scattering process intensifies. This scattering is attributed to the uneven surface and localized oxide charges.

Experimentally observed the effective surface mobility can be reasonably represented as a function of the surface field as

$$\mu_{ef} = \mu_0 (6 \times 10^4/K_S)^r \tag{4.2.6}$$

where μ_0 is mobility at small fields ($K_S < 6 \times 10^4$ V/cm) while K_S is the surface electric field. For a MOS transistor it is $K_S = -(V_G - V_T - V_D/2)/t_{ox}$ where t_{ox} is the oxide thickness. For N-channel transistors $r = 0.12$ while for the P-channel $r = 0.2$. A rough estimate says that the mobility on the surface is about twice less than inside the semiconductor at the same concentration of the main carriers.

Another phenomenon that is of interest when considering power transistors is the velocity saturation, which has been discussed so far. Namely, when the velocity of the electron in silicon approaches the thermal velocity, intense scattering occurs in which the charge carriers begin to lose energy very quickly. As a result, the electrons in silicon reach the saturation velocity $v_{sn} = 10^7$ cm/s at $T = 300$ K for $K_C = 2 \times 10^4$ V/cm. For the holes we have $v_{sp} = 7 \times 10^6$ cm/s at $T = 300$ K and $K_C = 5 \times 10$ V/cm. K_C is a critical field, i.e., the field in which the carrier velocity is saturated. The maximum drift velocity at the surface is lower than inside the semiconductor body and depends on the crystallographic orientation of the surface.

During the operation of a power component, situations often arise when the concentration of minority carriers is very high. This refers to the injected carriers in the base of the bipolar transistor. When the concentration of injected carriers exceeds the concentration of the majority carriers, it can be considered as secondary carriers in the cloud of the majority carriers. Due to drift or diffusion, there is a tendency for these two types of carriers to move in opposite directions, and due to electrostatic attractive forces, there is a tendency for the concentration of electrons and holes to be approximately (dynamically) equal, which means that the distributions of electrons and holes are not independent functions. If the injection is very large then it is necessary that the excess concentrations of the primary and secondary carriers (n' and p') to be equal. For such a situation, ambipolar mobility μ_A is introduced, which is given as

$$\mu_A = 2\mu_p\mu_n/\left(\mu_p + \mu_n\right) \tag{4.2.7}$$

This enables the influence of the majority carriers onto the mobility to be expressed in the continuity equation.

It should be borne in mind that in addition to scattering on the crystal lattice and impurities, in large injections, scattering between carriers is also important, so that

$$1/\mu = 1/\mu_r + 1/\mu_A + 1/\mu_{nos} \tag{4.2.8}$$

For Si, Ge, and GaAs the dependence of the electron mobility in an N-type semiconductor as a function of concentration can be well approximated with

$$\mu_n = \frac{\mu_0}{1 + (N_D/10^{17})^{1/2}} \tag{4.2.9}$$

where μ_0 takes different values depending on the semiconductor material.

This relation introduces a certain error (about 20%) for $N_D > 10^{17}$ cm^{-3}.

With large injections, in addition to reducing mobility, there is also a reduction in the lifetime of carriers. The main cause is considered to be the so-called Auger's recombination, which occurs when energy released during recombination is communicated to a third carrier who is therefore energized. This process can involve two electrons and a hole or two holes and an electron. In silicon, this type of recombination becomes observable at concentrations of about 10^{17} cm^{-3}. As a result of this process, recombination is more intense, so the effective lifetime is shortened (for silicon):

$$\tau_{\text{eff}} = \frac{\tau_0}{1 + a \cdot (n')^2} \tag{4.2.10}$$

where τ_0 is the lifetime at small concentrations (for silicon, of value of 10 μs), n' is the excess concentration, and $a = 2 \times 10^{-36}$ cm^6. For example, for $n' = 10^{18}$ cm^{-3} one gets $\tau_{\text{ef}} = \tau_0/3 = 3.5$ μs.

4.2.1.2 Energy Diagrams Revisited

As a consequence of the presence of the impurities the Fermi level of the material migrates. For n-type semiconductors with the rise of N_D raises E_F. In the opposite case, with the rise of N_A, E_F decreases. There is a special case, when the concentration of the impurities becomes of the order of (approximately) 0.1% (and above) of the density of the basic material, that E_F goes above E_c (in the n-type) or below E_v (in the p-type). Such semiconductors are considered degenerated. Now, in a degenerated n-type semiconductor there are electrons with energy below E_F while still in the conduction band. This is a property of conductors, too. These electrons are capable of tunneling into a metal which is tied to the semiconductor surface. That is the way how very low resistance contacts between metal and semiconductor are made.

To start with we may recall two facts that were mentioned above. First, the Fermi level in a semiconductor varies depending on the concentration of impurities. Second, in a solid body, the Fermi level is constant everywhere.

Its value, however, depends on the concentration of the impurities in the following way:

$$E_{\text{Fn}} = E_{\text{Fi}} + kT \cdot \ln(N_D/n_i) \tag{4.2.11a}$$

$$E_{\text{Fp}} = E_{\text{Fi}} - kT \cdot \ln(N_A/n_i). \tag{4.2.11b}$$

Namely, with the rise of the impurity concentrations the Fermi level migrates far from its intrinsic value. In a p-type semiconductor it decreases while in n-type rises. That is depicted in Fig. 4.2.5. Note the logarithmic scale for the abscissa.

Figure 4.2.6 depicts the position of the Fermi level in a semiconductor while changing the type and the number of impurities. First the case of intrinsic semiconductor is given. The Fermi level is approximately at the middle of the energy gap. Then a "normal" n-type semiconductor is shown as (b). A very high concentration of donor impurities (for silicon at about 10^{20} cm^{-3}) brings the Fermi level

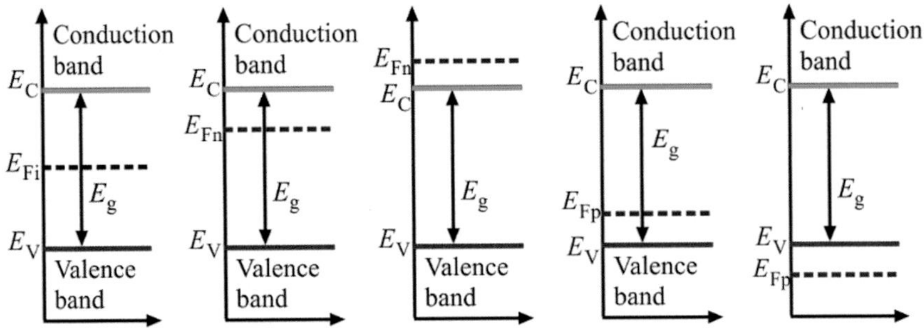

Fig. 4.2.6 Position of the Fermi level in a semiconductor. **a** Intrinsic; **b** n-type; **c** n-type degenerated; **d** p-type; and **e** p-type degenerated

into the conduction band as shown in (c). This kind of semiconductors is referred to as "degenerated." The opposite happens in p-type semiconductors as depicted in (d) and (e).

So far, we were considering a uniform distribution of the impurities in the semiconductor body. That, however, is not always the case. If the concentration in a slab is changed, the Fermi level is changing its relative position with respect to E_C and E_V. Since, however, its value is to be constant everywhere in the slab, the boundaries of the energy bands are changed (bent) as depicted in Fig. 4.2.7. There, in an n-type semiconductor, the concentration is decreased in direction opposite to the x-axis which bents upward the boundaries of the energy bands.

In such a situation, due to diffusion (migration) of the free carriers tending to make their own space-distribution uniform, in equilibrium, an electric field is created of a value

$$K(x) = -\frac{kT}{q} \cdot \frac{1}{n(x)} \cdot \frac{dn(x)}{dx}. \tag{4.2.12}$$

Its orientation is depicted in Fig. 4.2.8 while its value depends strongly on the relative change of the concentration in space meaning that abrupt changes of concentration, even with not changing the semiconductor type, may lead to large fields. Of course, it creates a force opposing the diffusion which stops further redistribution of charge. The corresponding voltage barrier created within the semiconductor in this

Fig. 4.2.7 Energy band diagram of a non-uniformly doped n-type semiconductor

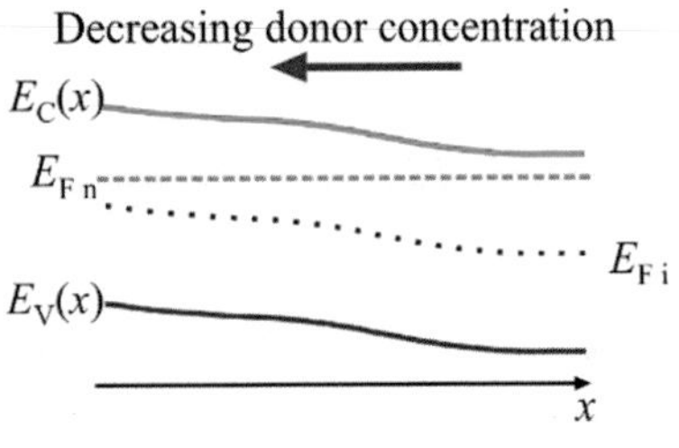

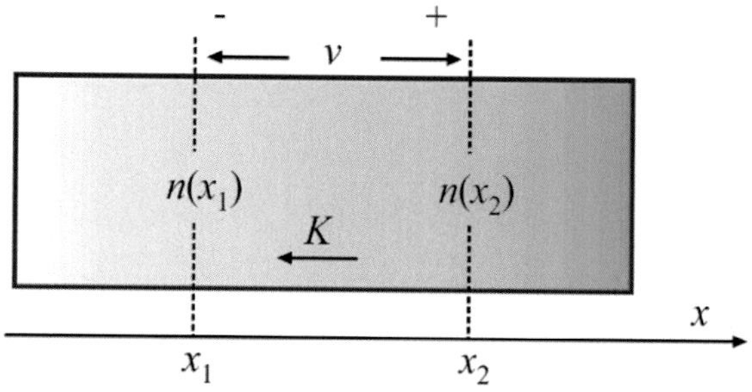

Fig. 4.2.8 Electric field and potential difference orientations within a non-uniformly doped n-type semiconductor slab. $n(x_1) < n(x_2)$

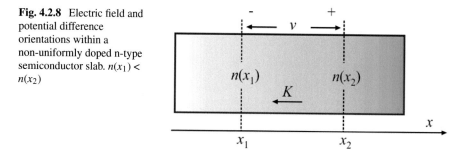

way is given by

$$v = \int_{x_1}^{x_2} K(x) \cdot \mathrm{d}x = V_\mathrm{T} \cdot \ln\left(\frac{n(x_2)}{n(x_1)}\right) \tag{4.2.13}$$

High concentration of the majority carriers has as a consequence other effect significant for the operation of the transistor. Namely, in this case the number of impurity atoms becomes so large that their presence has an impact on the energy structure of the semiconductor. Simply put, in the energy gap of a semiconductor, the energy band of impurities is created instead of the energy level of impurities. This is mainly due to the interaction of impurity atoms among selves since they cannot be considered isolated anymore. If the impurity band is wide enough, it will overlap with the adjacent band (conductive for electrons or valence for holes) so that the energy gap of the semiconductor narrows as a result. This situation is illustrated in Fig. 4.2.9. The increment of the width of the energy gap is usually given as

$$\Delta E_\mathrm{g} = \frac{2q^2}{16\pi\varepsilon_\mathrm{s}} \cdot \frac{q^2 n}{\varepsilon_\mathrm{s} kT} = \alpha\sqrt{n/n_0} \tag{4.2.14}$$

where for non-degenerated case we have $\alpha = 22.5$ meV and $n = 10^{18}$ cm^{-3} while for the degenerated one we have $\alpha = 162$ meV and $n_0 = 10^{20}$ cm^{-3}. For example, if $n = 5 \times 10^{17}$ cm^{-3} one gets $\Delta E_\mathrm{g} = 144$ meV.

Such significant changes in the width of the energy gap have direct consequences at the value of the carriers' intrinsic concentration n_i. Now we actually have

$$n_\mathrm{ie}^2 = n_\mathrm{i}^2 e^{\Delta E_\mathrm{g}/(kT)}, \tag{4.2.15}$$

where n_ie is the intrinsic concentration of electrons for a heavily doped semiconductor. It is easy to notice that due to the narrowing of the energy gap there is an increase of the intrinsic concentration. So, for $\Delta E_\mathrm{g} = 100$ meV we get $n_\mathrm{ie} = 6.8 \cdot n_\mathrm{i}$.

Having in mind that the product $n \cdot p$ is constant, an increase in n_i automatically means an increase in the concentration of the minority and a decrease in the

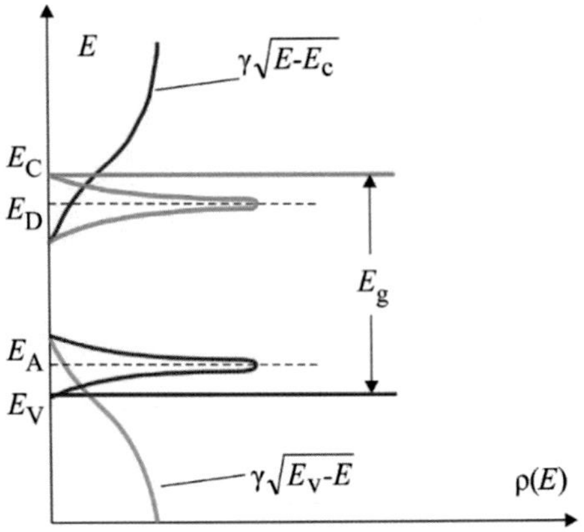

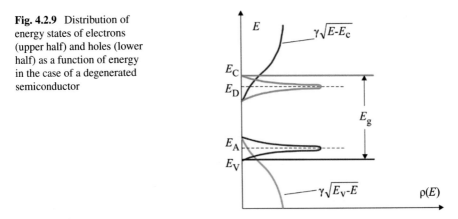

Fig. 4.2.9 Distribution of energy states of electrons (upper half) and holes (lower half) as a function of energy in the case of a degenerated semiconductor

concentration of the majority carriers, which partially neutralizes the effect expected to happen when increasing the concentration of the impurities.

4.2.1.3 The Breakdown Voltage and Measures for Its Increase

When considering the breakdown voltage of the p-n junction, the relation between the specific conductivities and the breakdown voltage for the abrupt junction is established in LNAE_ Book_1. It can be seen that the breakdown voltage is practically determined by the concentration in the region of lower conductivity. The surface of the junction was considered to be infinitely flat. When expressed over concentration, the breakdown voltage is approximately given by

$$V_B = 6 \times 10^{13} \cdot N^{-3/4}, \tag{4.2.16}$$

where N is the concentration on the side of lower conductance expressed in cm^{-3}. In the case of linearly graded p-n junction one gets

$$V_B = 9.3 \times 10^{9} \cdot a^{-2/5}, \tag{4.2.17}$$

where a is the gradient of concentration at the p-n transition. In the case of real components, however, the surface of the p-n junction is not flat but consists mainly of three parts: first, a flat part parallel to the surface of the semiconductor, then a cylindrical part extending to the surface of the semiconductor at right angle, and, finally, the curved part that connects these two areas. Of the greatest importance for reducing the breakdown voltage is the part that comes to the surface. At the ends of the p-n junction, the influence of the oxide (which covers the semiconductor) on the depleted area manifests itself. The positive charge in the oxide attracts electrons from the N-side of the p-n junction thus narrowing the depleted region on the N-side

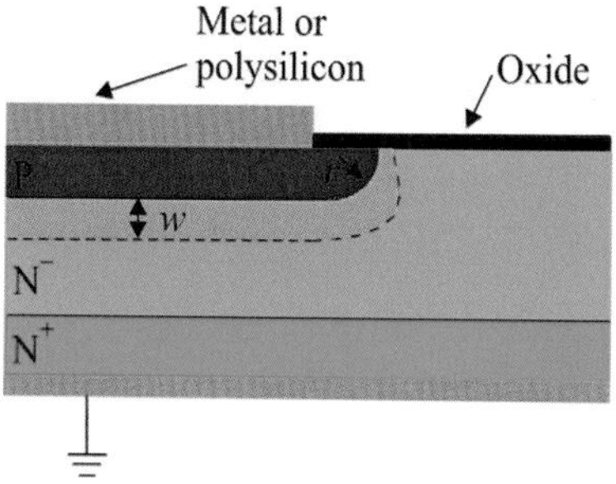

Fig. 4.2.10 Narrowing the depleted region below the oxide

(since when the concentration is higher the depleted region is narrower). This creates a larger electric field at the junction for the same external voltage, so it is easier to break down. This situation is shown in Fig. 4.2.10.

The bending of the junction influences the value of the breakdown voltage by itself. This is usually explained by the presence of larger electrostatic flux at the bending. The smaller the radius of the bending *r*, the smaller the breakdown voltage.

Procedures for increasing the breakdown voltage are aimed at eliminating these two main causes of its reduction. The most commonly used method of increasing the breakdown voltage is also the cheapest, and it consists in covering the p-n junction by the metallization of the electrode over the oxide as in Fig. 4.2.11. In this way, during the inverse polarization of the junction of interest, the surface of the semiconductor on the cathode side is depleted, thus achieving the opposite effect from the action of the charge in the oxide. This method carries the risk that, if the oxide is thinner or the voltage is too high, it will invert the cathode surface (in Fig. 4.2.11, a P-type channel would form on the surface). Therefore, the oxide at the junction itself is usually thicker. The anode electrode now has a recessed part at the point of contact and a raised part at the point of oxide—it has a plate shape.

The simplest way to increase the breakdown voltage by neutralizing the curvature of the p-n junction in shallow diffusions is shown in Fig. 4.2.12. Here, another diffusion of a much larger depth was simply embedded along the perimeter of the junction. In that way the radius *r* was significantly increased and thus the breakdown voltage.

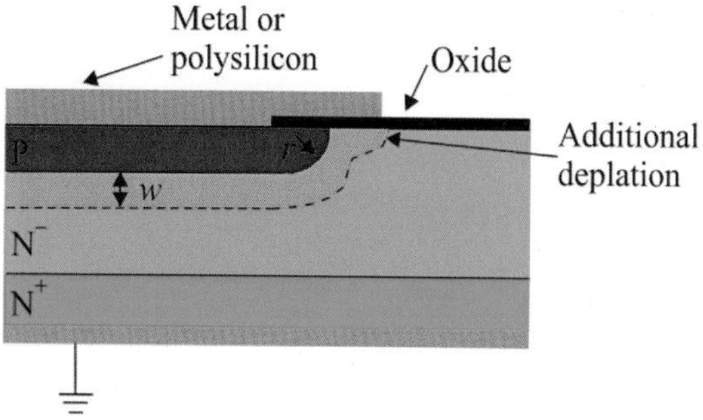

Fig. 4.2.11 Overlap of the p-n junction boundary by metallization (at the anode electrode side of the junction)

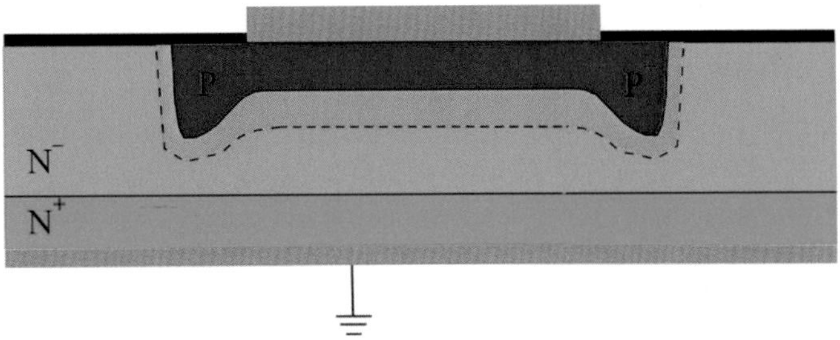

Fig. 4.2.12 Increasing the radius of the p-n junction by additional diffusion

Let us now consider somewhat more complex procedures for increasing the breakdown voltage. SIPOS (from semi-insulating-polycrystalline silicon) technology is often used for high-voltage discrete bipolar transistors. It consists of coating the non-oxidized surface of silicon with a thin layer of polycrystalline silicon (crystal bits surrounded by amorphous bits) which is very weakly doped with oxygen or nitrogen. Oxide is applied over this layer. In this way, the effect of charge in the oxide on the p-n junction is reduced, but other beneficial effects are also achieved. The resistance of this layer is very high but finite, so that, in the case of PNP transistors as in Fig. 4.2.13, for example, transfers the electrostatic influence of the negative potential of the collector to the N-type base surface. This impoverishes the surface of the base even more, and the breakdown voltage increases.

Figure 4.2.13 shows another procedure for increasing the breakdown voltage. Namely, a protective ring of the same type of conductivity as is the base is installed in the collector around the base, without being connected to any potential-floating. The use of such rings (they can be more concentric) allows increase of the radius of curvature of the p-n junction as in Fig. 4.2.14. Namely, at low inverse voltages the depleted region is bounded as in (**a**) while at high voltages the depleted regions of the main p-n junction and the ring merge so that the actual limit is now bounded as in (**b**). The N-band, the ring, and the P-band together can be considered as a voltage divider so that the excess voltage is consumed on the N-band-ring section, and thus the voltage at the main junction is reduced.

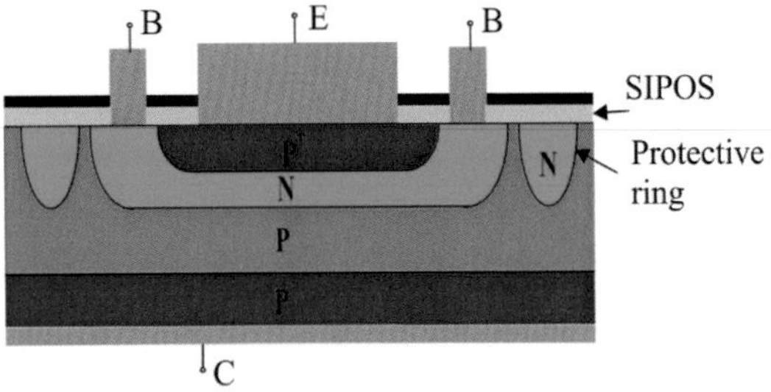

Fig. 4.2.13 Use of SIPOS and guard rings for increase of the breakdown voltage

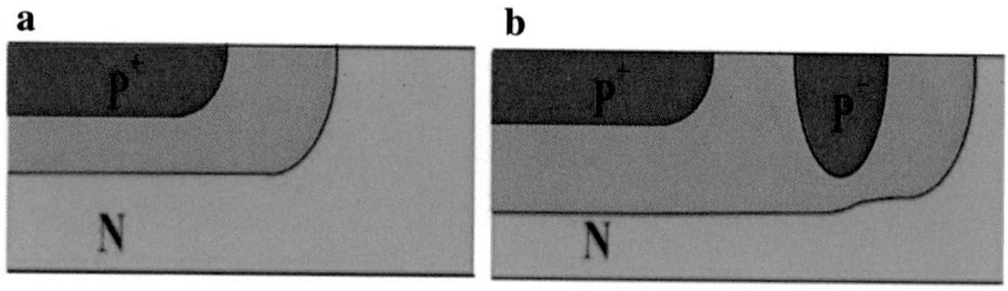

Fig. 4.2.14 Guard ring and the depleted region of the p-n junction. **a** For low voltages and **b** for large voltages

The breakdown voltage does not increase linearly with the number of rings. When insisting on extremely high breakdown voltages, a more expensive method must be applied, which usually consists of physically removing the curved and the vertical part of the junction. For junctions that need to work with voltages up to 400–500 V, deep trenching is used as in Fig. 4.2.15a. The disadvantage of this procedure is low reproducibility. For higher breakdown voltages, "sharpening" of the junction is used as in Fig. 4.2.15b. The greater the angle α, the higher the breakdown voltage will be. The reason for this is to increase the surface area of the junction so that the surface charges have a lower surface density. In Fig. 4.2.15c we have the so-called MESA etching. In this case, only a part of the semiconductor with a higher concentration of carriers is removed, but in such a way that a part of the depleted area is also taken away. The removal of part of the stationary spatial charge results in redistribution of charge at the boundary of the depleted area so that it extends to the part of the semiconductor with a higher concentration. The depleted area expands, and the breakdown voltage increases. It should be noted that a similar effect occurs when "sharpening" the junction by $\alpha < 0$. The depleted area, before and after etching, is shown in Fig. 4.2.15d. By using this procedure, breakdown voltages of the order of a thousand volts and more were reached.

The next type of activity that is important for increasing the breakdown voltage is passivation. Passivation is in fact coating of a component with an insulating layer so as to provide protection against surface contamination. The simplest passivation is silicon oxide. Silicon nitride is also used. However, these methods do not give satisfactory results for high-voltage components, so glass or organic polymers are often used. SIPOS also gives very good results by applying a layer of silicon nitride that prevents the dissolution of nitrogen from the atmosphere, and over it a layer of silicon dioxide is applied to prevent breakthrough in the nitride (as illustrated in Fig. 4.2.13).

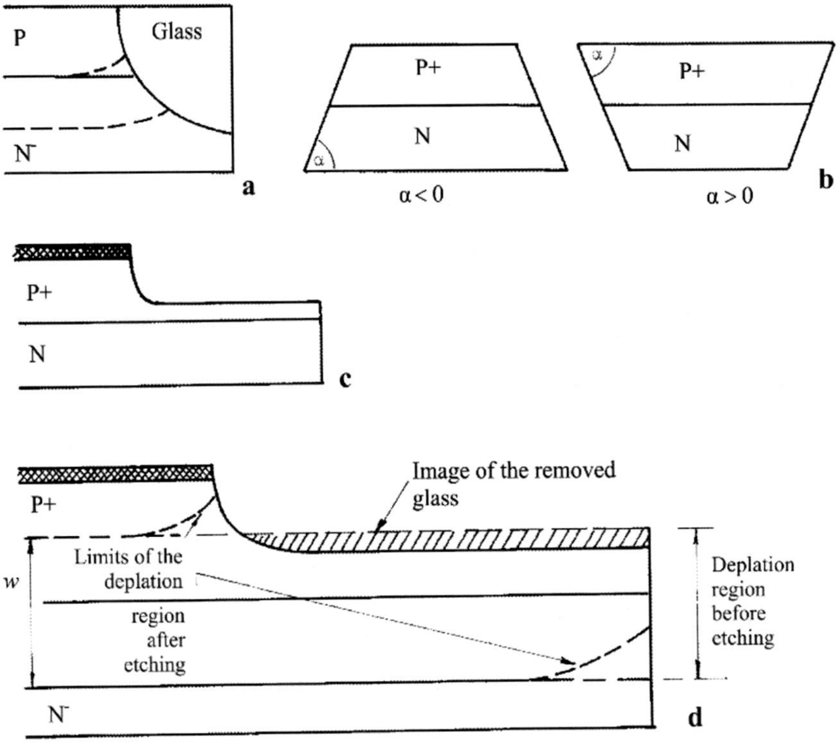

Fig. 4.2.15 Increasing the breakdown voltage by etching. **a** Etching of the trench, **b** sharpening of the junction, **c** the MESA process, and **d** description of the depletion region

4.2.2 Power Diodes

4.2.2.1 The p-n Junction and the Semiconductor Diode

The structure where semiconductors of n- and p-type are put together is of paramount importance. That is called a p-n junction. It is the base of all semiconductor electronics as such.

When two semiconductors of opposite types are layered one to another a structure arises known as p-n junction. Such a structure is depicted in Fig. 4.2.6a. Here only the acceptor (in the p-type semiconductor) and the donor (in the n-type semiconductor) atoms are depicted being the most important ones for the establishment of the junction.

Due to the diffusion force which is present on both sides of the slab the majority carriers from one side migrate to the other side leaving a depleted region at the border between the two types of semiconductors. This region contains only ions of the dopant materials which are not mobile. Together they are referred to as space charge. This is why the depleted region is named the space charge region. Since the charges in the sides of the space charge region are of opposite signs electrical field is created that is opposing the migration of the majority carriers. Eventually, equilibrium is established between the diffusion and the electrical forces over the space charge region and no migration is observed if no disturbance is applied to the junction.

The width of the space charge region, denoted as w, defines the difference of potentials of the p-type versus n-type semiconductor, the n-type being at higher potential. This difference (often referred to as barrier) is related to the difference of the Fermi levels of the semiconductors of both sides and may be expressed as

$$V_0 = V_{\mathrm{T}} \cdot \ln(N_{\mathrm{A}} N_{\mathrm{D}})/n_{\mathrm{i}}^2. \tag{4.2.18}$$

where N_{A} and N_{D} are the concentrations of acceptors and donors, respectively, $V_{\mathrm{T}} = kT/q$ and $q = 1.6 \times 10^{-19}$ °C. As can be seen this barrier is different for different semiconductor materials and is higher for the ones having low n_{i} i.e., having large E_{g}. For example, a silicon p-n junction with $N_{\mathrm{D}} = 10^{15}\ \mathrm{cm}^{-3}$ and $N_{\mathrm{A}} = 10^{18}\ \mathrm{cm}^{-3}$ at $T = 300$ K exhibits a barrier of $V_0 = 0.635$ V.

Note that this barrier is in favor for the minority carriers.

If a voltage is applied to the slab so that the p-type is brought to higher potential than the n-type, the barrier is reduced and the effective voltage over the space charge region is $v_{\mathrm{b}} = V_0 - v$. Now, the probability for the majority carriers to diffuse into the opposite side is risen and electrical current flows of the value (with second order effect neglected)

$$i = I_{\mathrm{s}} \cdot \left(e^{v/V_{\mathrm{T}}} - 1\right) \tag{4.2.19}$$

where I_{s} is the inverse saturation current (or leakage current).

Since both types of carriers are contributing to the main current of the device it is referred to as a bipolar component.

4.2.2.2 The On-Resistance

The so-called dynamic on-resistance of the p-n junction is obtained as

$$r_{\mathrm{on}} = \frac{1}{\frac{di}{dv}\big|_{v>V_\gamma}} = \frac{V_{\mathrm{T}}}{I}, \tag{4.2.20}$$

where V_γ is the built-in junction barrier seen from the outside terminals and, depending on the construction, material, and concentrations, ranges from 0.2 V to several volts. Note, at room temperature $r_{\mathrm{on}} = 0.0256/I$ (Ω). So, for example, if $I =$ 1 A the dynamic resistance of the junction becomes $r_{\mathrm{on}} = 25.6$ mΩ.

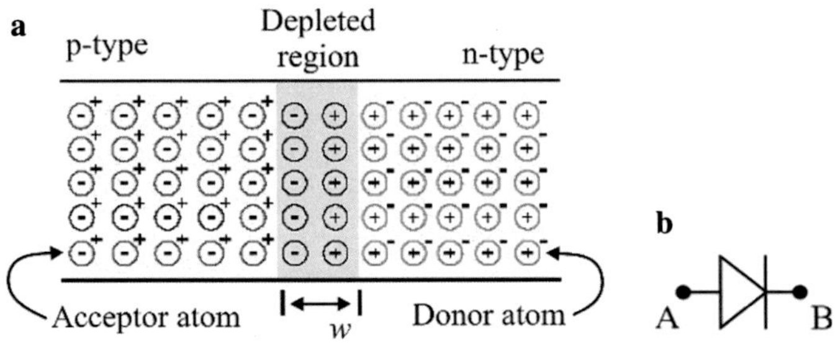

Fig. 4.2.16 **a** Formation of the p-n junction and **b** schematic symbol of a diode

4.2.2.3 Inverse Saturation Current

For $v < 0$, i.e., when the n-type semiconductor is on higher potential, $i \approx -I_s$ and is, for a large band of *inverse* voltages, very small and constant. Namely, in this case only the minority carriers are creating the current and, as shown earlier, their number is small. Depending on the basic material and the concentrations of the majority carriers the value of I_s belongs to the range $10^{-9} < I_s < 10^{-16}$ A. Note, a current of 10^{-16} A is in the range of 1000 electrons per second. Of course, in real devices, due to the so-called crystal defects and, consequently, generation-recombination centers, this current is much larger. I_s is highly temperature dependent which may be expressed as

$$I_s = A \cdot T^2 \cdot e^{-E_g/(kT)} \tag{4.2.21}$$

where A is a constant defined by the construction, the material, and the concentrations in the p-n junction. Note, A is proportional to the area of the junction (S).

As can be seen from (4.2.20) the p-n junction has very small resistance when forward biased while from (4.2.21) we see that for inverse biasing its resistance is very high. This is why one may say that the p-n junction exhibits rectifying effect, i.e., behaves as a switch controlled by its own voltage. The component having the p-n junction as the main functional part is then called p-n (or semiconductor) diode. Its schematic symbol is depicted in Fig. 4.2.16b.

4.2.2.4 Breakdown Voltage

The width of the depletion region is expressed as

$$w = w_0\sqrt{1 - v/V_0} \tag{4.2.22}$$

where w_0 is the width of unbiased junction and is defined by the impurity concentrations of both sides. In the so-called one-sided junctions where the concentration of one side is much larger than the other, the main part of w_0 is on the side where the impurity concentration is smaller.

There are components (diodes) where only one semiconductor material (of n-type) is used on which metal is deposited. In such a device the metal-semiconductor junction behaves similar to the p-n junction with V_0 halved. There are some other differences that will be mentioned later on. Such diodes are referred to as Schottky diodes.

From (4.2.22) we may conclude that if $v = V_0$ the depletion region disappears which means that the p-n junction behaves as a short circuit. The current of a real component is now defined by the bulk resistances of the semiconductors from the junction to the contacts and of the resistances of the contacts themselves.

For inverse bias, i.e., for $v < 0$, w rises with the rise of the inverse voltage so that the electric field over the depletion region is $K = (v - V_0)/w$. This field interacts with the atomic forces and at some level creates free electrons which are available for conduction. Now, these electrons may activate an avalanche effect so that the number of free electrons generated in the depletion region, and the proper inverse current, becomes very large. An avalanche breakdown occurs. Since, however, the voltage is high; the power dissipated at the junction is large leading to rise of the temperature which in turn raises n_i. This means that a thermal runaway is possible if precautions are not taken to limit the inverse current during breakdown.

The breakdown voltage of one-sided junction may be approximately expressed as (under the approximation of infinite area of the junction and for silicon)

$$V_B = \frac{\varepsilon \cdot K_C}{2 \cdot q \cdot N}. \tag{4.2.23a}$$

where N is the concentration at the lightly doped side and ε is the dielectric constant of the semiconductor. This expression takes the following forms for different semiconductor materials (being of interest here):

$$\text{For Si: } V_B = 2.96 \times 10^{17}/N \tag{4.2.23b}$$

$$\text{For 4H - SIC: } V_B = 135 \times 10^{17}/N \tag{4.2.23c}$$

$$\text{For GaN: } V_B = 99.4 \times 10^{17}/N. \tag{4.2.23d}$$

From the above expression for silicon with $N = 10^{15}$ cm^{-3} we get $V_B = 296$ V. Proportionally higher values may be calculated for the two wide gap materials. In real junctions, however, the edges of the junction are bent toward the semiconductor surface which in turn is populated by generation-recombination centers and is under influence of the trapped charges being above the surface (isolated by proper oxide or nitride). That reduces dramatically the value of the breakdown voltage which now becomes process dependent. Namely, in cases when high breakdown voltages are required special processing techniques are to be implemented to counteract these

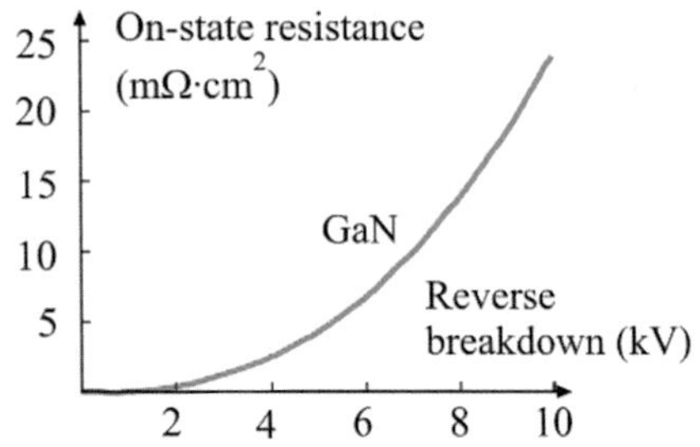

Fig. 4.2.17 On-resistance as a function of the breakdown voltage for the GaN Schottky diodes

influences as mentioned earlier. In a practical example for a 4H-SiC p-n junction with $N = 10^{15}$ cm^{-3} the expected value according to (4.2.23c) was 13.5 kV but when considering the second order effects 4.7 kV was obtained.

According to (2.23) the breakdown voltage may be increased by reducing the majority carrier concentration. That, however, leads to reduction of the forward current and consequently to rise the on-resistance. Figure 4.2.17 depicts the interdependence of the on-resistance and the breakdown voltage for a set of GaN Schottky diodes.

4.2.2.5 The Diode Forward Characteristic

As an illustration the characteristics of a forward biased 4H-SiC Schottky diode for different temperatures are depicted in Fig. 4.2.18. Note that the temperature coefficient of the current changes sign with clearly exposed a zero-temperature-coefficient point which (for this device) is at $v_F = 1.13$ V (approximately). Similar behavior may be observed for GaN Schottky diodes. Negative temperature coefficient of the current allows for parallel connection of diodes. Namely, if two (almost) identical diodes are connected in parallel the one conducting larger current will be heated more than the other and will have higher temperature. That will reduce its current.

Equilibrium is observed in real situation, and both diodes are conducting according to their properties. If the temperature coefficients were positive the first one would take over while the second would become redundant. Finally, one is to note that at high currents the characteristic becomes linear which means that the junction barrier is overridden and the device current is defined by the resistive properties of the semiconductor body and the contact resistance.

4.2.2.6 Junction (Space Charge or Transition) Capacitances

To finalize the description of the p-n junction properties its capacitances will be considered.

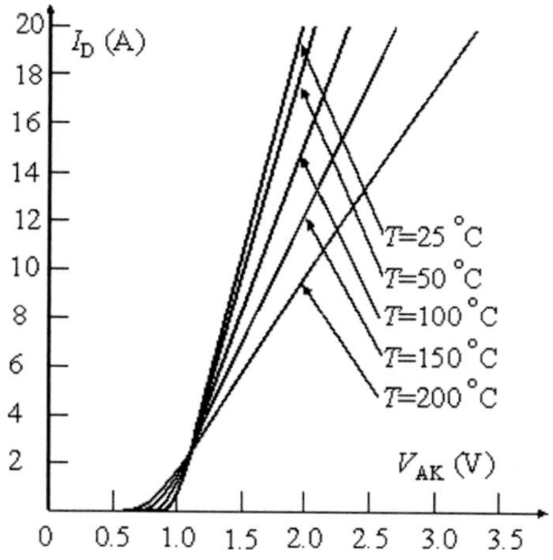

Fig. 4.2.18 Forward static characteristic of a 4H-SiC Schottky diode

Assuming the area of the junction is S and its width is w, since it represents a sandwich where the depletion region (insulator) is found between two semiconductors, the junction may be viewed as a capacitor. Its capacitance is

$$C_t = \varepsilon \cdot S/w. \tag{4.2.24a}$$

Here t stands for "transition," ε for the dielectric constant of the prime material, S for the junction area, and w for the space charge region width. Since w is voltage dependent one may write

$$C_t(v) = C_{t0}/\sqrt{1 - v/V_0}. \tag{4.2.24b}$$

Figure 4.2.19 illustrates this dependence for the diode of Fig. 4.2.17. $C_{t0} = C_t(0)$, in this case, is as large as nF. One may easily conclude that there is a residual capacitance to which the curve is converging for high inverse voltages. It represents the parasitic capacitances related to the mounting of the semiconductor chip to the component package.

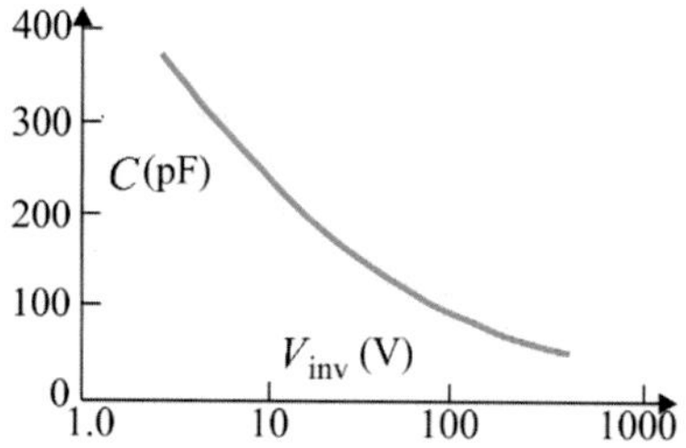

Fig. 4.2.19 Capacitance versus inverse voltage of a 4H-SiC Schottky diode

4.2.2.7 Diffusion Capacitance

The junction capacitance is dominating the dynamic behavior of the p-n junction and, accordingly, of the diode, at inverse polarization or, what is usually said, at negative bias. Since only the inverse saturation current is present at the junction, the "losses" of the capacitor which is now obtained are very small. That is why the p-n junction is used as a capacitor within integrated circuits and especially in high-frequency voltage-controlled oscillators.

When positive bias is applied, however, despite the fact that the diode is behaving as a small resistance it exhibits a large capacitance which is of a nature different than the junction capacitance. That new capacitance is called diffusion capacitance, and its value is calculated from

$$C_{\mathrm{d}} = C_{\mathrm{d0}} \cdot e^{v/V_{\mathrm{T}}} = \frac{\tau}{V_{\mathrm{T}}} \cdot I \tag{4.2.25}$$

where C_{d0} is its value for $v = 0$ and τ is the transit time of the minority carriers from the junction to the contact of the diode (one-sided junction is usually considered). As can be seen it rises exponentially with the rise of the forward biasing voltage of the junction.

The nature of this capacitance is related to the way how current is established for forward biased junction. With reference to Fig. 4.2.20a, when the p-n junction is forward biased the barrier at the junction is reduced by the value of the applied voltage. It is assumed that the diode is connected in a circuit as depicted in Fig. 4.2.20b. Due to the direct polarization a large number of carriers will be injected into the opposite side of the junction: electrons into the p-type and holes into the n-type. Now, these become minority carriers and, while migrating through the semiconductor, are gradually recombined with the majority carriers already existing on the corresponding side of the junction. The change of the concentration of minority carriers in the n-type semiconductor with time and space is shown in Fig. 4.2.20c. Of course, due to the migration of the carriers, current is established whose value will be determined by the gradient of the minority concentration. The gradient at a given space coordinate, however, as can be seen, changes through time until it reaches equilibrium which means that the value of the current will change, too. That is depicted in Fig. 4.2.20d. It looks as if we have the situation of the circuit depicted in Fig. 4.2.20e. Based on the analogy it is easy to conclude that a capacitance is acting in parallel to the on-resistance of the junction.

For direct polarization, since r_{on} is small, the importance of this capacitance comes forward only for signals of high frequencies. In switching, however, it becomes of crucial importance. Namely, in a forward biased diode the amount of excess space charge of minority carriers (the area below the curve for t_3 and above p_{n0} in Fig. 4.2.20c) may be large especially in cases of high currents. To switch off a diode, i.e., to reverse the voltage and reduce the current to $-I_{\mathrm{s}}$, it is necessary to first neutralize the space charge (by a current of value I_{r}) and then to rise the concentration of the minority carriers to their equilibrium value. These two phases are depicted in

Fig. 4.2.20 Explanation of the diffusion capacitance and the reverse recovery time

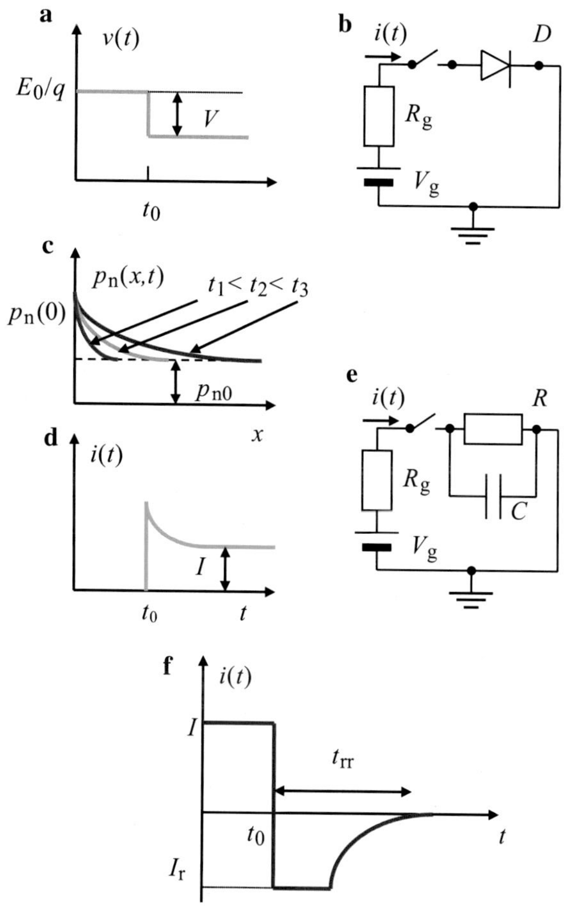

Fig. 4.2.20f. The overall time needed is called the reverse recovery time and denoted by t_{rr}.

Here comes the importance of Schottky diode for switching applications in power electronics. Namely, in this device, as mentioned, there is no p-type semiconductor and the metal is deposited on the surface of n-type semiconductor the voltage barrier being due to the difference of the Fermi levels of the metal and the semiconductor (as will be discussed later). The metal, of course, takes the role of the p-type semiconductor with the difference that in forward biasing it is not injecting holes into the n-type. The electrons injected into the metal, on the other side, are not minority carriers. There is no space charge of minority carriers, and there is no diffusion capacitance. This is why it is often said that the Schottky diode is a "majority carrier" semiconductor device.

So, apart from the halved forward voltage, the most important difference between the p-n diode and the Schottky diode is the reverse recovery time (t_{rr}). In a p-n diode,

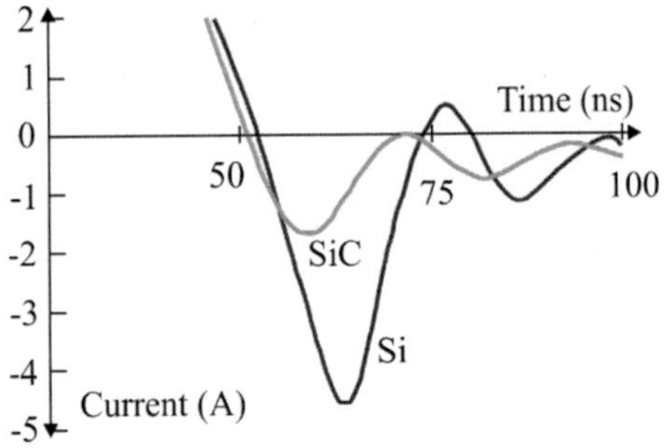

Fig. 4.2.21 Measured waveforms of switching off the currents in SiC Schottky and Si FRD diodes under equal conditions

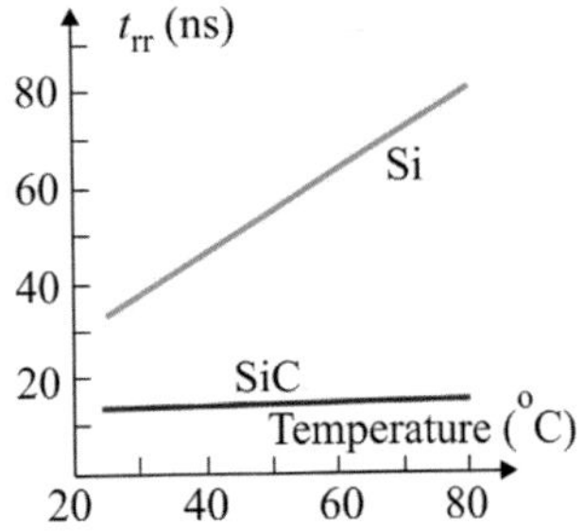

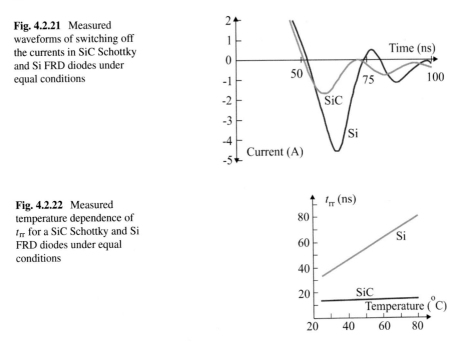

Fig. 4.2.22 Measured temperature dependence of t_{rr} for a SiC Schottky and Si FRD diodes under equal conditions

the reverse recovery time can be in the order of hundreds of nanoseconds to less than 100 ns for fast diodes. Schottky diodes do not have a recovery time, as there is nothing to recover from. The switching time is ~100 ps for the small-signal diodes and up to tens of nanoseconds for special high-capacity power diodes. For high-power Schottky diodes however, due to the additional structures needed to allow for very high breakdown voltages, additional p-n junctions are created around the main part of the component which define the inverse recovery time.

To illustrate the switching in a real Schottky SiC diode and to compare with its Si counterpart (here the so-called FRD is used which stands for fast recovery diode) under equal conditions, in Fig. 4.2.21 measured waveforms of the diode current are depicted. As can be seen the area under the SiC curve representing the charge that is to be neutralized (frequently denoted as Q_{rr}) is incomparably smaller. One is to add to that the temperature dependence of the reverse recovery time which is depicted in Fig. 4.2.22. It is evident that in the case of SiC diode t_{rr} is practically independent of temperature which is not the case when its Si counterpart is considered.

4.2.3 Power Active Devices

When comes to the power of an active electronic device there are three specific situations. Generally, one has in mind that both the voltage and the current of the device are large. The power is large, however, in the case when the device current is relatively small while its voltage is large (that is situation of an open switch) and in the opposite case when the voltage is small and the current is large (the situation of a closed switch). To withstand high voltages, high currents, and high powers the component has to have specific properties both looking from building materials point of view and from construction point of view.

The dissipated power has to be removed firstly from the part of the die which performs the fundamental function of the device and then from the die and finally from the package. For the first case the properties of the semiconductor material are in charge while for the second the implemented technology. The latter will be not discussed further here. Note, if cooling is not effective, the device gets hotter and the semiconductor within it eventually reaches its maximum temperature. That, at the beginning, affects the functionality of the device but if not taken care of it may lead to overheating and destruction of the device.

To get the feeling about the properties of the semiconductor materials from this point of view in Table 4.2.1 the thermal conductivity coefficients (Θ W/(cm K)) of the basic semiconductors for implementation in power electronic components are given.

As can be seen 4H-SiC has three times larger conductivity making it very attractive for power applications. Having in mind the almost equal dependence of the energy gap versus temperature of GaN and 4H-SiC (as depicted in Fig. 4.2.3), we may conclude that 4H-SiC is favorable from this point of view.

It is obvious that if a device is to be implemented as a switch the power in both states of low current and low voltage must be reduced. That means that if the switch is closed its voltage has to be as small as possible. The minimal voltage of the closed switch defines its power losses and may be of crucial importance when series-connected switches (intended to be in on-state for normal operation of the system) are used. The opposite stands for the open switch. A leakage current flowing through the open switch at high voltages may give rise to discharges which may jeopardize the functionality of the system.

The minimum voltage of the power device, as we will see soon, defines the efficiency of the power amplifier in which it is embedded. The lower the minimum voltage, the higher the efficiency.

Fast switching in power electronics means fast transition from one end (low voltage-high current) to the other (low current-high voltage) and vice versa, through the high-power region. That means that all problems that may be conceived in the use of a power device are probable to be met. Capacitances of all kinds are to be charged and discharged during the switching thereby affecting the speed. Here comes forward the construction of the device. Namely, since the current density is limited by the concentration of the carriers contributing to it, to get higher currents, larger area is needed. Larger area means larger capacitances. Two conflicting requirements impose additional limits to the applicability of the device.

When implemented in power amplifiers the transistors frequently perform as switches. In addition, in order to extend their application to high-frequency solutions reduced capacitances are of crucial importance. That stands for the reverse recovery time too.

4.2.3.1 Bipolar Power Transistors

We will start here with a very short repetitorium of the basic structure and functionality of the bipolar junction transistor which will be referred to by the short BJT in the rest of the book.

Figure 4.2.23a represents the cross section of one of the two variants of the BJT: the NPN transistor. Alternatively, one may use a PNP transistor which is not shown here. Figure 4.2.23b represents the schematic symbol of the NPN BJT. Three regions are seen: the emitter (E), the base (B), and the collector (C). The emitter and the collector are of the same type of conductivity while the base is of the opposite.

Figure 4.2.24 depicts the so-called normal polarization of an NPN BJT in a configuration with common base. Here the p-n junctions are marked. In this case the emitter junction is forward biased while the collector junction is inverse biased.

The functionality of this component is best understood via the analysis of the currents established in normal polarization. These are depicted in Fig. 4.2.25.

Starting with the emitter junction which is forward biased we have large drift current of the majority carriers in the emitter denoted by I_{nE}. These carriers are injected into the base where they become minority carriers. Part of them gets recombined with the majority carriers of the base to form the $(I_{nE} - I_{nC})$ current of the base. At the same time, the biasing of the emitter junction allows the minority carriers of the base to be injected into the emitter to form the I_{pE} component of the emitter current.

Now, for the remaining minority carriers in the base that came from the emitter, the collector junction is forward biased and these carriers are transferred to the collector

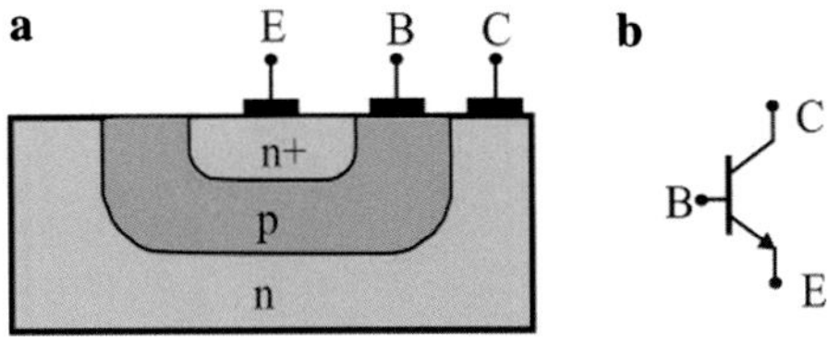

Fig. 4.2.23 NPN BJT. **a** Cross section and **b** symbol

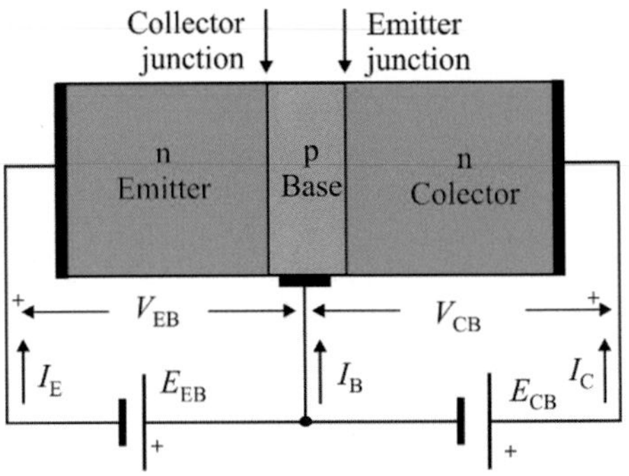

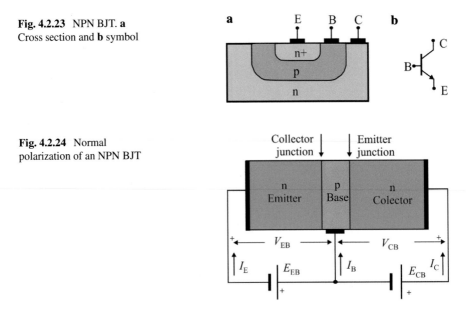

Fig. 4.2.24 Normal polarization of an NPN BJT

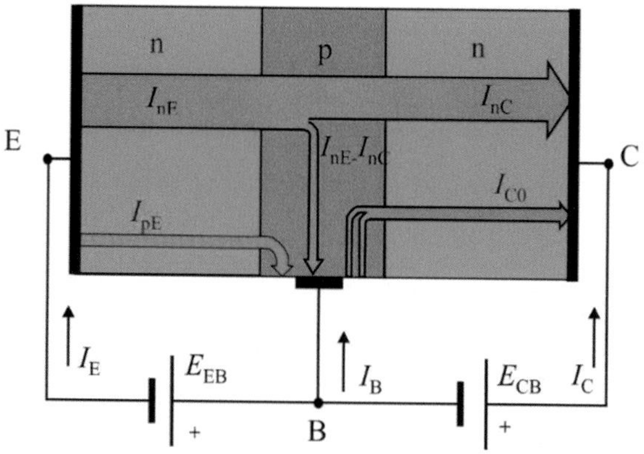

Fig. 4.2.25 Currents in NPN BJT

to form I_{nC} which is the main component of the collector current. In addition, minority carriers of both types are contributing simultaneously to the collector and the base current. These are denoted as I_{C0} and are referred to as inverse saturation current of the collector junction.

To reach the main functionality of the BJT which is: with small changes of the base-to-emitter voltage to get large changes of the collector current, it is necessary to be $I_{nE} \gg I_{pE}$, $I_{nC} \approx I_{nE}$, and $I_{nC} \gg I_{C0}$. That is achieved by controlling the concentrations in separate areas so that N_{DE}, $N_{DC} \gg N_{AB}$ and $N_{DE} > N_{DC}$, where N_{DE} and N_{DC} are concentrations of majority carriers in the emitter and collector, respectively, while N_{AB} is the concentration of the majority carriers in the base. In addition, the physical distance between the emitter and the collector (which is the base width) should be as small as possible to reduce the recombination within the base and to reduce the transit time of the minority carriers through the base.

The following is the Ebers-Moll model of the BJT

$$I_E = I_1 + \alpha_r \cdot I_2 \tag{4.2.26a}$$

$$I_C = I_2 + \alpha_d \cdot I_1 \tag{4.2.26b}$$

where

$$I_1 = I_{es}\left[\exp(V_{BE}/V_T) - 1\right] \tag{4.2.27a}$$

$$I_2 = I_{cs}\left[\exp(V_{CB}/V_T) - 1\right] \tag{4.2.27b}$$

and I_{es} and I_{cs} are the inverse saturation currents of the emitter and the collector junctions, respectively, while α_d and α_r are the direct and reverse current gain coefficients, respectively. Figure 4.2.26 depicts the schematic representation of the Ebers-Moll model of an NPN BJT. Note, all constants mentioned above are functions of the material properties, concentrations, and dimensions of the transistors. Details were given in LNAE_Book 1.

For normal polarization, having in mind the values of the inverse saturation currents, the model reduces itself into

$$I_E = I_1 \tag{4.2.28a}$$

$$I_C = \alpha_d I_1 + I_{C0}. \tag{4.2.28b}$$

In the case of CE configuration, the (NPN) transistor model is usually expressed as

$$I_B = I_{es}\left[\exp(V_{BE}/V_T) - 1\right] \tag{4.2.29a}$$

$$I_C = \beta \cdot I_B + I_{C0} \tag{4.2.29b}$$

where

$$\beta = \alpha/(1 - \alpha) \tag{4.2.30}$$

with $\alpha = \alpha_d$.

The symbol β denotes the current gain coefficient of the BJT in CE configuration. While α is smaller than unity (but very near to it) β is a very large number.

To further understand the functionality of the BJT Fig. 4.2.27 depicts the potential distribution in a PNP BJT. In both parts (a and b) the dashed line represents the potential distribution in a non-biased transistor. The difference in the potentials of the emitter and collector is a consequence of difference in concentration of the majority carriers and will be compensated by the difference in the voltage drops on the corresponding semiconductor-metal junctions at the emitter and collector terminals. The full lines represent the potential distribution after biasing. As can be seen the forward biasing at the emitter reduces the barrier while the inverse biasing at the collector increases the barrier.

What is very important to notice is the fact that the depletion region at the collector junction extends itself mostly in the base so the effective base (denoted by w) gets reduced when the inverse voltage at the collector is raised (as in Fig. 4.2.27b). That phenomenon, named Early effect, has multiple consequences. First, the variation of

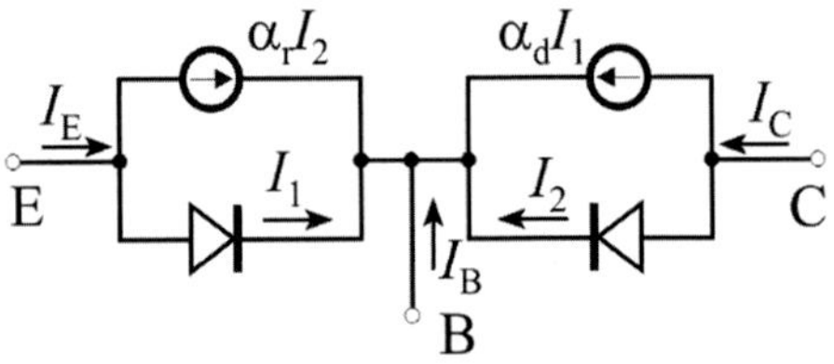

Fig. 4.2.26 Schematic representation of the Ebers-Moll model

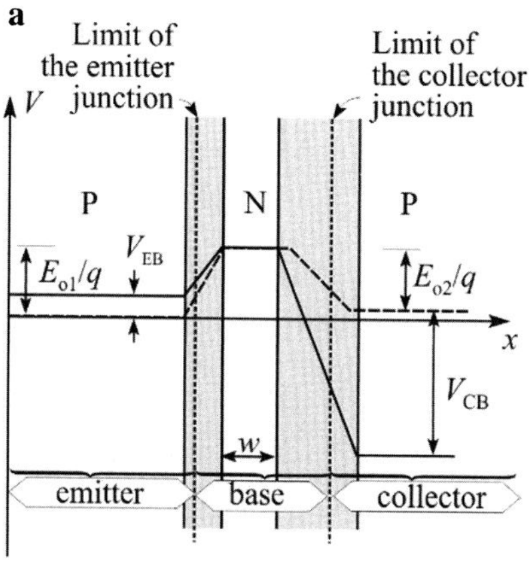

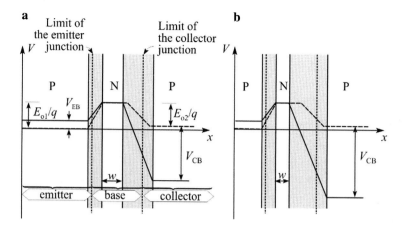

Fig. 4.2.27 Distribution of the potential in a PNP BJT. **a** Low collector voltage and **b** higher collector voltage

the effective width of the base influences the transfer of the minority carriers in the base so affecting the current gain coefficients and inverse saturation currents. Equally important is the phenomenon of base punch-through which happens for large inverse collector voltages. In that case the depletion region from the collector side covers the whole base and a "short circuit" is established between the emitter and the collector. This phenomenon is referred to as punch-through of the BJT and is illustrated in Fig. 4.2.28.

Figure 4.2.28 depicts the run of the current of a BJT when high voltages are present. The value V_{CBo} corresponds to the (avalanche) breakdown of the collector

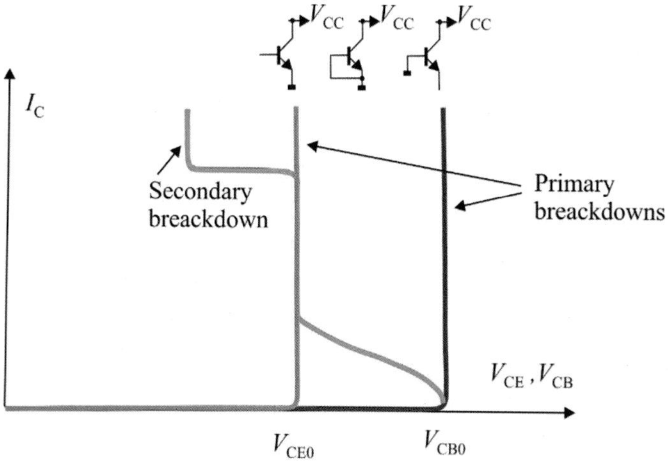

Fig. 4.2.28 Breakdowns in a BJT

junction and is characteristic for CB configuration as shown by the magenta line. In CE configuration the voltage V_{CE0} is the value which corresponds to the break down as shown by the orange line. In the case when a short circuit (as in Fig. 4.2.28) or resistive connection exists between the base and the emitter, after starting the breakdown, the current follows the green line. It reaches the orange line so that the voltage V_{CE0} is referred to sustainable voltage. Finally, after exposed to large currents and the sustainable voltage, the transistor goes into second breakdown shown by the blue line. One claims that local heating in the base-to-emitter junction causes increase of the current density and consequently leads to a current runaway.

Having all that in mind one may conclude that for getting high-voltage transistors one needs a BJT with a wide base as shown in Fig. 4.2.29. In this case the depletion region of the collector junction needs much higher voltage to reach the emitter.

The problems with this solution are manifolds. Above all, the current gain coefficient is significantly reduced which means the power gain is reduced, too. To avoid that new structure is in use in which in fact new collector region with reduced majority concentration is inserted between the base and the collector as depicted in Fig. 4.2.30. Now, the base width is preserved while the depletion region extends on both sides in the base and (mostly) in the collector. This structure is referred to a homotaxial transistor.

Significant improvement of the transistor's characteristics occurs with the component whose structure can be designated as N^+-P-P^--N^-N-N^+ (Fig. 4.2.31). With this structure, a higher maximum punch-through voltage, higher avalanche breakdown voltage, and higher resistance to secondary breakdown are obtained. It is usually made in the epitaxial area while base epitaxy is possible too, so that only the emitter is formed by diffusion.

Some features of bipolar power transistors will be discussed in the sequel. Let us first consider the current gain coefficient at moderate currents. In this situation the current gain coefficient is determined by the emitter efficiency I_{nE}/I_E. There are two

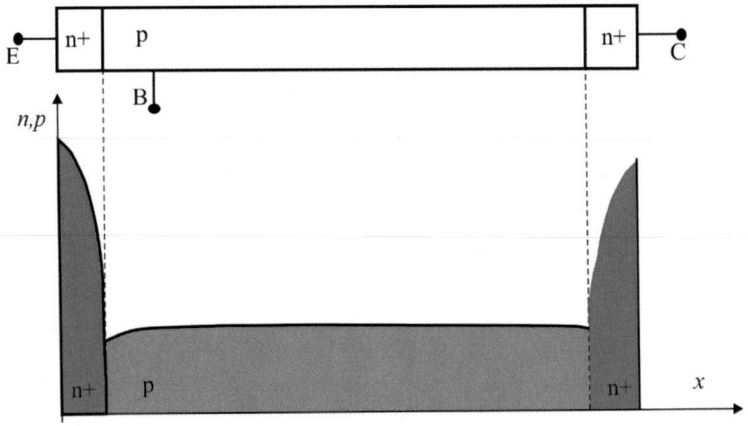

Fig. 4.2.29 BJT with a wide base

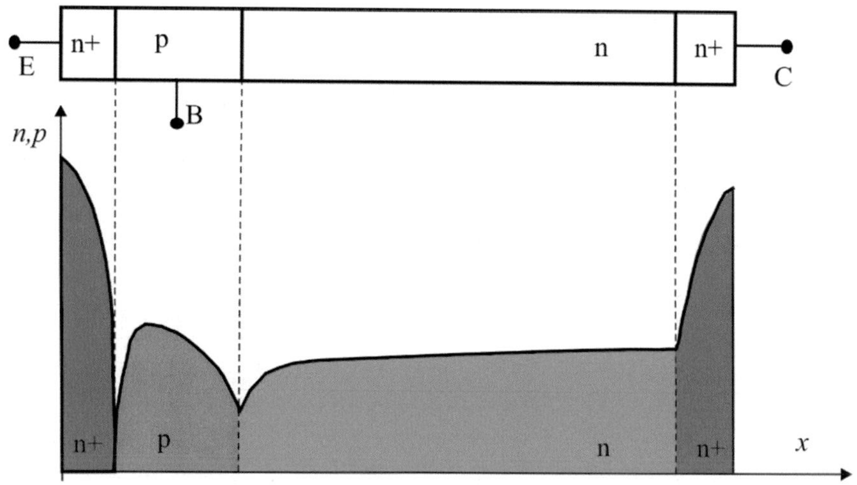

Fig. 4.2.30 Homotaxial BJT (with extended collector)

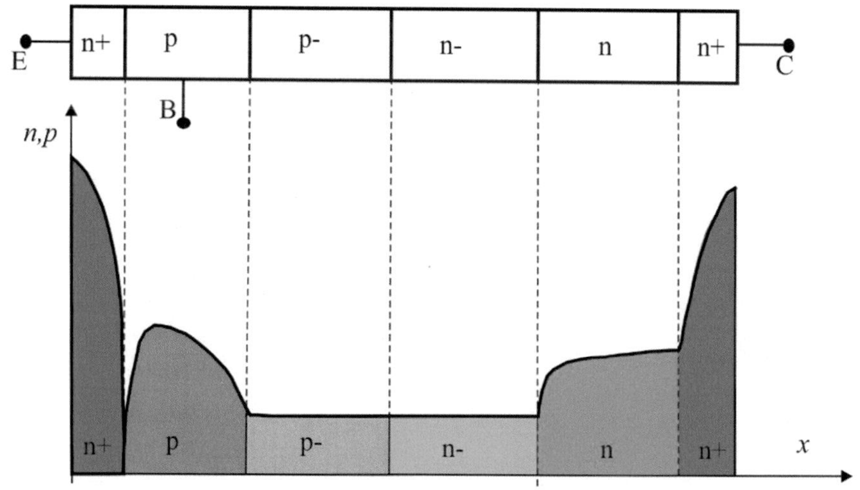

Fig. 4.2.31 The N^+-P-P^--N^-N-N^+ (epitaxial) power transistor

problems with increasing the efficiency of the emitter by increasing the concentration of the main carriers in it.

First of all, due to the narrowing of the energy gap and the increase of n_{iE} (and p_{iE}), the effective value of the concentration of the majority carriers is significantly lower than the concentration of impurities, which gives less emitter efficiency than would be expected. Then, due to Auger's recombination at high concentrations of impurities, the effective concentration decreases, but also the lifetime of the main carriers is shortened, and thus their diffusion length. This also has the consequence of reducing the efficiency of the emitter, i.e., the current gain.

On the other hand, the presence of a layer of low concentration (N^- for an NPN transistor) in the collector leads to the expansion of the base of the transistor, which also results in a reduction of the gain. Namely, for reasonably high voltages on the collector, the mobile spatial charge that makes up the collector current in the depleted region at the P-N^- junction has a higher concentration than the stationary charge of positive donor ions in the N^- region, which has the same effect as if the base boundary (of the P-region) has moved toward the collector—there was an effective expansion of the base. The magnitude of the base expansion depends on the emitter current and is not expressed at low currents.

Note that the temperature coefficient of the collector current is positive. That prevents parallelizing BJTs. Namely, with two BJTs connected in parallel due to the natural non-identity of the transistor; the current in one of them is larger. As a result, since the voltages on both components are equal, the dissipation of the first one increases, and thus its temperature (positive coefficient), which means that it draws even more current and thus becomes the dominant component in the parallel connection. The non-dominant will be out of function. That imposes the need of creation of power BJTs with large emitter-base junction area. That gives rise to the following.

Of particular importance is the effect of current accumulation (congestion). This effect is easily explained by the finite resistance of the base body, which is not small given the low concentration in the base. The base current in planar transistors extends laterally, which means that there are different distances, and thus the resistance between the base connections and individual points on the emitter. Where there is greater resistance, there will be a greater voltage drop across it and a smaller voltage drop across the emitter junction.

In the area of the junction where the voltage is lower, the current density will be lower so that there is an increase (accumulation) of the current density at the edges of the emitter that are close to the base connection (this is usually cylindrically symmetric with respect to the emitter). This results in non-uniform current gain along the length of the emitter junction and affects the local expansion of the base where the current is higher, thereby reducing the current gain. This effect can be significantly mitigated if a special emitter and base geometry is performed so that, seen from above, they have the shape of combs that fit into each other. This situation is shown in Fig. 4.2.32.

Finally, let us consider the influence of the construction of a bipolar power transistor on its output characteristics. In addition to the fact that the current gain is less, the effect of quasi-saturation is also observed. Namely, between the current saturation which determines the minimum voltage on the transistor and the voltage saturation which represents the active operating region, there is an area with a lower current gain coefficient according to Fig. 4.2.33. This phenomenon is attributed to the base expansion and is all the more pronounced if the resistance of the N^- region in the N^+-P-P^--N^--N^+ structure is higher (i.e., the concentration is lower). When the voltage on the collector is large enough, the effect of base expansion ceases so that

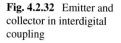

Fig. 4.2.32 Emitter and collector in interdigital coupling

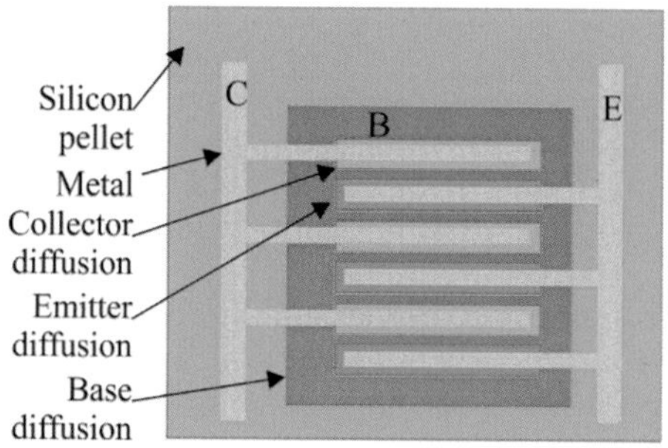

the characteristics turn into saturation. The reciprocal value of the slope of the line separating the region of quasi-saturation and voltage saturation is determined by the conductivity of the N^- region and denoted as $1/R_C$. Transistors that do not have an N layer do not exhibit quasi-saturation.

It should be noted here that there are a number of effects that are inherent in bipolar transistors, regardless of the amount of power dissipated on them, and which are not commented on here. Among them, we single out the decrease in gain at high currents, which is even more important for power transistors since the gain at moderate currents is also reduced. Then, due to the increased base width and the lower diffusion constant of the secondary carriers in the base, the upper cut-off frequency is significantly reduced. The temperature dependence of the characteristic is of importance, too.

All these are partly illustrated in Figs. 4.2.34, 4.2.35, and 4.2.36. The input characteristic for two temperatures is depicted in Fig. 4.2.34. The output characteristics are given in Fig. 4.2.35. Figure 4.2.36 depicts the dependence of the CE current gain coefficient on the collector current. Finally, Fig. 4.2.37 depicts the safe operation area of a power BJT as a log-log diagram. Note the difference in the limits of the working area for pulsed and DC mode of operation. When pulsed, time is available for the collector junction to be cooled. The green line represents the maximum power hyperbola.

Fig. 4.2.33 Illustration of quasi-saturation

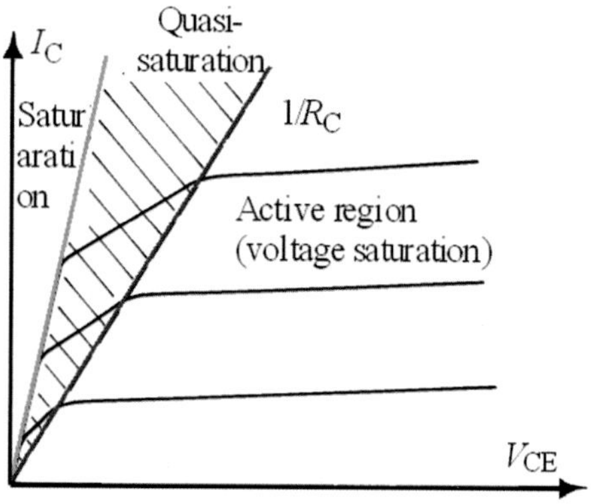

Fig. 4.2.34 Input characteristics of a power BJT

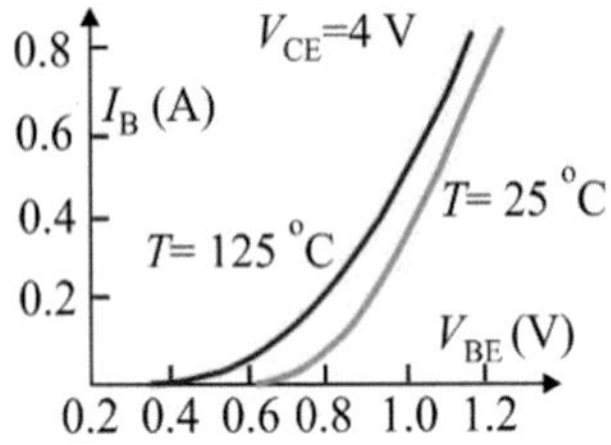

Fig. 4.2.35 Output characteristics of a power BJT

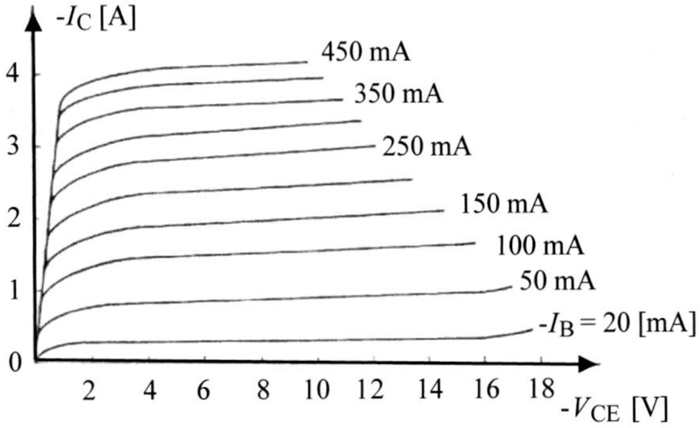

Fig. 4.2.36 CE current gain of a power BJT

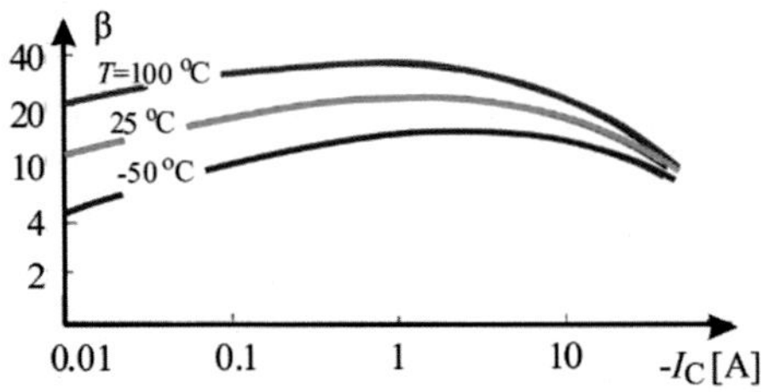

Fig. 4.2.37 Safe operating area of a power BJT

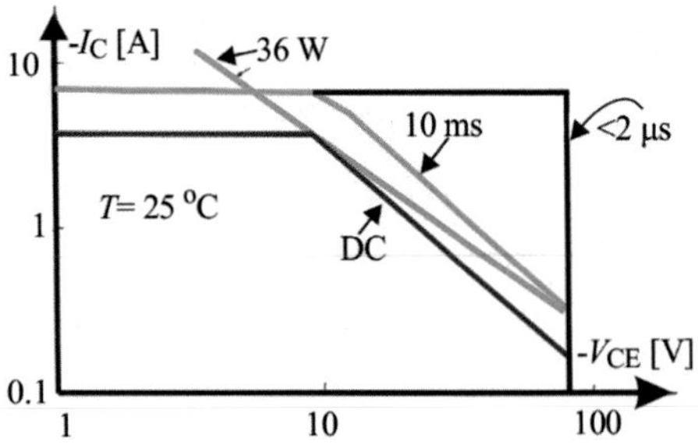

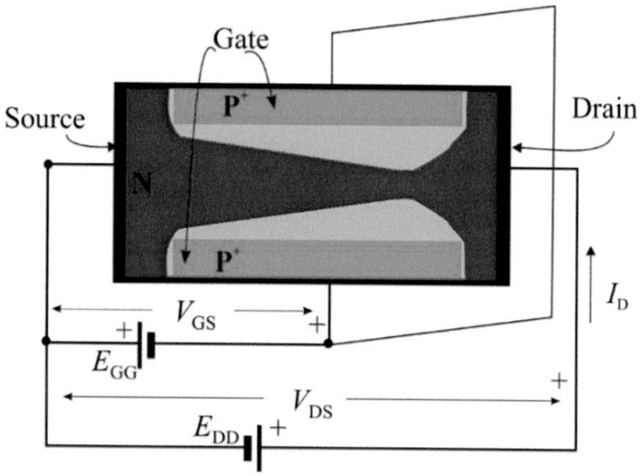

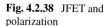

Fig. 4.2.38 JFET and polarization

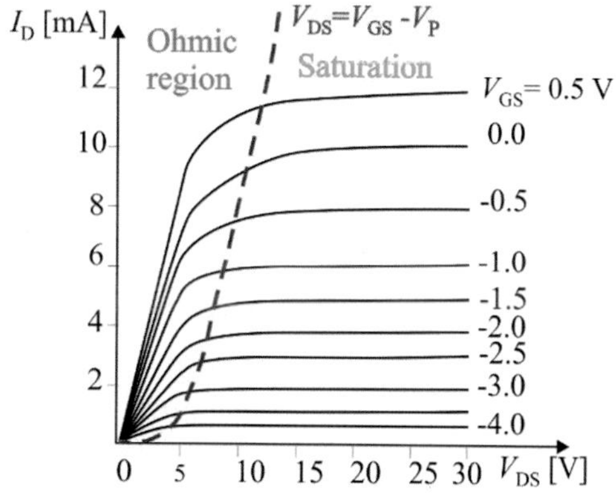

Fig. 4.2.39 Output characteristics of an N-channel JFET

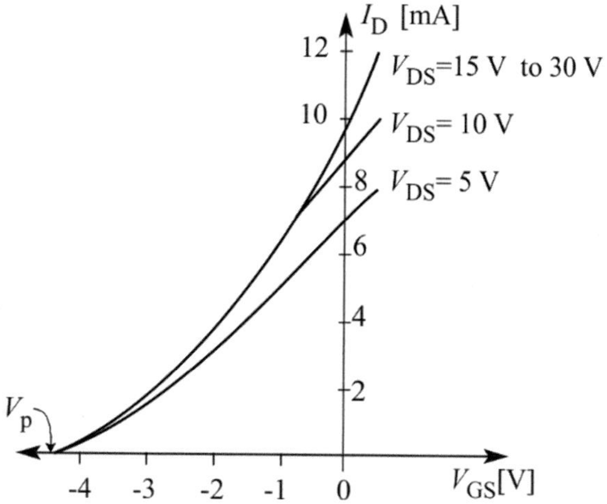

Fig. 4.2.40 Transfer characteristics of an N-channel JFET

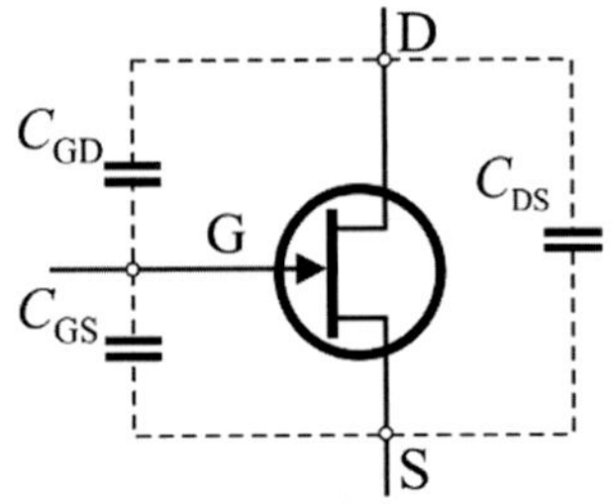

Fig. 4.2.41 Capacitances of an N-channel JFET

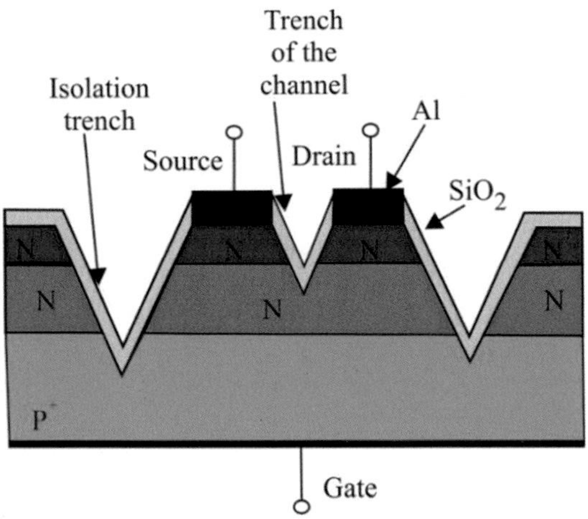

Fig. 4.2.42 Cross section of a power JFET with V-groove

4.2.3.2 Junction Field Effect (JFET) Power Transistors

JFET, since when it is built into an amplifier it produces smaller gain than a bipolar transistor, is less commonly used. However, it is characterized by two very important features that make it suitable for use in certain situations. First of all, its high input resistance is of great importance, so that the excitation circuit of the power amplifier does not have to generate a large power but should be a voltage amplifier ($I_{in} = 0$). In addition, the temperature coefficient of the drain current is negative, which allows the construction of a single component in the form of a parallel connection of a series of individual ones. If the temperature coefficient is negative, the component that has a higher current in the first moment heats up more, which results in a reduction of the current so that both components become equally influential.

In the next a short repetitorium on the JFET will be given. The main theory was developed in LNAE_Book_1.

The structure (in principle) and of the polarization of an N-channel JFET is depicted in Fig. 4.2.38. Its output characteristics are given in Fig. 4.2.42 and are modeled as follows.

The model of the JFET is expressed by several expressions characterizing parts of its characteristics as depicted in Fig. 4.2.39.

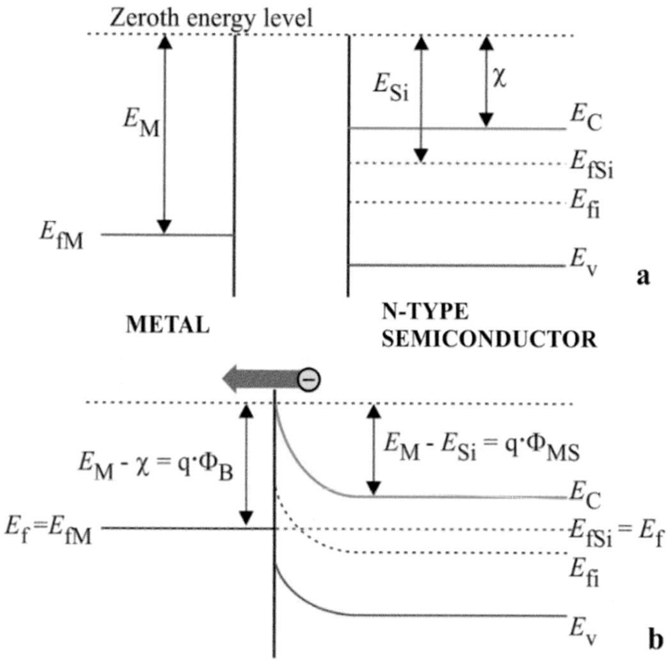

Fig. 4.2.43 Energy bands in a metal-semiconductor contact. **a** Before contact and **b** after contact

Fig. 4.2.44 Energy band diagram of a p-n junction on a single semiconductor: homojunction. **a** Before and **b** after junction

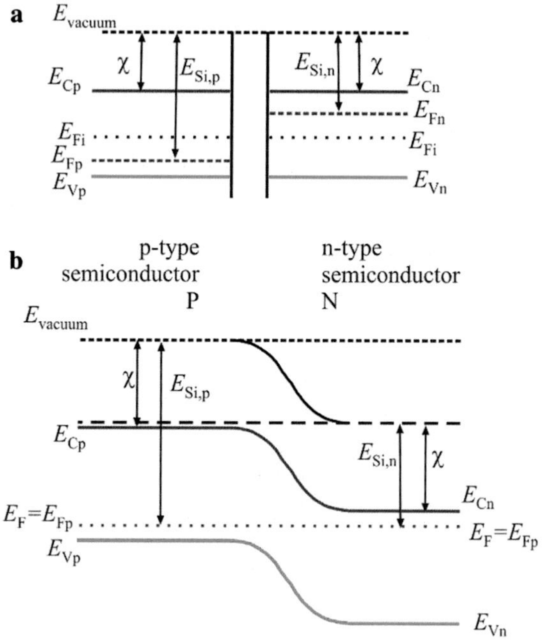

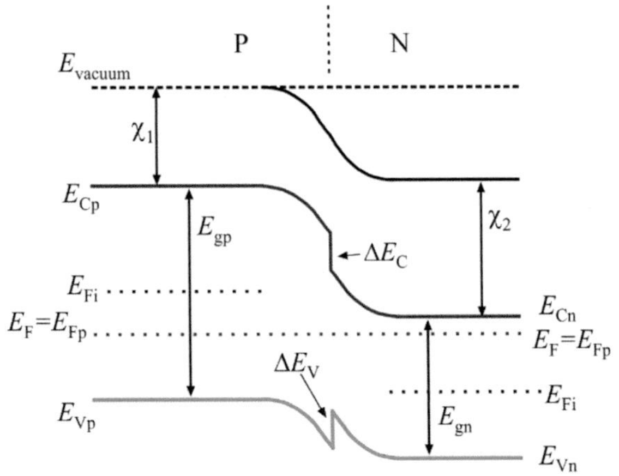

Fig. 4.2.45 Energy band diagram of a p-n junction of two semiconductors with different energy gaps (heterojunction)

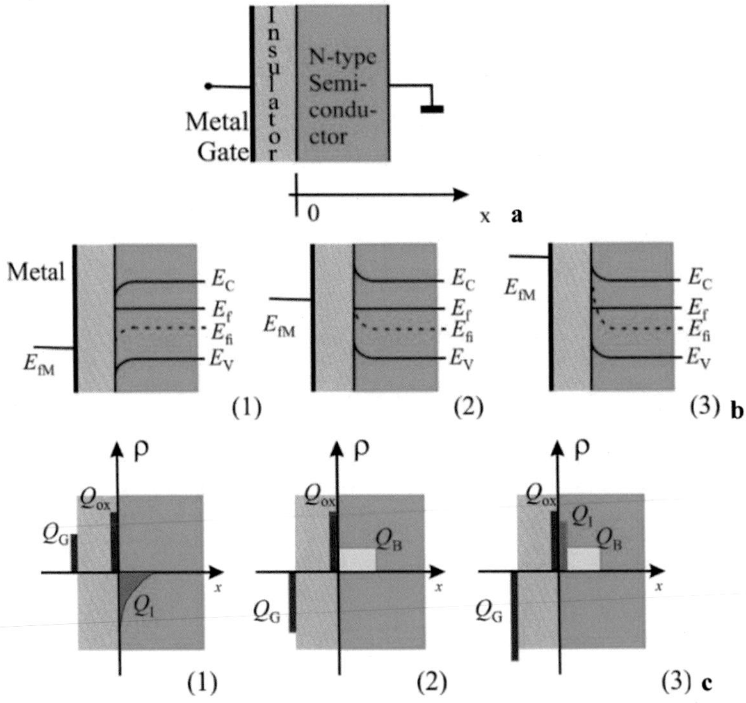

Fig. 4.2.46 MOS structure. **a** Cross section through the structure, **b** energy diagrams, **c** charges, (1) accumulation, (2) depletion, and (3) inversion

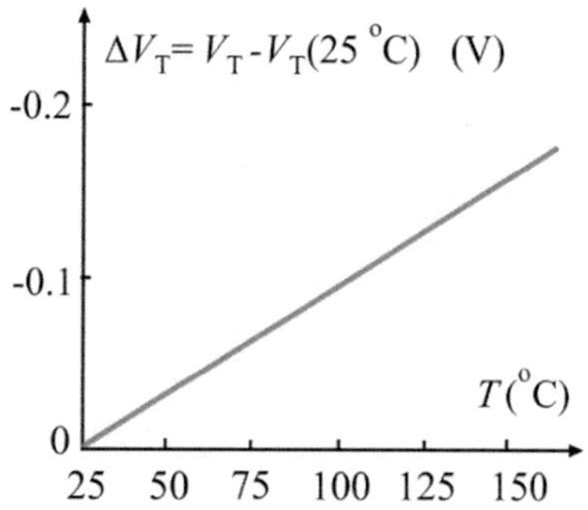

Fig. 4.2.47 Temperature dependence of the threshold voltage

For the ohmic region one use

$$I_D = I_{DSS}\left\{2\left(1-\frac{V_{GS}}{V_p}\right)\cdot\frac{V_{DS}}{-V_p}-\left(\frac{V_{DS}}{V_p}\right)^2\right\} \tag{4.2.31}$$

where V_p stands for the pinch-off voltage explained by the JFETs transfer characteristic given in Fig. 4.2.40 while I_{DSS} is the saturation current for $V_{GS} = 0$ which can be estimated from the same figure. This expression is valid for both N- and P-channel devices provided that V_p is negative for the N-channel and positive for the P-channel component.

Then, for the saturation region one use

$$I_{DS} = \begin{cases} I_{DSS}\left(1-\frac{V_{GS}}{V_p}\right)^2 & \text{za } |V_{GS}| < |V_p| \\ 0 & \text{za } |V_{GS}| > |V_p| \end{cases}. \tag{4.2.32}$$

To complete the picture, Fig. 4.2.42 depicts the capacitances of the JFET. Here, C_{GD} and C_{GS} are the junction capacitances of the gate-to-channel junction while C_{DS} is a very small parasitic capacitance. When used in amplification purposes the most important is the input capacitance expressed as

$$C_{in} = C_{GS} + (1-A)\cdot C_{GD}, \tag{4.2.33}$$

where A is the local gain. This makes C_{GD} the most important one.

Due to the negative temperature coefficient of the JFET, the power version of it may be formed as parallel connection of several low-power JFETs. There are several structures that allow the creation of JFET as a parallel connection of several transistors with a short channel. One of these structures is VJFET, i.e., JFET with V-groove.

This component is not planar - the gate is located on the opposite side of the pellet. The cross section is shown in Fig. 4.2.42. The drain current is established laterally. An N-type epitaxial layer (channel) first grew on the P^+ substrate (gate). In order to apply the metal contacts of the source and drain, a thin N^+ layer was applied on top of it. Two types of grooves were formed by chemical etching. The first (shallower) defines the area of the channel (separates the source and drain), and the second isolates one individual structure from the others. Seen from the top, source and drain look like combs whose teeth enter each other similar to the emitter and base interdigital structure of Fig. 4.2.32.

One such component, for example, has a total surface area of 0.2 mm^2 and consists of thirty channels 3 μm long. The bottom of the channel groove is only 1 μm away from the P^+ area. A breakdown voltage between the source and the drain of 37 V was obtained. The saturation current of the transistor was 270 mA. The cut-off frequency of this transistor was 12 GHz, and the maximum power was 4 W.

In order to obtain a power JFET, it is necessary to have a high breakdown voltage between the drain and the gate and as short a channel as possible, i.e., small resistance of the channel. There are some problems with this. Let us first consider the problem of gate breakdown. In a normal transistor, the inverse saturation current of the gate is less than nano-amperes. When multiple transistors are connected in parallel, it increases, which is not critical. In addition, however, the increase in the gate current is influenced by an additional component of the current of the secondary carriers from the channel, which can significantly increase the inverse current of the gate junction and thus significantly increase the probability of avalanche breakdown at the gate-drain end of the junction. The current of the secondary carriers in the channel is formed at high voltages on the drain and at short channels due to impact ionization. In the N-channel transistor, the electrons formed during ionization participate in the drain current while the holes are accepted as secondary carriers at the gate-channel junction, thus increasing the gate current.

When the gate is inverse biased, the depleted area expands into the channel until it "gets pinched." Then saturation occurs.

If the width of the channel in the absence of external voltage at the gate narrows to zero, then in order for the drain current to flow, a certain drain voltage is required so that the output characteristics, initially, have a slope equal to zero. After overcoming this barrier, JFET acts as a resistor so that the characteristic has the shape of a straight line with a positive slope. If the voltage at the gate is higher, the drain current starts flowing later (the characteristic moves translationally to the right).

In the next, starting again with fundamental physics, we will try to come to the properties of the MOS transistors and to its model.

4.2.3.3 Isolated Gate Field Effect (IGFET) Power Transistors

MOS (metal–oxide–semiconductor)-based devices are dominant for switching purposes nowadays which primarily came to use in digital electronics. Their use in power switching is based on two important properties:

1. No current is needed to keep the transistor in ON or OFF state, and

2. The device current has a negative temperature coefficient which allows for very large number of MOS transistors to be connected in parallel. So, in fact, a power MOS transistor is a large integrated circuit. The second property is appreciated also when the MOS transistors are implemented as linear components in power amplifiers since continuous operation is present and heating is intensive.

To understand the mechanisms and the properties of this device, however, deeper understanding of the device functionality is needed. One of the fundamental specifics of this device is the CIS (conductor-isolator-semiconductor) structure which is more frequently referred to as the MOS and is among the most influential discoveries in the electrical world of all times.

4.2.3.3.1 Metal-semiconductor Junction Revisited

The valuable properties of the Schottky diode were already discussed above. Its function is based on a metal-semiconductor junction and here we will discuss the energy diagram of it as the first example of a structure consisting of two materials with different values of their own Fermi levels.

Figure 4.2.43a depicts two separate energy diagrams. On the left is metal. E_{FM} stands for its Fermi level. E_M and E_{Si} (on the right) are work functions, i.e., the minimum energies needed for an electron to be removed from the metal or semiconductor, respectively, to a point in the vacuum immediately outside the solid surface. On the right is semiconductor (here n-type which is important for creation of a Schottky diode). χ is the affinity of the semiconductor which differs from the working function (E_{Si}) since in the interval between E_{Fn} and E_{Cn} there are no free electrons.

In Fig. 4.2.43b the metal and the semiconductor are brought together. The Femi level now is the same on both sides. Note that, however, at the surface of the semiconductor it is much lower than it was earlier. Now the semiconductor surface behaves almost as a p-type. To accomplish that, redistribution of the free majority carriers (electrons) in the semiconductor is due to happen. They migrate into the metal leaving a depleted region at the surface. A potential barrier is created of a height $\Phi_{MS} = (E_M - E_{Si})/q$ which stops further migration of electrons from the semiconductor into the metal. On the opposite side another barrier is formed of a value $\Phi_B = (E_M - \chi/q)$ which opposes the migration of the electrons from the metal into the semiconductor. While these two barriers are different in height, opposite is the difference in concentration of electrons in the metal and in the semiconductor, so that, statistically speaking, equal amount of current is contributed by both sides.

4.2.3.3.2 Semiconductor-Semiconductor Homojunction Revisited

The contact between two semiconductors having the same Energy gap is referred to as the homojunction. We will now discuss the energy diagram of the p-n homojunction which is usually implemented in semiconductors and especially in diodes.

Figure 4.2.44b depicts two separated semiconductors of different types having the same energy gap. The only difference is in the value of the Fermi level. When brought together (as in Fig. 4.2.44b), the Fermi level becomes equal on both sides and, consequently, bending of the energy levels at the borders of the gap is observed. A potential barrier is formed. The diagram presented here implies rather symmetrical junction. That means both sides have equal concentration of the majority carriers. That, however, is rarely the case. Usually, in order to achieve electrical symmetry (remember the difference in the electrical properties of the electrons and holes), the concentration on the p-side is much larger to compensate for the lower mobility of the holes. That, of course, is not the only reason to build non-symmetrical junctions. In many cases it is necessary to make a shallow junction (looking from the surface downward to the semiconductor) which imposes very high concentration in the top layer for the depletion region not to be extended in it.

Parts of the electronic device where concentration is very high are usually denoted by "+." In the opposite case, when the concentration is purposely made very low the proper part of the semiconductor is labeled "–." So, we have N^-, N, and N^+, and consequently P^-, P, and P^+, regions.

4.2.3.3.3 Semiconductor-Semiconductor Heterojunction

Now, going back to the energy diagrams, Fig. 4.2.45 depicts a hetero p-n junction. As can be seen the p-type semiconductor has larger energy gap than the n-type. As a consequence, at the junction border discontinuity of the energy diagram may be observed.

The value of the bandgap of a material is directly related to its dielectric constant while the latter is defining the light speed in a semiconductor. This means that the light passing from one to the other side will be diffracted or even refracted (depending on the incoming angle). This is the basic phenomenon that was implemented in lasers and other optoelectronic devices. For discovery of the semiconductor superstructures in the year 2000 the Nobel Prize in physics was awarded jointly to Kroemer, H. and Alferov, Z. I. The importance of the heterojunction for power transistor development will be discussed later on. Note, the heterojunction depicted in Fig. 4.2.45 is referred to as the "straddling gap" one. Depending on the mutual positions of the energy bands two more versions of heterojunctions may arise which are not of importance at this stage of the proceedings.

4.2.3.3.4 MOS Structure

Figure 4.2.46 depicts the fundamentals of the MOS structure.

In Fig. 4.2.46a the cross section is depicted first. The structure consists of a layer of conducting material (usually metal but frequently also a highly doped polycrystalline silicon), an insulator (usually an oxide of the basic semiconductor), and the very semiconductor (here n-type silicon) which is usually referred to as the substrate.

In Fig. 4.2.46b the energy diagrams while in Fig. 4.2.46c charge distributions are depicted. Note that due to the nature of the structure of the lattice of the oxide, a positive charge is inherent to it. This charge, here denoted as Q_{ox}, is usually located near to the semiconductor surface (its value is of the order of 10^{11} positive ions per cm^3. Note, all charges hereafter and capacitances will be expressed per unit area.).

Three special cases are presented. With (1) the state of accumulation is denoted. In this case, for an n-type semiconductor, positive charge is brought to the conducting electrode here denoted Q_G as depicted in Fig. 4.2.46c (in a transistor based on this structure the conductor is named gate hence the letter G in the index.). It is attracting electrons from deep of the semiconductor, and they become accumulated at the surface hence the term accumulation. The accumulated charge is here denoted as Q_I and is negative. In this state the semiconductor surface has much higher conductivity than the rest of it while the carriers are of the same type. Note, Q_{ox} helps accumulation in an n-type substrate and hinders it in a p-type.

When negative charge of a proper value is brought to the gate the majority carriers will be repelled from the surface of the semiconductor and it may be depleted to the level of an intrinsic semiconductor. That is shown in case (2) in Fig. 4.2.46. This state of the surface is called depletion. Note, in an n-type semiconductor Q_{ox} hinders the depletion. In the state of depletion positive (in n-type substrate) ions of the donor atoms are left and an immobile charge is created at the surface. That is denoted as Q_B in Fig. 4.2.46c. The resistivity of the surface is now very high.

Finally, if large amount of negative charge is brought to the gate [Fig. 4.2.46c case (3)] minority carriers are attracted from deep of the substrate to change the type of conductivity of the surface. The surface becomes inverted. In such a case, for n-type substrate, the surface behaves as a p-type semiconductor. The accumulated charge, Q_I, is now positive. The larger negative (for n-type substrate) Q_G, the larger Q_I and the conductivity of the surface since larger number of minority carriers will be available for conduction. A very important side effect may be observed in this case. Namely, since the surface became a p-type semiconductor it forms a p-n junction with the substrate. So, a diode is created as a consequence of inversion. If inverse biased, this diode will keep the conducting p-type surface isolated from the substrate.

4.2.3.3.5 Threshold Voltage

So far, we were dealing with energies. It is usual in microelectronics to have voltage equivalent of the energy obtained by division with the electron charge, i.e., $V = E/q$.

It is usual to pronounce the surface inverted when the Fermi level at the surface [see Fig. 4.2.46b case (3)] is as much above its intrinsic value as it is below deep in the substrate, i.e., if $(E_{Fi} - E_F)_{surface} = (E_F - E_{Fi})_{substrate}$. This is referred to as the strong inversion. The gate voltage (with reference to the substrate (see Fig. 4.2.46a) needed to achieve strong inversion is called threshold voltage and denoted by V_T. As may be deduced from the Fig. 4.2.46 its value has several components. First, one is to overcome the usual barrier arising at the contact of two materials with different work function (like in the case of the Schottky barrier). This part of the threshold voltage will be denoted by Φ_{MS} (potential between metal and semiconductor). According to Fig. 4.2.46 and Fig. 4.2.43 its value is $\Phi_{MS} = (E_M - E_{Si})/q$. For n-type silicon with normal concentration of donor impurities its value is approximately −0.3 V. The second component of V_T is related to Q_{ox}. The part of V_T needed to overcome Q_{ox} is equal to $-Q_{ox}/C'_{ox}$, C'_{ox} being the capacitance of the sandwich per unit area ($C'_{ox} = \varepsilon_{ox}/t_{ox}$, ε_{ox} being the dielectric constant of the oxide and t_{ox} being its thickness.). Further on, to achieve strong inversion a voltage of the value of $\Phi_B = -2 \cdot \Phi_F = -2 \cdot (E_F - E_{Fi})/q$ is to be applied. Finally, depletion of the surface is to be achieved by implementing a voltage of $-Q_B/C'_{ox}$. Its value depends on the concentration of the majority carriers in the substrate, and using half of the width of the depletion region under no voltage condition and with a voltage barrier equal to Φ_B one may get

$$Q'_B = \sqrt{2 \cdot \varepsilon_{Si} \cdot q \cdot N_D \cdot (-\Phi_B)}. \tag{4.2.34}$$

So, the complete expression for the threshold voltage for an n-type substrate under the condition of no voltage applied (between the surface and the substrate) is given by

$$V_T = \Phi_{MS} + \Phi_B - Q_{ox}/C'_{ox} - Q'_B/C'_{ox}. \tag{4.2.35}$$

In the case when a potential difference of V (V) exists between the surface and the substrate one uses the following

$$V_T = V_{T0} + B \cdot \left(\sqrt{|V - 2 \cdot \Phi_F|} - \sqrt{|2 \cdot \Phi_F|}\right) \tag{4.2.36a}$$

where, for n-type substrate,

$$B = \frac{\sqrt{2 \cdot q \cdot \varepsilon \cdot N_D}}{C'_{ox}}, \tag{4.2.36b}$$

$$V_{T0} = \Phi_{MS} - 2 \cdot \Phi_F - Q_{ox}/C'_{ox} - Q'_{B0}/C'_{ox}, \tag{4.2.36c}$$

and $Q'_{B0} = Q'_B$ (0) (as given by (4.2.34)).

Note, for n-type substrate the voltage needed to produce inversion is negative so is the threshold voltage.

It is important to have in mind that the value of V_T discussed so far was obtained under the condition of strong inversion. However, inversion starts with lower (absolute value) voltages and just after depletion starts there are minority carriers under the surface which are available for conduction. This state of the surface is referred to as sub-threshold, and, consequently, other quantities related to it bear that attribute.

An important property of the threshold voltage is its temperature dependence. V_T decreases with rise of temperature. An example of this dependence is given in Fig. 4.2.47 for a silicon structure.

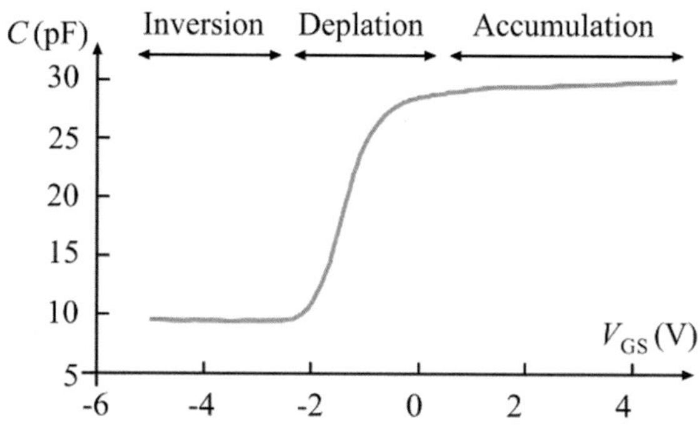

Fig. 4.2.48 Capacitance of a MOS capacitor at low frequencies

4.2.3.3.6 MOS Capacitance

The MOS structure depicted in Fig. 4.2.46a may be viewed as a capacitor since an insulator is sandwiched by two conducting materials. Note, the oxide film can be as thin as $t_{ox} = 1.5$ nm. We already mentioned the oxide capacitance as a representative of this structure. Its value per unit area is $C'_{ox} = \varepsilon_{ox}/t_{ox}$, e.g., 34.5×10^{-5} pF/(μm^2) for a structure with $t_{ox} = 100$ nm. By changing the potential of the gate, however, the conductive state of the surface is changed. So, if the surface is going from accumulation into depletion, the plate of the capacitor on the substrate side is moving farther from the oxide (deeper into the substrate) and the capacitance is decreased. With formation of the inverted layer, we may observe two capacitors connected in series in between the gate and the substrate. The first one is the capacitance of the oxide (its electrodes are the gate and the inverted layer) while the second is the capacitance of the p-n junction formed between the inverted layer and the substrate. The latter is voltage (and frequency) dependent as explained earlier for the junction capacitance (in LNAE_Book_1). The overall dependence at low frequencies (the high-frequency case will be not discussed here) is depicted in Fig. 4.2.48 for a structure with $N = 4 \times 10^{15}$ cm^{-3}, $t_{ox} = 80$ nm, and area $S = 10^{-4}$ in^2.

4.2.3.3.7 MOS Transistor

A silicon MOS transistor, usually named Metal–Oxide–Semiconductor Field Effect Transistor (MOSFET - to emphasize its functionality which is based on the effect of the electric field established between the gate and the semiconductor), is obtained when at the ends of the MOS structure (looking parallel to the surface) additional diffused parts are added to serve as transition to the contacts. In that way a four terminal device arises as depicted in Fig. 4.2.49.

The two new parts are named source (S) and drain (D) despite the fact that the component is fully symmetrical. Namely, by proper reconnection of the terminals or proper biasing of the device the role of these two parts becomes different. It is important to have in mind that these two parts are serving as "transition" between

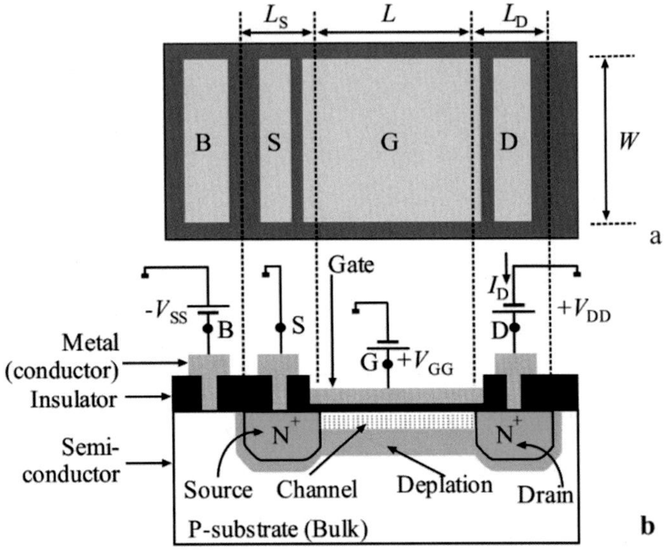

Fig. 4.2.49 N-channel enhancement type MOS transistor with normal polarization shown. **a** Top view and **b** cross section

the inverted surface (named channel) and the metal deposited at the surface which is enabling wiring of the device. To that end these two areas are to be degenerated to allow for the contacts between the metal and semiconductor to have very low resistance (ohmic contact).

The figure shows the state of normal polarization. That means that the potential at the gate is large enough to provoke inversion of the surface beneath, and both source and drain junctions are inversely biased. To that end the substrate (frequently named bulk and denoted by B), for an N-channel device, has to be the part brought to the lowest potential of all. That, in addition, inversely biases the junction between the channel and the substrate denoted in the figure as depletion. Now we may conclude that the structure source-channel-drain, which is forming a path for current conduction, is fully isolated from the bulk which is achieved without additional technological measures. This property is unique to MOS transistors and allows for many mutually isolated transistors to be built in the vicinity to each other or, in other words, allows for high packing density which is of crucial importance for integrated circuit production.

Note, only one type of carriers is contributing to the current in the channel so that these devices are referred to as unipolar.

Figure 4.2.49a depicts the top view and the dimension notation. L is the channel length while W is its (and transistors) width. The area defining the MOS capacitance is $L \cdot W$.

In addition to the MOS capacitance there are more as depicted in Fig. 4.2.50. First, in order to fight tolerances in the production process, there is overlap between

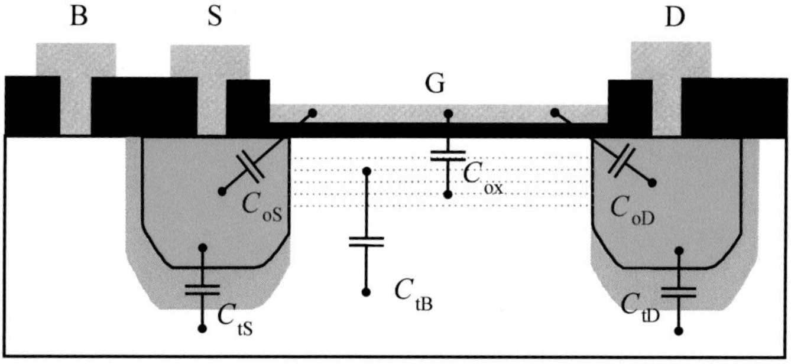

Fig. 4.2.50 Capacitances of the MOS transistor

the gate conducting material and both the source and the drain. These are named overlap capacitances C_{oS} and C_{oD}. Then, there are the depletion capacitances of the source and the drain denoted as C_{tS} and C_{tD}.

Note, when discrete transistors are made for the market, usually the source is short-circuited to the bulk so a three-terminal device is delivered. In such cases only C_{tD} is of importance. Finally, there is a depletion capacitance between the channel and the bulk (C_{tB}) which adds to the MOS capacitance as shown in Fig. 4.2.50. When the channel and the source are short-circuited so is this capacitance which is one of the reasons for the two to be brought together. The other reason is the fact that the threshold voltage, as given by (4.2.34a), is no longer voltage dependent and may be calculated simply from (4.2.35).

Similar to the difference of importance of C_{tS} and C_{tD} is the difference between C_{oS} and C_{oD}. Namely, C_{oD} establishes a feedback path from the drain (being the output terminal of the device) to the gate (being the input terminal of the device). This capacitance is now seen from the gate multiplied G times, G being the incremental gain of the circuit in which the transistor is embedded. This is called the Miller effect, and C_{oD} is the Miller capacitance. Since the gain is signal dependent this capacitance is seen to be highly nonlinear, and since the value of G may be large, it may go through large values even though its nominal value is very small (of the order of fF).

Before proceeding to the current of the MOS transistor we will note that the one shown in Fig. 4.2.49 is attributed as enhancement type. Namely, in this type of devices, in order to create a conducting path from the source to the drain (channel) the gate voltage has to reach a substantial value, in some cases of several volts. To overcome this an additional process is introduced in the MOS technology, namely implantation. By this process before creating the thin oxide a shallow implantation with impurities creating a semiconductor of opposite type to the bulk is performed so that the surface is inverted. Now, depending on the "dose" of implantation the threshold voltage may be "programmed." This is referred to as depletion type of a device.

Note, to disable conduction between the source and the drain the gate voltage has to be lower than the threshold voltage (absolute values are considered) no matter the type of the device (depletion or enhancement).

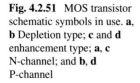

Fig. 4.2.51 MOS transistor schematic symbols in use. **a**, **b** Depletion type; **c** and **d** enhancement type; **a**, **c** N-channel; and **b**, **d** P-channel

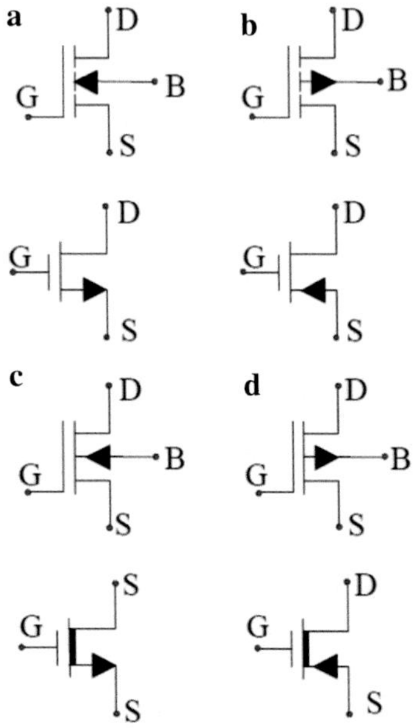

Fig. 4.2.51 depicts the schematic symbols in use for MOS transistors.

When voltage is brought to the drain with respect to the source (positive for N-channel device as depicted in Fig. 4.2.49 current is established. We will consider here only the case when the source is short-circuited to the bulk since discrete devices are of interest. In that case ($V_S = V_B = 0$) for the current at low voltages between the drain and the source one use (for a P-channel device)

$$I_D = 2 \cdot \beta \cdot \left\{ \begin{array}{l} \left(V_{GS} - V_{T0} - \frac{1}{2} V_{DS}\right) \cdot V_{DS} \\ -\frac{2}{3} \cdot B \cdot \left[(-V_{DS} - \Phi_B)^{3/2} - (-\Phi_B)^{3/2}\right] \end{array} \right\} \tag{4.2.37a}$$

where

$$\beta = \frac{\mu_p}{2} C'_{ox} \cdot \frac{W}{L} \tag{4.2.37b}$$

and V_{T0} and B are given in (4.2.36). Note, for convenience, the letter β is frequently substituted by A.

It is easy to find out from this expression that the derivative of the current with respect to the gate voltage (the transistor's transconductance) is a function of the drain voltage and of β. In some references instead of (4.2.37b) one use $\beta = \mu_p \cdot C'_{ox} \cdot W/L$ and (4.2.37a) is corrected accordingly. For example, for an N-channel device with $t_{ox} = 10$ nm, $\mu_n = 350$ cm^2/(Vs), and $V_T = 0.7$ V, one gets $\beta = 60 \cdot (W/L)$ μA/V^2. This means that the transconductance (and the current) for a given biasing is larger for wider transistors and smaller for the longer ones. That means that it is possible to

Fig. 4.2.52 Transfer characteristics of two MOS transistors

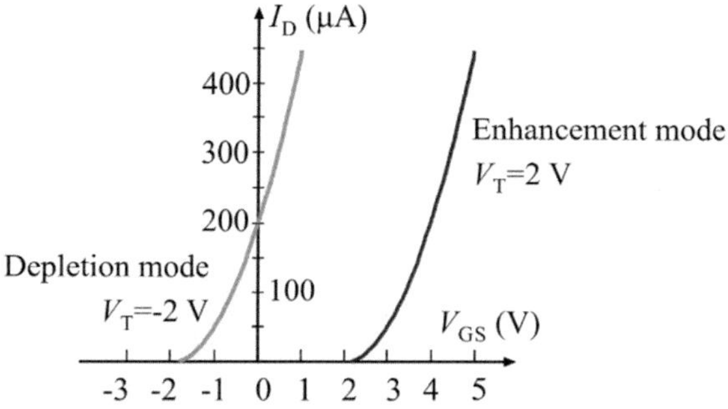

keep the gate capacitance constant by keeping the area $W \cdot L$ constant while changing the transconductance in a wide interval. So, for high-current devices large W/L is needed while devices which should act as high resistances will use small W/L. If one is after a high-current device bearing high voltages, however, one is to apply additional measures to reach the goal in order to avoid very large gate capacitances.

If the charge in the depleted region (Q_B) is considered constant, a simplification of (4.2.37a) may be produced as (Fig. 4.2.52)

$$I_D = 2 \cdot \beta \cdot \left\{ (V_{GS} - V_{T0}) \cdot V_{DS} - \frac{1}{2} V_{DS}^2 \right\}. \tag{4.2.37c}$$

The mobile charge in the channel may be expressed as

$$Q_I' = -C_{ox}'(V_{GS} - V_T - V)) \tag{4.2.38}$$

where V is the potential at a point in the channel measured from the source (which is connected to the bulk). $V(0) = 0$ V and $V(L) = V_{DS}$. This means that with the rise of the potential starting from the source the mobile charge is reduced. At the drain, if $V_{DS} = V_{GS} - V_T$, it disappears. The channel is pinched-off, and for higher drain voltages a depletion (of very high resistance) is seen in series between the channel and the drain. Due to that resistance the current stops to rise, i.e., saturates. The corresponding drain voltage is called the saturation voltage $V_{DSsat} = V_{GS} - V_{T,}$ and the drain current is called the saturation current (V_{DSsat} substituted in (4.2.37c) in place of V_{DS}):

$$I_D = A \cdot (V_{GS} - V_T)^2 = I_{DSS} \cdot (1 - V_{GS}/V_T)^2 \tag{4.2.39}$$

with $A = 2 \cdot \beta$ and $I_{DSS} = A \cdot (V_T)^2$. Equation (4.2.39) now relates an input and an output variable which is why it is referred to as the transfer characteristic of the MOS transistor. Figure 4.2.52 depicts transfer characteristics of an enhancement mode and a depletion mode transistor.

The above developments may be stated as basic. There are many secondary effects which come in fore in different situations. Here we will discuss some of them directly related to the drain current dependence on the drain-to-source voltage.

With the rise of the drain-to-source voltage above V_{DSsat} the width of the depletion region between the drain and the channel continues to rise so shortening the channel and consequently rising the field seen between its ends. That means that the velocity of the carriers is risen and consequently the current rises. To avoid complex expressions, it is usual this effect to be inserted into the expression for the drain current (above V_{DSsat}) in the following way (for N-channel device):

$$I_D = A \cdot (V_{GS} - V_T)^2 (1 + \lambda \cdot V_{DS}) \tag{4.2.40}$$

where λ (V^{-1}) is a parameter. Note $I_D = 0$ for $V_{DS} = -1/\lambda$ no matter the value of V_{GS}. This means all lines representing the characteristic beyond saturation converge into a single point on the negative V_{DS} axis. That is the so-called Early voltage for MOS devices. The value of λ is of the order of 10^{-2} V but may be larger for power components.

With further increase of the drain-to-source voltage avalanche breakdown of the drain-to-bulk p-n junction may be expected. This limits the maximum voltage between the source and the drain. This part of the characteristic is frequently approximated so that the drain current (say Eq. (4.2.40)) is *divided* by the following quantity:

$$(1 - V_{DS}/BV_{DB})^{2 \cdot m} \tag{4.2.41}$$

where BV_{BD} is the breakdown voltage (of the same sign as V_{DS}) and m is an even natural number describing the steepness of the characteristic.

Having all these considerations in view Fig. 4.2.53a depicts the so-called output characteristics of a MOS transistor. In the field of output characteristics, one may recognize the following regions.

- *Cut-off*. The region of zero valued currents below the threshold voltage. One is not to forget, however, the sub-threshold current which may become of importance in some real-life situations.
- *Active region*. Part of the field where saturation occurs. When in this region the transistor is "normally biased." It is characterized by high transconductance and hence with high gain. One is to be aware that in this region the product $P_D = V_{DS} \cdot I_{DS}$ is large and is the cause of heating the device. Its maximal value is, of course, limited by P_{Dmax} which is an important parameter of any power transistor.
- *Ohmic region*. This part is characterized by strong dependence of the current of both gate and drain voltages.
- *Breakdown region*. Here the drain current becomes as large as allowed by the outside circuitry.

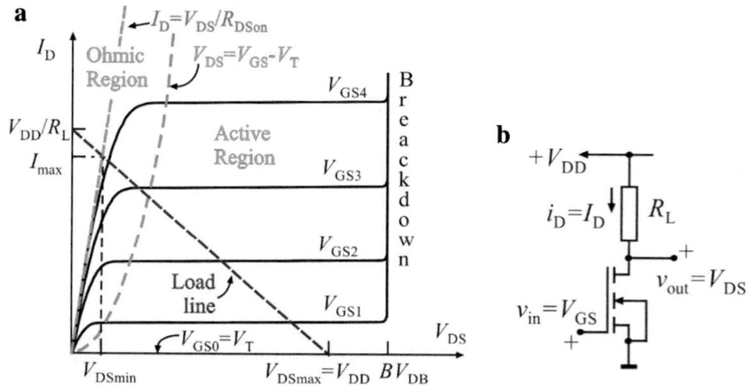

Fig. 4.2.53 **a** Output characteristic of a MOS transistor. In case of N-channel $V_{GS4} > V_{GS3} > V_{GS2} > V_{GS1} > V_T$. **b** Inverter with an NMOS enhancement type transistor

The Ohmic and the active regions are separated by a line on which $V_{DS} = V_{GS} - V_T$, as shown.

One is to note the dashed (blue) line marked as $I_D = V_{DS}/R_{DSon}$. It is the tangent with a slope of $\frac{dI_D}{dV_{DS}}\big|_{V_{DS}=0}$. The part of the field beyond this line is not accessible. The resistance R_{DSon} is an important parameter of the power MOS transistor, as will be explained in the next chapter.

Figure 4.2.53b depicts the simplest circuit built of an NMOS transistor of enhancement type and a load resistor (R_L). It is frequently referred to as inverting amplifier (for small-signal applications) or, in general, as an inverter. The instantaneous value of the output voltage of this circuit may be expressed as

$$v_{out} = V_{DD} - R_L \cdot i_D \tag{4.2.42a}$$

where i_D and v_{out} are instantaneous values. For steady state DC conditions, it becomes

$$V_{DS} = V_{DD} - R_L \cdot I_D. \tag{4.2.42b}$$

This expression is known as the *load line* for resistive loads (the dashed red line). The instantaneous values of the gate and drain voltages and the drain current are found at the interception of this line and the transistor's characteristics. It defines two important quantities for the circuit of Fig. 4.2.53b: the maximum and the minimum drain voltages. The maximum drain voltage in this case is simply $V_{DSmax} = V_{DD}$. It is reached when the transistor is cut-off. The minimum drain voltage is obtained at the interception of the load line and the R_{DSon} line as shown in Fig. 4.2.53a. Its value is

$$V_{DSmin} = \frac{V_{DD}}{1 + R_L/R_{DSon}}. \tag{4.2.43}$$

We may conclude now: the smaller $R_{DSon,}$ the smaller V_{DSmin}. Small V_{DSmin} is of great importance for the efficiency of power amplifiers and, of course, defines the minimal value of the transistor voltage when it is operated as a closed switch. Since high current is experienced at the point of interception (denoted as I_{Dmax}) the value of the product $P_{CS} = I_{Dmax} \cdot V_{DSmin}$ may be prohibitive for the implementation of the device (CS here stands for "closed switch").

4.2.3.3.8 The Power MOS Transistor

Transistors with an insulated gate are characterized by the unique property that their current gain, in stationary mode, is infinite, which means that the drain current can be large, and at the same time the input current (gate current) is equal to zero. However, when the transistor is transferred from off-state to conducting state and vice versa, it is necessary to manipulate the charge at the gate in order to form a channel or to make it disappear. Therefore, in short time intervals (during the transient mode), the charging and discharging current of the gate capacitor flows.

This feature of IGFET, at first glance, should give MOS technology a significant advantage over bipolar when high currents need to be sustained. Namely, when BJT is used to deliver high currents in stationary regime, it is necessary to provide, according to the magnitude of the current gain of the transistor, a high base current. For example (see Fig. 4.2.36), if $\beta = 20$, to produce a current of $I_C = 3$ A, $I_B =$ 150 mA is required. Based on this, we conclude that it is desirable to have a BJT with high current gain which means that smaller input current will be required.

The first idea to solve this problem is Darlington's pair, whose current gain is large. However, this coupling still has incomparably less current gain than IGFET.

Unfortunately, the disadvantage of IGFET is the large minimum voltage V_{DSmin} compared to the BJT saturation voltage $V_{CEmin} = V_{sat}$. As a consequence, IGFETs have significantly higher dissipation when it should act as a closed switch, i.e., when the drain current is high. Although some structures have been developed that, to some extent, alleviate this shortcoming (e.g., VMOS), BJT remains incomparably more successful in this respect.

To get high power, as already discussed, both high voltages and high currents are to be enabled.

Enhancement mode N-channel components (rarely P-channel) are usually used to make power transistors. The usual MOS structure we have considered so far can be used for low drain voltage transistors. Power components require high breakdown voltages and high currents [the constant $\beta = \mu_0 \varepsilon_{ox} W/(2L \cdot t_{ox})$ which multiplies the expression for the current to be large]. These requirements dictate the use of components with a short channel, with a built-in additional drain area of lower conductivity and with a vertical (instead of horizontal) structure.

As for the high currents several ideas are usually implemented. First, as already stated, the power MOS transistor is in fact a parallel connection of many single transistors to form a large integrated circuit. This allows for adding the currents while at the same time the gate capacitances are added. To achieve parallel connection of the transistor easier, usually a vertical structure is used in which the drain is common to all cells in the integrated circuit. In such a case the letter V may precede the shorthand for the transistor (e.g., VMOS).

Then, the dimensions of every single transistor are manipulated. To increase the channel width a cylindrical structure is used in which the source is of a circular (or hexagonal) shape surrounding from outside the channel, the drain being in the center.

Finally, the channel length is reduced. That is usually achieved by *double* diffusion from the same opening in the surface protective coating: first for the bulk and second for the source. That is why the device is dubbed DMOS. Channels (and bulks) shorter (thinner) than 1 μm are usually obtained. That means very high fields are experienced along the channel rising the velocity of the carrier to extreme values.

At this point two important phenomena may be observed. First, when high fields are applied additional scattering of the carriers near the surface takes place which reduced their mobility. One may approximate the new value of the mobility under these conditions by

$$\mu_{\text{eff}} = \mu_0 \cdot (6 \times 10^4 / K_{\text{surf}})^r, \qquad (4.2.44)$$

where K_{surf} is the surface electric field, μ_0 is the mobility at low fields ($K_{\text{surf}} < 6 \times 10^4$ V/cm), and r is a constant. For a MOS transistor $K_{\text{surf}} = -(V_{\text{G}} - V_{\text{T}} - V_{\text{D}}/2)/t_{\text{ox}}$. For silicon N-channel $r = 0.12$ and for P-channel, $r = 0.2$. A rough calculation shows that the surface mobility is twice as small as the bulk mobility for the same concentration of the majority carriers.

The other phenomenon is related to the saturation of the carrier velocity at high fields. Namely, the highest value of the average velocity of the carriers in a solid body is defined by their thermal movement and (to simplify) no field can force them to move faster. To keep the velocity constant with the rise of the field, the mobility is to decrease. The saturation velocity value is approximately $v_{\text{sat}} = 10^7$ cm/s for electrons for silicon (2.2×10^7 cm/s for SiC and 2.5×10^7 cm/s for GaN). The fields needed to reach these electron velocities at room temperature ($T = 300$ K) are about $K_{\text{s}} = 2 \times 10^5$ V/cm (see Table 4.2.1 and Fig. 4.2.54). Here comes the advantage of the device dabbed HEMT (high electron mobility transistor) which will be discussed next.

Due to the velocity saturation and the decrease of the mobility at high fields the expression for the drain current is changed. Instead of (4.2.40) one use

$$I_{\text{D}} = \frac{\varepsilon_{\text{ox}}}{2 \cdot t_{\text{ox}}} \cdot W \cdot (V_{\text{GS}} - V_{\text{T}}) \cdot (1 + \lambda \cdot V_{\text{DS}}). \qquad (4.2.45)$$

In other words, the transfer characteristic becomes linear which is of great importance for application in power amplifiers.

For better characterization of the power MOS transistors based on silicon, measured transfer characteristics of a representative component are depicted in Fig. 4.2.55.

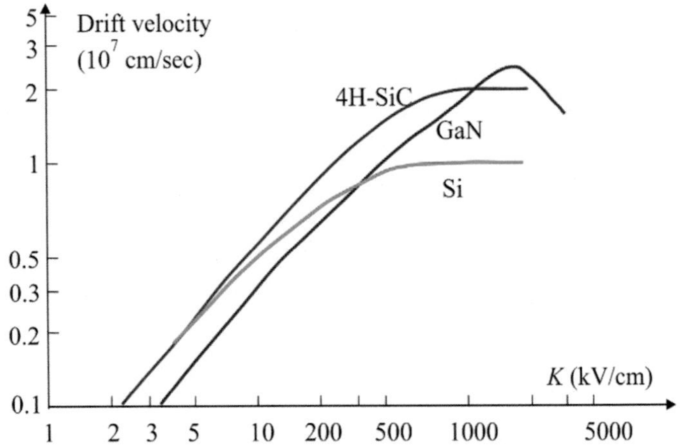

Fig. 4.2.54 Carrier velocity as a function of the electric field in a semiconductor slab

Fig. 4.2.55 Transfer characteristics of a power MOS transistor

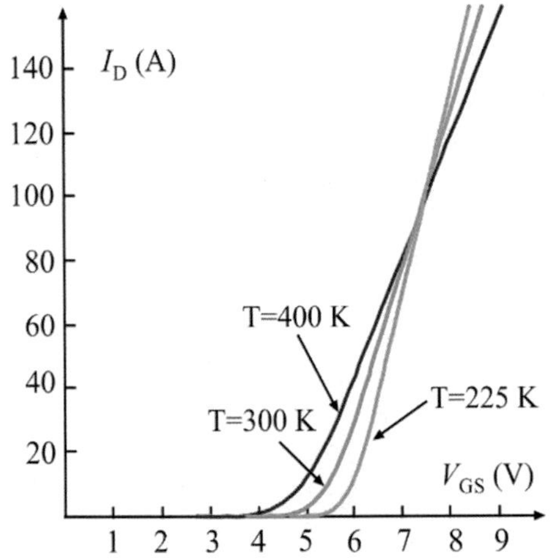

Let us consider now the need for a high-voltage device. To allow for high voltages at the drain terminal, the depletion region at the drain-to-bulk junction is to be forced to broaden toward the drain. In the opposite it would soon cover the bulk (the so-called punch-through effect normally present in bipolar transistors) and short circuit between the drain and source would happen. To achieve this the drain has two layers obtained by epitaxial growth. The N^+ layer needed to form the metal–semiconductor contact and the N^- layers (in which the rest of the component is embedded) allowing for unilateral junction so that the P (bulk)-N (drain) depletion region stretches itself toward the drain only.

Due to the high resistivity of the drain region λ in (4.2.45) is decreased meaning that the characteristics are nearer to horizontal.

To complete the picture on the properties of the power MOS transistor its gate will be addressed with some more details.

First, one is to bear in mind that the oxide thickness is extremely small which produces extremely high field on it. One volt on the oxide as thick as 10 nm produces electric field strength of 10^8 V/m which is rarely met in the "macro" world. Collecting charge to produce one volt at a capacitance of the order of magnitude of 10 pF is not a difficult thing to do. So extremely high care is to be applied when bringing signals to the gate to avoid gate-to-channel breakdown. Such a failure manifests itself as a punctual short circuit which allows a small current to flow from the gate to the channel so degrading or even fully disabling the function of the transistor. The datasheet rating for the gate-to-source voltage is usually between 10 and 30 V for most power MOS transistors.

As a first realization of this type of components we will shortly consider the lateral doubly diffused MOS (LDMOS) transistor depicted in Fig. 4.2.56. The term doubly diffused comes from the sequence of diffusion of the P- and N^+-type regions. These are created from the same opening at the surface, hence "double." This technological action enables very short channels to be created which is of crucial importance for getting higher current for the same drain voltage. Note that higher currents may be obtained by both enlarging the transistor's width and parallelizing several transistors. Both actions are feasible with the drawback that the input capacitance of the component is increased. The laterality here is mostly related to the possibility for this device to be integrated in an amplifier circuit—it is planar which means all contacts are from the same side. To allow for higher voltages, however, the drain is to be displaced. So that we have to have a structure similar to the collector of the power BJT. Namely a low concentration N^--epitaxial layer is used as a high resistive part of the drain. To facilitate the transport of the carriers in the drain (to reduce its surface resistivity) the gate metallization is extended over this N^- region.

There are several structures that realize a short-channel vertical MOS transistor. Among them, the most important are VMOS and DMOS transistors.

Alternatively, vertical components are in use. These, of course, are meant to be implemented as discrete devices. In relation to the horizontal (lateral) it has better temperature properties, better utilization of the silicon surface, and therefore a lower price. The channel must be as short as possible in order to obtain as much current as possible at the same voltages. Increasing the channel width, of course, achieves

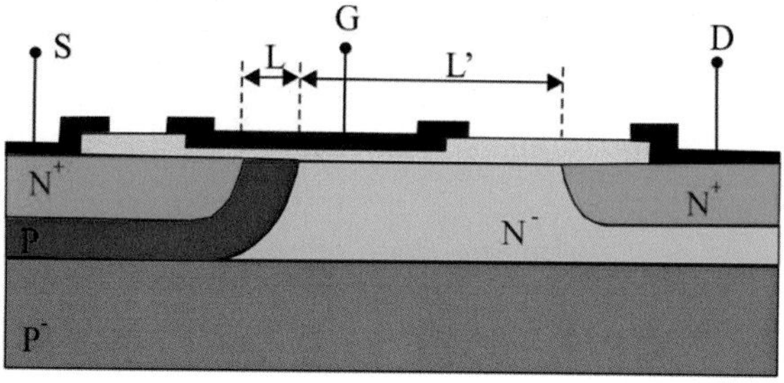

Fig. 4.2.56 Lateral doubly diffused MOS (LDMOS)

the same effect, but at the same time increases all the capacitances of the transistor. The drain region whose conductivity is lower (N^-) allows the depleted drain region to expand mostly toward the drain, which, similarly to bipolar transistors, increases the breakdown voltage. Source and substrate (here it is more convenient to use the name body instead of substrate to distinguish between epitaxial substrate) of this component are (again) diffused from the same opening of the mask (therefore the component is called DMOS—double diffuse MOS). In this way, a very short channel is created. Finally, the body is usually short-circuited for the source. Increasing the current is also achieved by reducing the threshold voltage, so that polycrystalline silicon is usually used as a gate.

As the first alternative we will here mention the VVMOS depicted in Fig. 4.2.57. It is an alternative to DMOS power transistors being V-shaped, hence VMOS. We distinguish between VVMOS—vertical VMOS and VUMOS—vertical UMOS, which means that the bottom of the groove is cut, so instead of V-groove we have U-groove. The structure of VVMOS is shown in Fig. 4.2.57. It is created similarly to DMOS, i.e., the N^- drain grows epitaxially on N^+ substrate. Then, by double diffusion, the body (P^-) and the source (N^+) are formed. Now, in the middle of the diffusion, a V-shaped groove is created by etching deep enough to fit into the N^- region. The surface of the groove is covered with oxide and then with the gate. As can be seen from Fig. 4.2.57, the channel is formed in the P region along the surface of the groove. In this way, a component with a very short channel can be obtained, the length of which can be determined precisely.

The most frequently used power MOS transistor nowadays is the VDMOS which is in fact a double diffused vertical MOS structure depicted in Fig. 4.2.58. Apart of the groove the structure is equal to the one of the VVMOS.

To illustrate, a typical VDMOS would have the following performances: maximum voltage $V_{DSmax} = 450$ V, maximum current $I_{Dmax} = 6$ A (higher current is allowed in pulse mode), maximum gate voltage $V_{GSmax} = 40$ V, and maximum dissipation $P_{Dmax} = 90$ W.

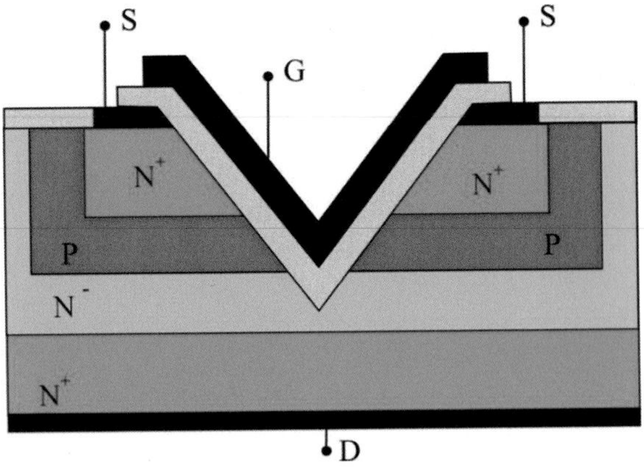

Fig. 4.2.57 Vertical V-groove MOS transistor (VVMOS)

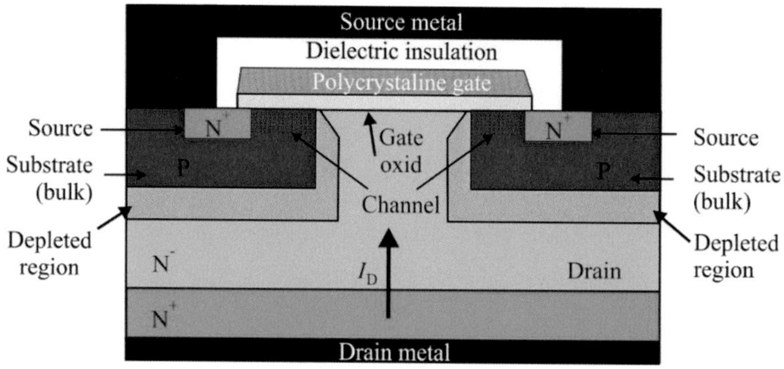

Fig. 4.2.58 Cross section (simplified) of an enhancement type N-channel VDMOS under normal polarization. Note, the source metal covers the bulk to provide for the short circuit between the source and the substrate

The characteristics of one power MOS transistor are shown in Fig. 4.2.59. In Fig. 4.2.59a output characteristics are given, and in Fig. 4.2.59b transfer characteristic (given for completeness). Figure 4.2.59c represents the dependence of thermal resistance on time (single pulse operation) for different ratios of pulse duration and signal period (duty cycles). It should be noted here that the thermal resistance is less than 1 K/W. Note, $D = 1$ here means DC regime of operation. Finally, Fig. 4.2.59d represents the safe operation area of the MOS power transistor.

Figure 4.2.60 depicts the fact that R_{DSon} depends on the characteristics on which the tangent is drawn for its definition. Namely, characteristics with smaller value of V_{GS} have lover slop at the origin leading to higher values of R_{DSon}. That may be deduced from (4.2.37a) after taking the proper derivative. A typical value of 20 mΩ may be observed here while transistors with a lower value may be frequently encountered. The figure also shows that R_{DSon} is a function of temperature, too.

With reference to Fig. 4.2.50 the following interterminal capacitances are present in the MOS transistor:

- C_{GS}—the gate-to-source capacitance. It is mainly formed of two components C_{oS} and C_{ox}. Namely, in normal operation the channel at the drain side is pinched-off which disconnects the capacitance from the drain. So, the full value of C_{ox} (which is almost constant) is connected to the source. Note, when in ohmic region the channel is complete and C_{ox} is shared between the source and the drain (one half to each of them).
- C_{GD}—the gate-to-drain capacitance. In normal polarization it is only C_{oD}. Half of C_{ox} is added when in ohmic region.
- C_{DS}—is the drain-to-bulk (being connected to the source) depletion capacitance.

In device datasheets, however, quantities obtained from these are evaluated and reported. Namely the input capacitance C_{ISS} ($= C_{GS} + C_{GD}$), the reverse transfer capacitance (or Miller capacitance) C_{RSS} ($= C_{GD}$), and output capacitance C_{OSS} ($=C_{DS} + C_{GD}$) are given.

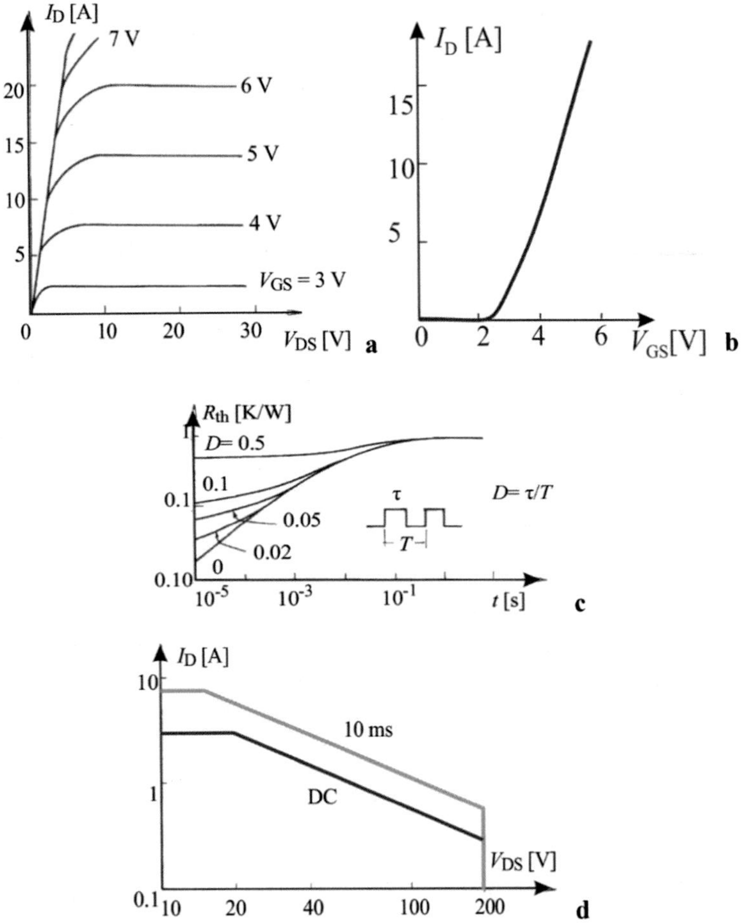

Fig. 4.2.59 Characteristics of a power MOS transistor. **a** Transfer characteristic, **b** output characteristics, **c** time dependence of the thermal resistance, and **d** safe operation area

Table 4.2.3 depicts measured values of the transistor capacitances for Si, GaN, and SiC-based power transistors. The measurement being performed for different working conditions, the values from this table should be taken for qualitative comparison only. As can be seen the values differ in favor to GaN and SiC.

Among the capacitances discussed C_{ISS} is of special importance since it contains the Miller capacitance C_{GD}. Namely, for the current of the Miller capacitance of the circuit of Fig. 4.2.61a (assuming C_{GD} is voltage independent) one may write

$$\begin{aligned} i_C &= C_{GD} \cdot \frac{d}{dt}(v_G - v_D) \\ &= C_{GD} \cdot \frac{dv_G}{dt}\left(1 - \frac{dv_D}{dv_G}\right) = C_{GD} \cdot (1 - G) \cdot \frac{dv_G}{dt} \end{aligned} \quad (4.2.46)$$

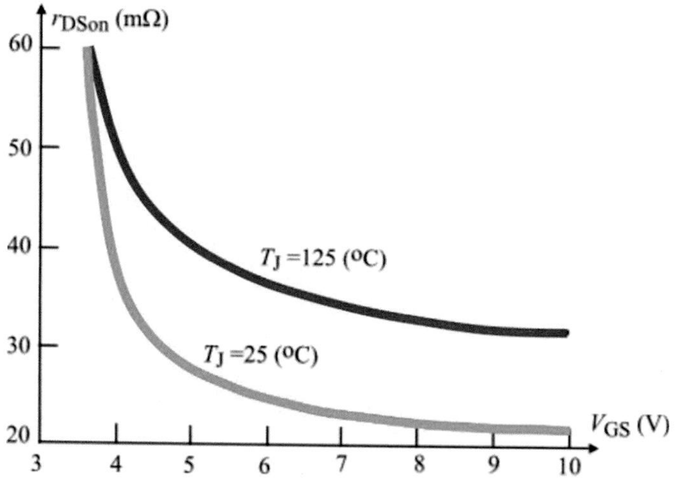

Fig. 4.2.60 R_{DSon} versus V_{GS} for two temperatures

Table 4.2.3 Power transistor capacitances

Type	C_{ISS} (pF)	C_{RSS} (pF)	C_{OSS} (pF)
SiC MOS (V_{DS} = 600 V)	660	4	60
GaN HEMT (V_{DS} = 360 V)	180	0.02	23
Si MOS (V_{DS} = 25 V)	2300	70	1000
Si IGBT (V_{AK} = 20 V)	5290	100	124

where v_D and v_G are node voltages of the drain and the gate terminal, respectively, and G is the incremental gain which is here negative. From this expression two important conclusions may be drawn. First, G is strongly dependent on the input voltage. Following the load line, it starts with $G = 0$ for $v_G < V_T$, goes through its maximum value for values of v_G and v_D belonging upper edge of the active region, and falls again to zero in the saturation region. That is depicted in Fig. 4.2.61b (bottom line) for the circuit of Fig. 4.2.61a with no reactive elements included, i.e., for static or very slow changes of the input voltage. Consequently, the equivalent capacitance $C_{GD} \cdot (1 - G)$ is also strongly nonlinear.

Second, the value of the input current (for a given capacitance) is strongly dependent on the slope of the input signal or, in other words, on the speed of switching. This means that highest speeds will rise the input currents. To illustrate, in Fig. 4.2.62a, b the (overall) input current of the circuit of Fig. 4.2.61a as a function of time is depicted for $dv_G/dt = k = 100$ V/ms and $k = 500$ V/ms, respectively. Significant rise of the peak value of the input current may be observed.

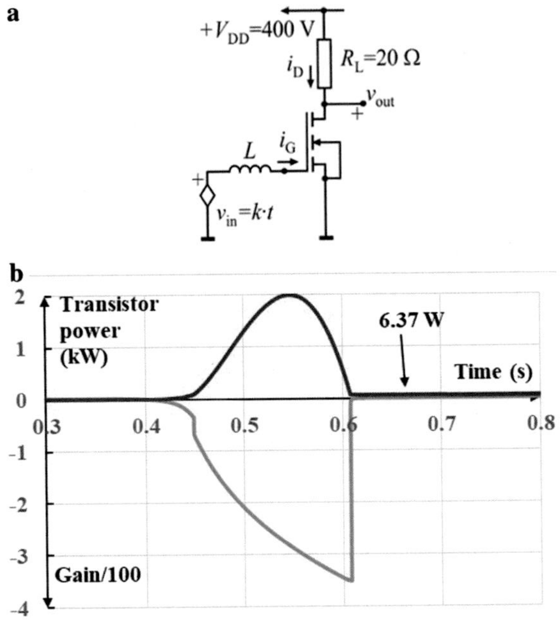

Fig. 4.2.61 **a** Test circuit. **b** Static (no reactive circuit elements included) incremental gain (G) of the inverter and power consumption of the transistor

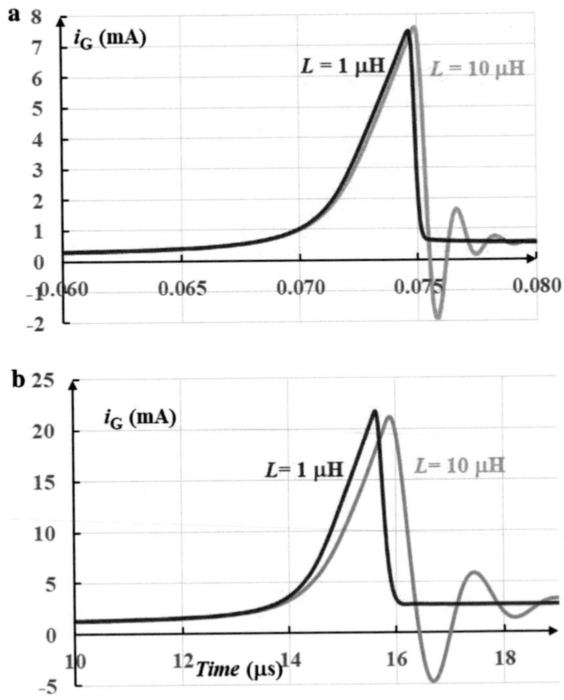

Fig. 4.2.62 Influence of the input capacitance and a parasitic inductance to the input current of an inverter. **a** k = 100 V/ms and **b** k = 500 V/ms

In the same figure the effects of a presence of a parasitic inductance in the input circuit are investigated. As can be seen larger values of parasitic inductances may lead to oscillating behavior of the switching.

To conclude, Fig. 4.2.61b depicts the power dissipation (top line) at the transistor during switching. One may be seen that while the transistor's working point is traversing the active region the dissipated power reaches high values. Here, care is to be taken not to exceed the rated pulsed maximal power of the transistor. It may be seen from the same figure also that when switched on the transistor is still consuming some power albeit small. That is the smallest power which it would consume if used as a closed switch.

4.2.3.3.9 Switching Properties of the Power MOS Transistor

Switching off and on the electronic device is of major concern in modern power conversion applications. To get a basic notion on the problem, here the circuit of Fig. 4.2.63a was simulated. It is the same as the one depicted in Fig. 4.2.61a with difference that the input voltage source is substituted by a driving circuit modeling the output of an opto-coupler. When the switch is on the gate voltage becomes reduced to the on-voltage of the diode which leads the transistor in cut-off.

The simulation results for $L = 5\ \mu\text{H}$ are depicted in Fig. 4.2.63b. The switch was turned on at $t = 10\ \mu\text{s}$ with his own transition time of 1 ps. Delay time of approximately 270 ns along with a rise time (or transition time) of approximately 80 ns may be observed. In order to check the origin of the shape of the waveform an additional simulation was conducted with the diode short-circuited. No noticeable changes were found confirming that the transistor is the one which fully defines the switching.

Figure 4.2.63c depicts the same time domain response but with $L = 15\ \mu\text{H}$. As can be seen, due to the transient in the gate circuit, as depicted earlier in Fig. 4.2.62, a negative pulse is added to the output voltage and consequently a positive pulse to the drain current. This phenomenon is more pronounced when L is risen in a way that the amplitude of the negative pulse is risen, and additional pulses are observed due to oscillations in the gate circuit. This, again, confirms the importance of keeping the gate inductance as low as possible.

Finally, from Fig. 4.2.63b one may find out the minimum value of the drain-to-source voltage of the transistor being approximately 3.22 V (under the condition of $R_L = 20\ \Omega$). This is an important information since it represents the voltage drop on a closed switch.

This is to be checked for compliance with the maximum pulsed ratings of the transistors which may be read from the safe operating area (SOA) as depicted in Fig. 4.2.64.

This 100 A, 500 V transistor allows for dissipation of 12,000 W, as can be read from the top right-most line (hyperbola) of the diagram. This value is chosen since the transient in our circuit lasts less than 100 μs which is marked in Fig. 4.2.64 for the hyperbola.

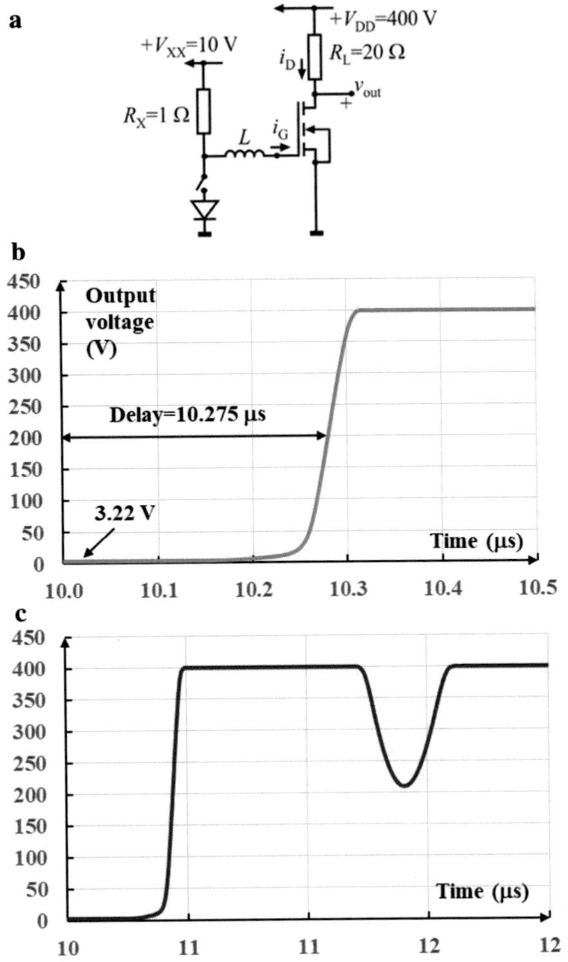

Fig. 4.2.63 **a** Test circuit for switching off, **b** output voltage ($L = 5\ \mu H$), and **c** output voltage ($L = 15\ \mu H$)

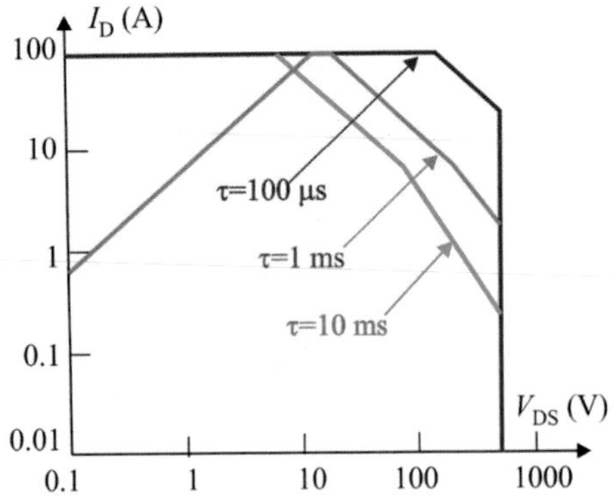

Fig. 4.2.64 Safe operation area of a VDMOS

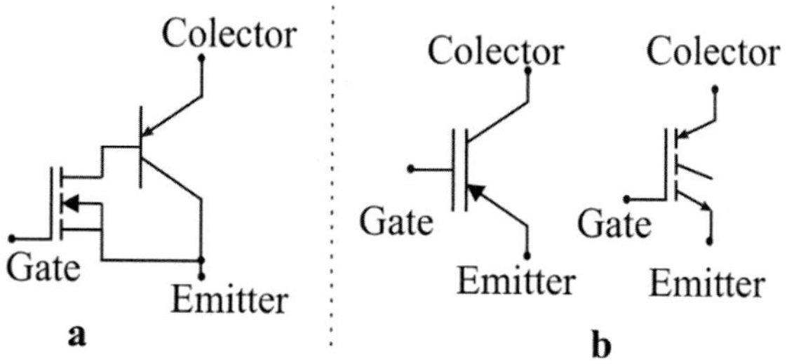

Fig. 4.2.65 **a** Basic structure of an N-channel IGBT and **b** symbols for IGBT

4.2.3.3.10 IGBT

As a component that accepts the IGFET property related to the input current and the BJT property related to the output voltage, the IGBT (from: insulated gate bipolar transistor = BJT with isolated gate) has been developed.

The electrical circuit schematic that basically explains the structure of the IGBT is shown in Fig. 4.2.65a while Fig. 4.2.65b shows two symbols that are alternatively used in the literature. The floating skewed line on the right of Fig. 4.2.65b represents the substrate connection.

As one can see, this is a combination of MOSFET and BJT. At the input is an N-channel MOSFET (there is also a dual component) which starts to conduct current when the input voltage exceeds the threshold voltage. The MOSFET's drain excites the base of a PNP transistor whose emitter is the output terminal. This terminal is marked here as the IGBT *collector*, and it is interchangeably named the *anode* (analogous to thyristors which will be discussed soon) or *drain* (analogous to MOSFET). The voltage V_{DS} of the MOSFET is equal to the voltage V_{CB} of the bipolar transistor, and the current of the BJT's base is in fact the drain current of the MOSFET. The voltage at the component's output is the BJT's V_{CE} voltage.

The structure of the IGBT is shown in Fig. 4.2.66a, and its equivalent circuit is given in Fig. 4.2.66b. As can be seen, the structure of the IGBT resembles the structure of the VDMOS depicted in Fig. 4.2.58, with the P area (which served as the substrate of the MOS transistor) now contacted and short-circuited for the N^+ area that served as the source. The essential difference is in the additional P^+ layer introduced between the lower metallization and the N^+ region that served as the drain. Of course, here too, the component has a cylindrical symmetry.

In the equivalent circuit given in Fig. 4.2.66b we recognize three transistors. The drain of the MOSFET is the region of N^- which is associated with the resistance of the semiconductor body marked with R_D. The MOSFET channel is formed on the surface of the P^- region, under the gate, and the source is the N^+ region, which is also located under the gate. The N^- band, however, is at the same time the base of the PNP transistor T_P, which is specific in that it has a built-in layer with a high concentration of N^+ toward the emitter (lower P^+ band), which enables high inverse voltages on the emitter of T_p.

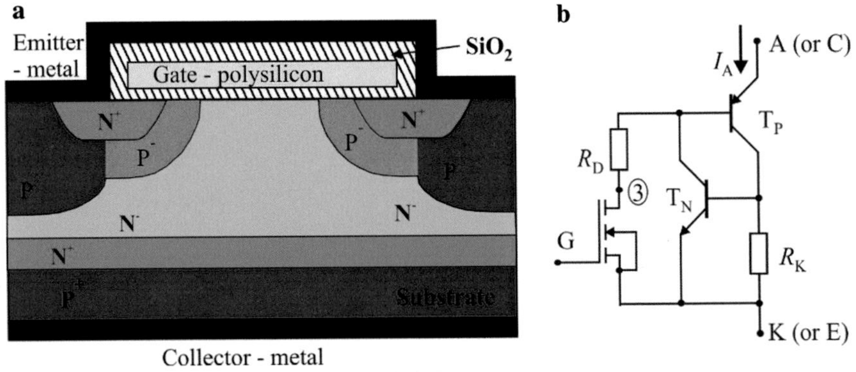

Fig. 4.2.66 **a** Structure of the IGBT and b equivalent circuit

The structure is cylindrical so that the upper N^+ regions are in fact the same ring that makes up the NPN emitter of the transistor T_N. The base of this transistor is the P^+ region just below the emitter, and the P^- region is expressed as a resistor R_K.

The output characteristics of one IGBT are shown in Fig. 4.2.67a. They represent the dependence of the output (collector) current on the output voltage (between the collector and the emitter) with the voltage between the gate and the emitter as a parameter. The specificity of these lines is related to the fact that the collector current cannot be established before the input voltage reaches the threshold voltage of the MOS transistor, and at the same time, its output voltage must be higher than the conduction threshold of the PNP transistor.

One interesting thing is related to the behavior of this component. Namely, it was said that the temperature coefficient of the MOSFET's drain current is negative, and the temperature coefficient of the BJT's emitter current is positive. Thanks to that, the base current of the PNP transistor has a negative temperature coefficient and has a stabilizing effect on the emitter current of the PNP transistor, i.e., on the current of the IGBT collector. The total effect is shown in Fig. 4.2.67b. It can be seen that at low currents the negative temperature coefficient dominates, and at high currents the positive one. In any case, the fact that there is a point of zero temperature coefficient at relatively high currents (slopes) speaks favorably about the temperature properties of this component.

Next, we will show the temperature dependence of the minimum output voltage (the on-voltage here denoted V_{AKsat}). It is depicted in Fig. 4.2.68. Here we get a confirmation of the previous discussion since for smaller currents ($I_A = 15$ A) the current decreases with temperature while for large currents ($I_A = 60$ A) the opposite happens. Finally, in order to complete the picture on the static characteristics of the

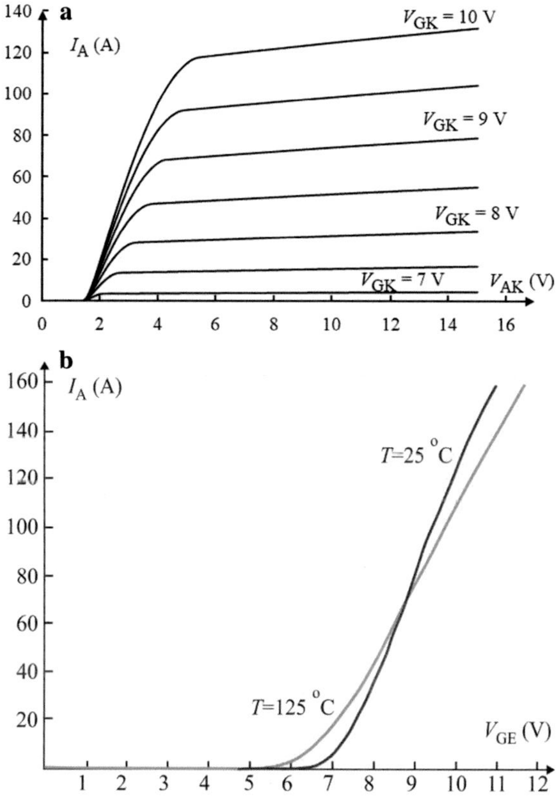

Fig. 4.2.67 **a** Output characteristic of an IGBT and **b** transfer characteristics measured for two different temperatures

IGBT and on the improvements as compared with the VDMOS Fig. 4.2.69 depicts the on-voltages as functions of the output currents. It is noticeable that at large currents, what is of prime importance, the voltage of the IGBT is several times smaller than the one of the VDMOS.

As can be seen from the part of the characteristics for low V_{AK} the on-voltage, which was expected to be almost equal to V_{CEsat}, is much larger due to the fact that $V_{AKsat} = V_{EBsat} + I_D \cdot R_D + V_{DKmin} = V_{CEsat} + I_C \cdot R_K$, I_D and I_C being the

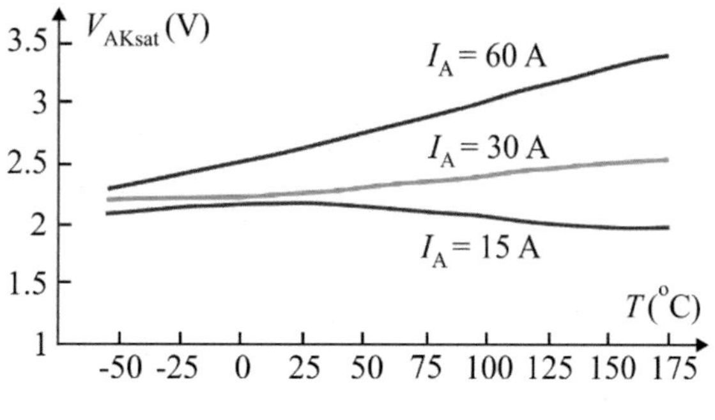

Fig. 4.2.68 Temperature dependences of the minimum voltage of an IGBT for different quiescent points

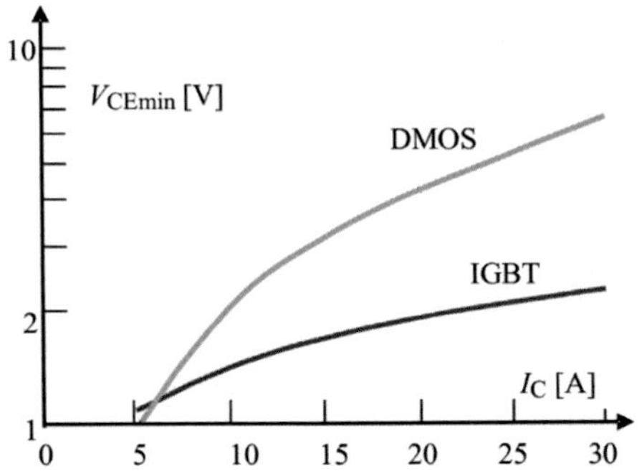

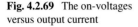

Fig. 4.2.69 The on-voltages versus output current

currents of the MOS and bipolar transistor (T_P), respectively. Its value rises for larger anode currents. Nevertheless, compared to a power MOSFET of the same die size and operating at the same temperature and current, an IGBT can have significantly lower on-state voltage.

A specific phenomenon may be ignited at very high currents I_A and voltages V_{AK}. That is a breakdown between the two terminals due to "firing" of the pnpn (thyristors) structure present between the terminals. Namely, the voltage drop $I_C \cdot R_K$ increases with the increase of the current and may reach the built-in potential of the emitter junction of the parasitic npn transistor denoted T_N. If that happens, a positive feedback loop is activated between the two BJT since the collector current of each of them is a base current to the other one. At some temperatures, the loop gain may become larger than unity and surge of the collector currents may happen while drastically reducing V_{AK}. The problem is in that the state is latched, and it may be "extinguished" only by reducing the current below the value which keeps the loop gain equal to unity or by reducing V_{AK} to zero.

This phenomenon together with the punch-through phenomenon (here less influential since the base is really wide), and the secondary breakdown, is defining the ratings of an IGBT at high voltages.

Figure 4.2.70 depicts the SOA of an IGBT for room temperature. This diagram is to be consulted when checking for power limits of the implementation. Note, for pulses shorter than 50 μs, at room temperature, power of 800 W is allowed. If, in such a case, currents as large as 90 A are to be allowed, the voltages have to be lower than 100 V. At the opposite end, if voltages as large as 1000 V are to be allowed, the currents have to be lower than 10 A. For (short pulsed) voltages of the order of 400 V, the currents are to be smaller than 20 A.

As for the on-voltage, when 2 V are requested, currents not larger than 20 A are allowed.

In the next the dynamic properties of the IGBT will be discussed.

IGBT is a complex device with all sorts of capacitances present within it. There are capacitances of the MOS transistor (gate-to-channel, gate-to-source, gate-to-drain, and drain-to-substrate) and the capacitances of the bipolar structures where additional three p-n junctions (each of them having two capacitances) may be recognized. To

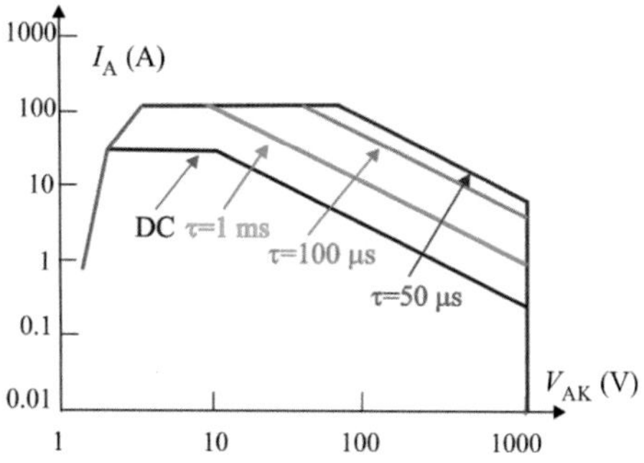

Fig. 4.2.70 The SOA of an IGBT

avoid complex explanations, we are referring here to Table 4.2.3 for the capacitances seen from the terminals of the IGBT. These values should be taken as reference only since they are voltage and current dependent. Nevertheless, none of them is small even not smaller than their counterparts.

As it was done for the devices discussed previously, we will here again look for the transient behavior of circuits where IGBT is the switching component.

When, instead of the MOS, in the circuit of Fig. 4.2.61a IGBT is used it exhibits transients of the input current as depicted in Figs. 4.2.71 and 4.2.72. Note the value of the inductance was in the first case $L = 10$ nH and in the second $L = 100$ nH. Again, two values of the slope of the input voltage were implemented in order to expose the capacitive behavior.

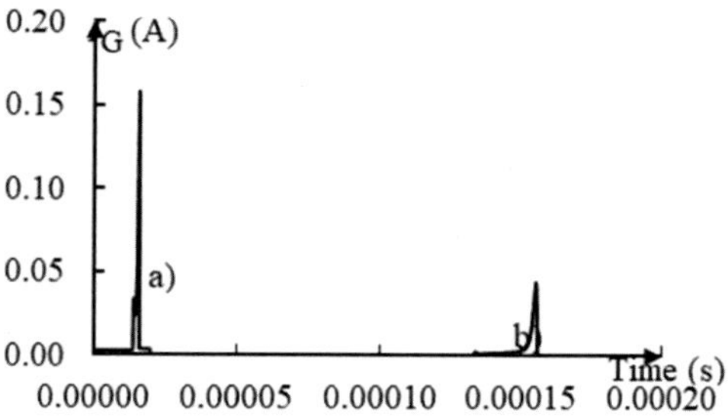

Fig. 4.2.71 The input current of the circuit of Fig. 4.2.61a with an IGBT and $L = 10$ nH. The slope of the input ramp voltage is 500 V/ms (**a**) and 50 V/ms (**b**)

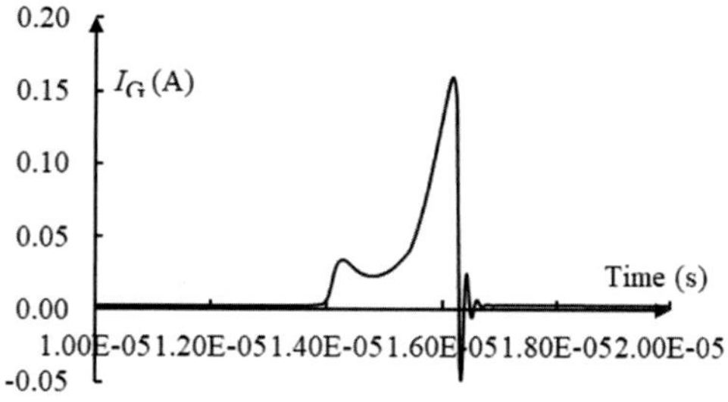

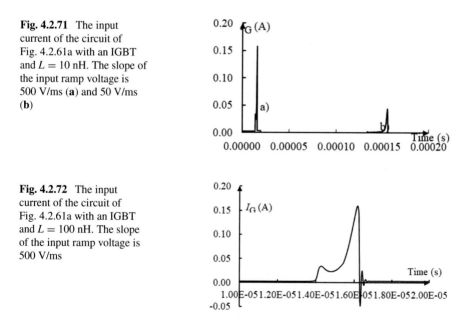

Fig. 4.2.72 The input current of the circuit of Fig. 4.2.61a with an IGBT and $L = 100$ nH. The slope of the input ramp voltage is 500 V/ms

As can be seen from Fig. 4.2.71 larger slopes (faster signals) will produce more expressed transients with relatively large input currents during the switching on of the transistor. The amplitudes of the input current in the transient for a larger value of the inductance, as seen from Fig. 4.2.72, are not much affected while now, oscillations may be observed. These become even more pronounced for larger values of L.

The switching off the IGBT is of prime importance. These are investigated with the help of the circuit of Fig. 4.2.62 with the MOS transistor substituted by an IGBT. $V_{XX} = 15$ V was used.

Two simulation results are depicted in Figs. 4.2.73 and 4.2.74.

In the first case the inductance value was $L = 0.4$ μH. Delay time of approximately 300 ns may be read from the diagram of Fig. 4.2.73a.

It is worth noting that the on-state voltage (before switching) produces a voltage drop $V_{AKmin} = 2$ V only (Fig. 4.2.73b).

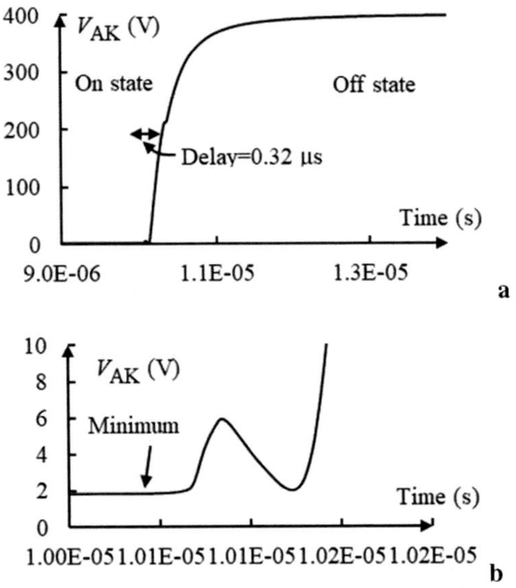

Fig. 4.2.73 Switching off the circuit of Fig. 4.2.62 with IGBT. $L = 0.4$ μH. Complete response (**a**) and the start of the transient (**b**)

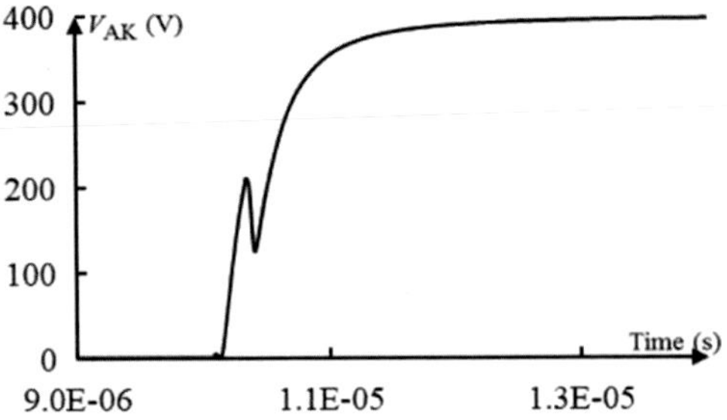

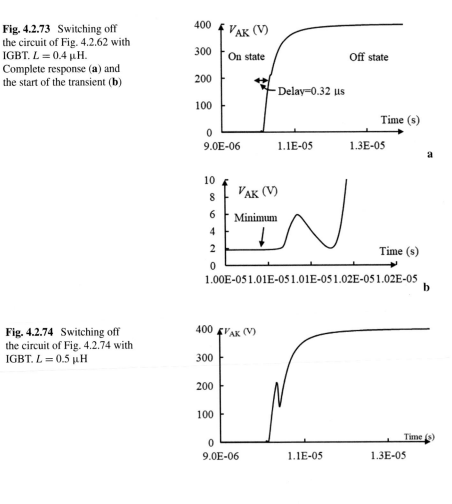

Fig. 4.2.74 Switching off the circuit of Fig. 4.2.74 with IGBT. $L = 0.5$ μH

In the second case L was incremented just for 25%. A serious deterioration of the switching-off response may be observed as depicted in Fig. 4.2.74. For larger values of L, a long train of oscillations occurs.

4.2.3.3.11 GaN HEMT

Since its emergence in the early 1990s, GaN has attracted attention as highly promising material system for both optical and electronic applications. The devices developed since then, AlGaN/GaN high electron mobility transistors (HEMTs), have been a subject of intense recent investigation and became attractive candidates for high-voltage, high-power operation. Other commonly used names for HEMTs include MODFET (modulation doped FET), TEGFET (two-dimensional electron gas FET), and SDHT (selectively doped heterojunction transistor). An overview of the processes in the III-nitride technology may be found in literature.

To start with let us get reminded the fundamental advantages of GaN as semiconductor material over silicon. That would be as follows:

- According to Table 4.2.1, GaN has much higher mobility of the electrons or, what is the same, a slab of GaN under the same electrical field will exhibit higher velocity of the carriers and, consequently, will have larger current than a silicon one. That is important since the resistivity of parts of the component being on the path of the main current is reduced leading to lower losses, i.e., R_{DSon}.
- Larger value of the breakdown field together with lower intrinsic concentration (Table 4.2.1) leads to higher breakdown voltages.
- Smaller capacitances (Table 4.2.3) lead to faster switching.
- Larger maximal velocity (see section: Power MOS transistor) allows for larger currents at high fields.

Figure 4.2.75 summarizes these claims. It depicts the limiting curves for the minimum R_{DSon} for a given breakdown voltage of Si, GaN, and 4H-SiC devices. A dot on any of these curves represents one device with limiting properties. The advantage of GaN and SiC is obvious. And, the margin is large. As for the space applications, due to higher exposition to cosmic rays and ultraviolet light, a specific advantage of GaN over silicon is its inherent radiation hardness.

In order to be more specific when making these comparisons in Fig. 4.2.76 the simplest structure of a GaN-based HEMT is depicted.

Here, looking from the bottom to the top center, on a SiC or sapphire (high resistance – insulating substrate) a buffer layer of AlN (or other, as depicted) is grown first. Its role is partly to allow for transition in the difference between the crystal lattice constants of the substrate and the materials which will come on the top and partly to rise the breakdown voltage of the vertical structure. Then the GaN layer is grown on the top of which the heterojunction is built by further growing $Al_xGa_{(1-x)}N$. It is at the bottom side of this heterointerface where the 2DEG is formed as depicted in the figure. It will serve as the conducting channel of the device. The top $Al_xGa_{(1-x)}N$

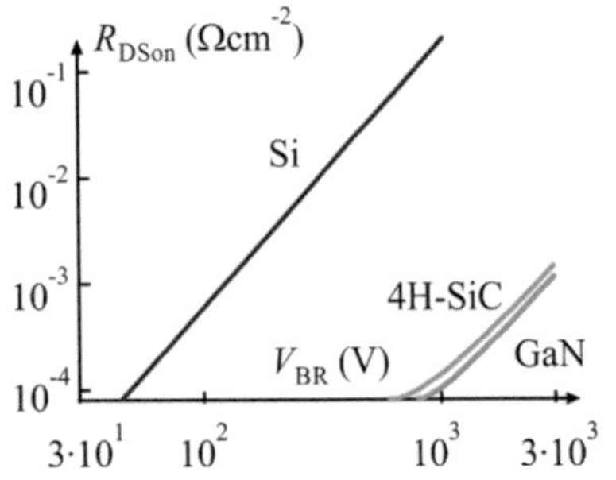

Fig. 4.2.75 R_{DSon} as a function of the breakdown voltage for Si, GaN, and SiC-based devices

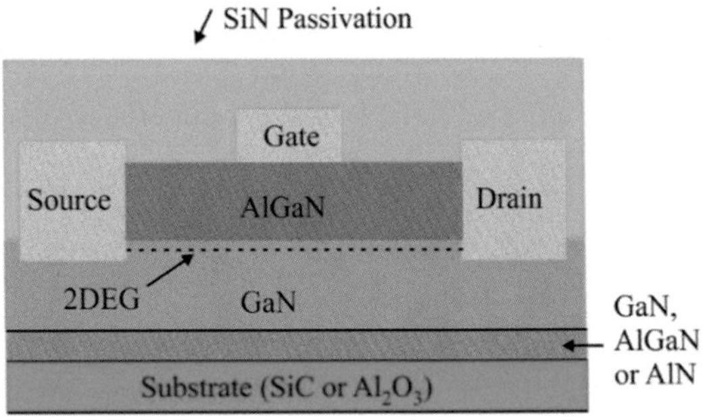

Fig. 4.2.76 Cross section of a GaN-based HEMT

layer may be doped to become n-type semiconductor. Finally, a metal layer is added to serve as the gate of the transistor forming a Schottky diode with the $Al_xGa_{(1-x)}N$.

Looking laterally, metal is deposited on the left and the right of the 2DEG to form the source and the drain terminals, respectively. Note the absence of degenerated areas below the metal-semiconductor interfaces which were unavoidable in silicon-based devices to make ohmic contacts.

So, in this structure we recognize several fundamental differences as compared with the silicon-based MOS transistor.

- The existence of 2D high mobility electron gas, which will be discussed in this paragraph, allows for much higher currents for a given voltage which may lead to reduction of the sizes. There is no need for inversion of the surface. Current is flowing between the source and drain for negative and positive voltages. The 2DEG behaves as in the case of depleted mode MOS with the difference in the higher mobility of the carriers. The device is switched fully off by large negative voltages at the gate which should deplete the 2DEG.
- The gate is not isolated. It is rather CS (conductor-semiconductor) than CIS (conductor–insulator-semiconductor) device. In fact, a Schottky diode is formed between the metal gate and the $Al_xGa_{(1-x)}N$. At certain positive voltages (larger than the barrier of the metal–semiconductor junction) this diode will conduct. This is depicted in Fig. 4.2.77. This fact may be of importance in biasing the gate circuit of high-voltage and high-current switches. If large positive voltages are

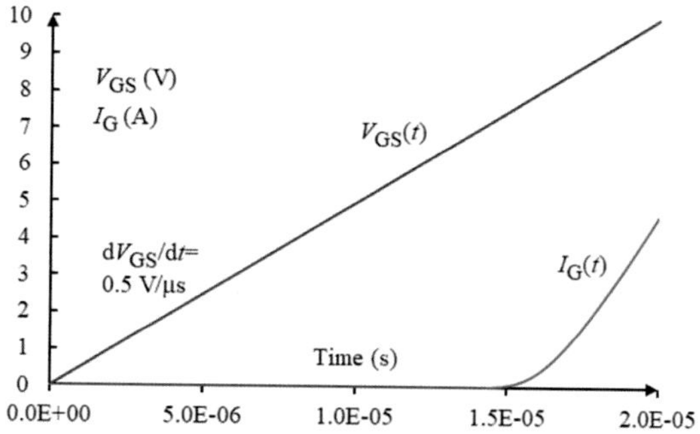

Fig. 4.2.77 The characteristic of the gate-to-source diode of a HEMT

brought between the gate and the source, large gate current will be established possibly overloading the gate driving circuit.

- This device does not exhibit the self-isolation property. Namely, there are no p-n junctions at all here which is unique for this device. As a consequence, no parallelization is possible on the same substrate despite the negative temperature coefficient. If another device were to be built on the same substrate the neighboring device terminals would be short-circuited via the GaN layer the surface of which may be inverted by presence of positive charge in the passivation SiN. This limits the drain current of this (obviously) lateral device. To avoid that HEMTs are designed in a so-called interdigital structure in which several transistors are placed in parallel but each consecutive one is alternatively sharing its source or drain with the neighboring one as depicted in Fig. 4.2.78.
- The breakdown mechanism which defines the maximum drain voltage has nothing to do with avalanche breakdown of a diode which was the case in the MOS transistor. Here a specific mechanism may be activated at high drain voltages leading to drain-to-gate breakdown for which leakage current coming from the gate into the channel is responsible. This phenomenon will be discussed here later on. In addition, the breakdown voltage is defined by the breakdown electric field of GaN and AlN but a phenomenon known as the insufficient confinement of the 2DEG in the channel allows for the 2DEG to overflow from the potential well into the GaN buffer layer (to be discussed later on), thus causing a buffer layer punch-through effect. This is avoided by improving the device structure by adding additional layers under the 2DEG leading to the so-called double heterojunctions.

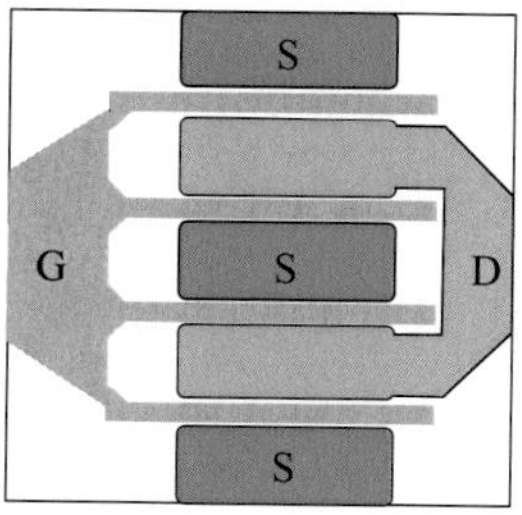

Fig. 4.2.78 Top view of a HEMT

In the next the physics and properties of the GaN HEMT will be discussed in a little more detail.

4.2.3.3.12 Properties of $Al_xGa_{(1-x)}N$

The $Al_xGa_{(1-x)}N$ layer situated between the metal and the GaN layer is of fundamental importance for the functionality of HEMT. On its top side a Schottky diode is formed while on its bottom side a heterojunction creating the 2DEG is present. Hence, its properties deserve more attention.

$Al_xGa_{(1-x)}N$ is an alloy in which the number of Al and Ga atoms in the crystal lattice adds to 100% of the N atoms. Therefore, x stands for a number less than 1 and is equal to the percentage divided by 100. For every single value of x a different material is obtained. Here a short review of the main properties of these materials will be given.

The relative dielectric constant of $Al_xGa_{(1-x)}N$ may be computed from

$$\varepsilon(x) = 10.4 - 0.3 \cdot x. \tag{4.2.47}$$

For $x = 0.3$ it would be $\varepsilon = 10.31$ ($\varepsilon(0.3) = 8.8$ was reported, however).

If Ni is used as a metal for the effective Schottky barrier potential one may write

$$q \cdot \Phi_B = 1.3 \cdot x + 0.84(\text{eV}). \tag{4.2.48}$$

For $x = 0.3$ it would be 1.23 eV.

The band offset is given by

$$\Delta E_C = 0.7 \cdot \left[E_g(x) - E_g(0)\right]. \tag{4.2.49}$$

For the band gap one may write

$$\begin{aligned} E_g(x) &= 6.13 \cdot x + 3.42 \cdot (1 - x) - x \cdot (1 - x) \\ &= 3.42 + 1.71 \cdot x + x^2(\text{eV}). \end{aligned} \tag{4.2.50}$$

For $x = 0.3$ we have $E_g = 4.023$ eV and $\Delta E_C = 0.4221$ eV.

These data will enable to calculate the proper part of the 2DEG charge density.

4.2.3.3.13 The $Al_xGa_{(1-x)}N$/GaN Heterojunction

The energy band diagram of this heterojunction is depicted in Fig. 4.2.79. Note, to follow the explanation given for the MOS transistor the sides (material_with_higher/material_with_lower energy gap) were switched as compared to

Fig. 4.2.45. In this case doped $Al_xGa_{(1-x)}N$ is considered. To illustrate, as above, $x = 0.3$ was used.

As can be seen, accommodation of the edges of the energy bands (E_C and E_V) take place on both sides in order for the Fermi level to be constant in the heterojunction as a whole. The crucial consequence of that is the formation of a shallow layer on the GaN side of the heterojunction in which the semiconductor (which is undoped and normally having intrinsic concentration) behaves as degenerated. The Fermi level here is above E_C. This phenomenon is frequently referred to as the quantum well. The concentration of free electrons here is high as if it were highly doped while without a presence of impurity atoms which would create positive ions which, in turn, would define the mobility of the electrons due to scattering (diverting the moving electrons by their attractive forces and so reducing their average velocity). So, their mobility is much higher than usually. This is unique and, one may say, extraordinary situation. The free electrons here create a very thin sheet of charge available for conduction which is named 2DEG. Their number is not influenced by the drain-to-gate voltage and may be reduced by high negative voltages at the gate which will lead to switching off the transistor.

Two additional mechanisms are active at the surface defining the total charge available for conduction.

Namely, due to electronic charge redistribution inherent to the crystal structure of the group III-N semiconductors, they exhibit exceptionally strong polarization. This polarization is referred to as spontaneous or pyroelectric. In other words, because of

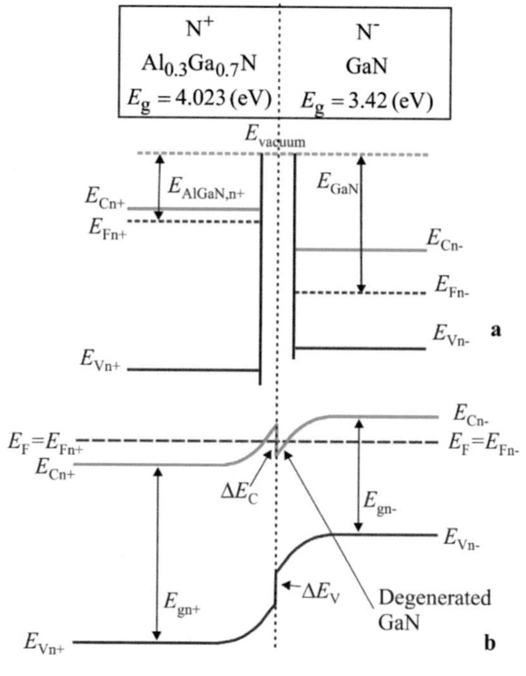

Fig. 4.2.79 The $Al_xGa_{(1-x)}N$/GaN heterojunction, **a** before and **b** after junction

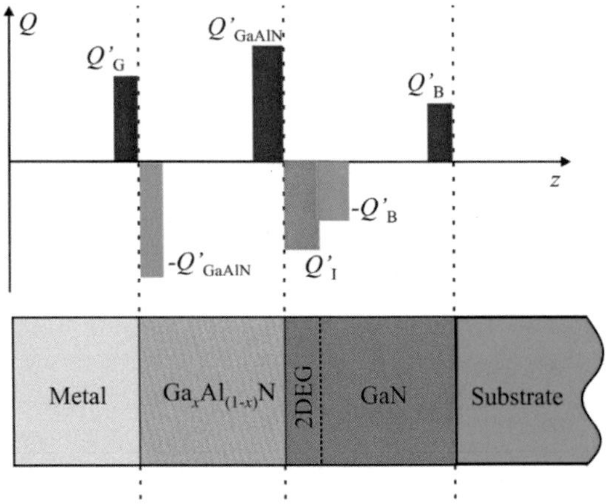

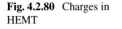

Fig. 4.2.80 Charges in HEMT

their specific crystal structure GaN-based and group III-N-based semiconductors can have different polarities resulting from uneven charge distribution between neighboring atoms in the lattice. As a consequence, charges of opposite signs are formed on the top and the bottom side of the GaN layer of the device. The number of these charges is denoted as Q'_B in Fig. 4.2.80 the prime symbol standing for "per unit area," usually per cm^{-2}, as it did for the MOS structure. This quantity is also referred to as sheet concentration.

Furthermore, again due to the difference in the parameters of the crystal lattice, since the GaN substrate is comparatively thicker, it is formed in a relaxed state while the AlGaN is under mechanical stress. When stress is applied along the <0001> direction to the group III-N semiconductors' lattice, the ideal lattice parameters of the crystal structure will change to accommodate the stress. Therefore, the polarization strength will be changed. This additional polarization in strained group III-N crystals is called piezoelectric polarization. Thus, in addition to spontaneous polarization, the strain developed in the AlGaN results in a piezoelectric polarization in AlGaN. When AlGaN is grown over the GaN substrate, due to the difference in their polarization, a net polarization charge develops at the interface denoted by Q'_{GaAlN} in Fig. 4.2.80. This charge, being positive at the GaN side, induces free electrons in the GaN surface enhancing the 2DEG the charge of which is denoted by Q'_I in Fig. 4.2.80.

It is worth noting that there is piezoelectric effect in the GaN layer as it is spontaneous polarization in the GaAlN layer, but these are usually of smaller extent and may be considered negligible in comparison with the main effects.

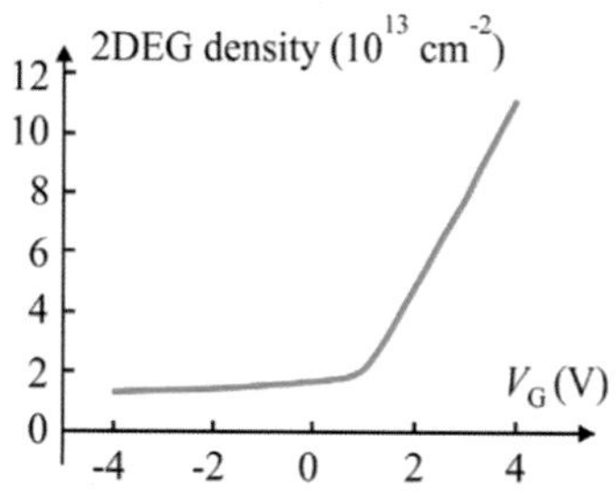

Fig. 4.2.81 Q'_s as a function of the gate voltage

The overall number of free electrons available for conduction at the top surface of GaN is usually expressed as

$$Q'_s = Q'_{ind} - \frac{\varepsilon}{q^2 d(x)}(q \cdot \Phi_B + E_F - \Delta E_C) \qquad (4.2.51)$$

where Q'_{ind} represents the charges related to both polarizations and the rest of the formula represents the quantum well. Formulae considering more details may be found in the literature. For $Al_{0.3}Ga_{0.7}N$ value of 1.1×10^{13} cm^{-2} was reported while a value of 1.2×10^{14} cm^{-2} may be found elsewhere.

We would like to stress here that in the HEMT literature a "creative" notation is used for the left-hand side of (4.2.51), namely n_s, with apparent intention to make a distinction with notations used in silicon technology alluding that here we are meeting something "completely new." The symbol n, however, in semiconductor physics, is used to denote concentration, which is number of electrons per unit *volume*. From that point of view using n_s which would stand for number of electrons per unit *area* is considered confusing and not accepted here.

As mentioned, the presence of a gate voltage may influence the number of electrons available for conduction by induction, i.e., by presence of charge in the gate electrode (denoted by Q'_G in Fig. 4.2.80). If positive enough the gate may increase significantly the number of electrons while, if negative enough, may fully deplete the 2DEG and cut-off the transistor. The dependence of Q'_s on the gate voltage is depicted in Fig. 4.2.81.

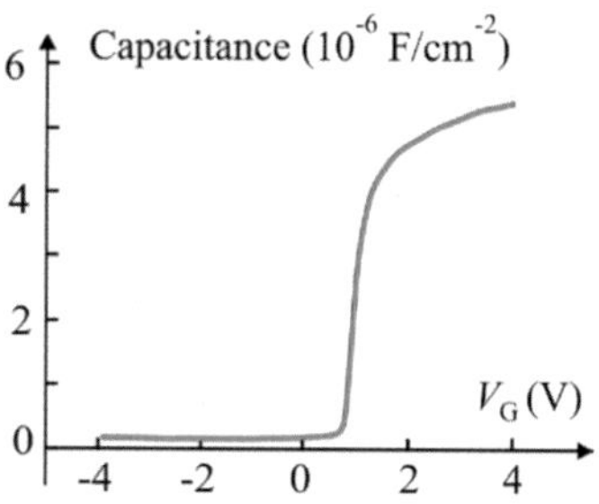

Fig. 4.2.82 Gate-to-2DEG capacitance of a HEMT

4.2.3.3.14 The HEMT Input Capacitance

Similarly, to the MOS transistor HEMT is characterized by its terminal capacitances for which numerical examples are given in Table 4.2.3. As may be seen the capacitances are smaller than those of Si MOS transistor. The main difference is here, however, in the value of the input capacitance C_{ISS} which is more than ten times smaller.

This is an important advantage of the HEMT. To illustrate, on Fig. 4.2.82, we reproduce the gate-to-source capacitance (capacitance of the Schottky diode) as a function of the gate voltage. It resembles the depletion capacitance of a Schottky diode before conduction. No data on the barrier were given. Note, when the Schottky barrier is approached the gate diode becomes conducting and gate current starts flowing. It forms a voltage drop on the (summary) resistance in the gate-to-source loop reducing the increment of the effective voltage on the capacitance (on the junction), so the curve flattens.

The capacitance between the gate and the 2DEG at zero gate voltage may be estimated as

$$C_{GB0} = \varepsilon \frac{L \cdot W}{d} \quad (4.2.52)$$

where ε and d stand for the dielectric constant and the thickness of the AlGaN layer, respectively, while L and W are the length and the width of the transistor (in fact of the 2DEG under the gate), respectively. For a transistor with $\varepsilon = 10.31 \cdot 8.854 \times 10^{-12}$ F/m, $d = 10$ nm, $L = 250$ nm, and $W = 100$ μm from (4.2.52) one gets $C_{GB0} = 0.23$ pF.

To appreciate the importance and the influence of the input capacitance to the transient behavior of a HEMT-based switch the circuit of Fig. 4.2.61a was simulated anew with a HEMT in place of the MOS transistor. Note $L = 0$ H was used in these simulations. A truncated ramp signal was used with two different values of the slope and adequate reduction of the flat level. The nonlinearities described for the MOS transistor are here present too, but as can be seen, much more dependent on the switching speed. This is mostly due to the smaller gain-independent part of C_{ISS} allowing the nonlinearity of the gain to dominate the shape of the response. Note, however, that the nonlinearity comes in fore for large positive gate voltages (high drain currents) albeit below conduction of the Schottky diode (which, according to Fig. 4.2.77, happens for $V_{GS} \approx 7$ V). The simulation results are depicted in Fig. 4.2.83.

4.2.3.3.15 Breakdown Voltage of HEMT

Breakdown in HEMT may be observed when the device is off ($V_{GS} < V_{TH}$) and at high drain-to-source (and drain-to-gate) voltages. As mentioned earlier two mechanisms may be considered responsible for the occurrence of a breakdown in a HEMT. We will here consider the explanation of the phenomenon attributed to the "vertical" drain-GaN-AlN-substrate path given above as complete (with suggestion to consult

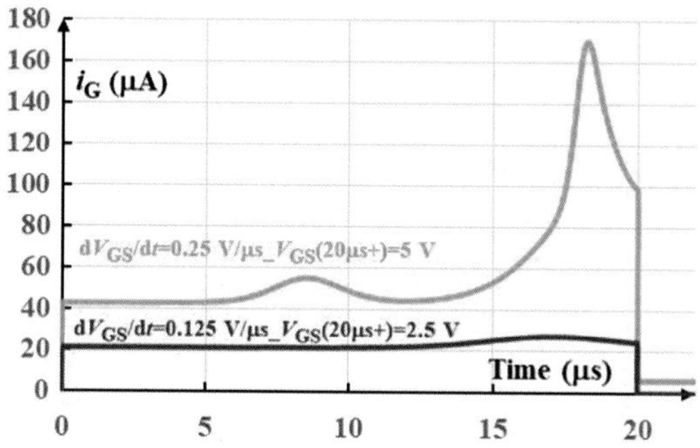

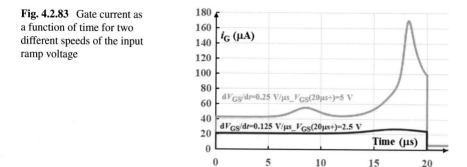

Fig. 4.2.83 Gate current as a function of time for two different speeds of the input ramp voltage

the literature cited) and will concentrate a bit more to the lateral drain-GaN channel-GaAlN-gate path. Namely, here, similar to the edges of the p-n junction, at the gate edge on the drain side extremely high field arises leading to "thermionic field emission" and "field-assisted tunneling." In other words, leakage occurs at the edge and electrons are injected into the normally fully depleted channel. Once electrons are present the rest of the breakdown process is quite the same as the avalanche in the p-n junction: By increasing the drain bias voltage, the electric field at the drain-edge of the gate increases and so the electrons (which have leaked from the gate) will acquire enough energy to cause impact ionization which leads to an avalanche effect and a sharp rise in drain current. Thermal avalanche is initiated, too, due to high dissipation present.

Various techniques were used to increase the breakdown voltage; the first one of them is to increase the distance between the gate and the drain. That would rise the resistance of the part of the channel between these two electrodes, i.e., will rise the R_{DSon} of the transistor. That is why within that datasheet of the component the R_{DSon}-versus-V_{BR} characteristic is always given such as in Fig. 4.2.75. As can be seen device with breakdown voltages as high as 3×10^3 V may be found. Additional techniques are applied in order to reduce R_{DSon}. By one of them the so-called metal field plates are used as an effective method as they do in SiMOS technology. A cross section of HEMT in which field plates are built is depicted in Fig. 4.2.84. The field plate consists of an additional metal layer located over the gate metal and extending into the region between the gate and drain. The field metal provides an electric field termination layer that reduces the magnitude of the field at the gate edge, thereby suppressing the gate leakage current. However, field plates also introduce additional capacitance. Therefore, a degradation in switching performance occurs. For this reason, a minimum field-plate length should be employed.

Two different configurations for field plates have been experimented: gate connected and source connected. The FP-to-channel capacitance in the case of source connected field plate becomes the drain-source capacitance, which could be absorbed in the output circuitry. Hence the drawback of additional C_{GD} (Miller capacitance) which is encountered in gate connected source plates is eliminated.

It is interesting to note that, as opposed to the situation in Si-based MOS, the temperature coefficient of the breakdown voltage is positive. That is attributed to

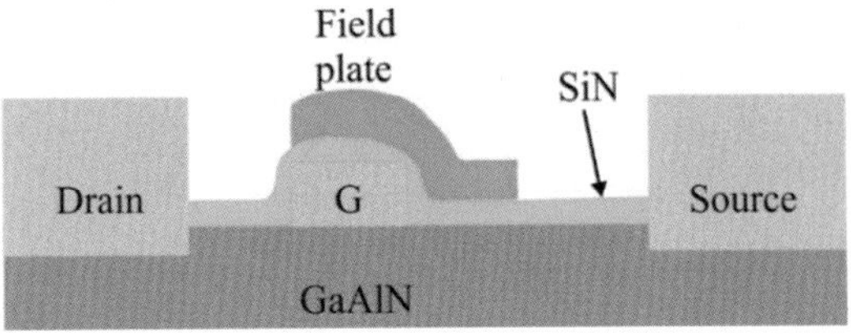

Fig. 4.2.84 Field plate in a HEMT

the fact that less electrons are created in the channel by thermal excitation while, in addition, the impact ionization requires higher energy due to a reduced electron transit time.

4.2.3.3.16 Threshold Voltage of HEMT

We will define here the threshold voltage similarly to the definition given for the SiMOS transistors. Namely, we will consider that the (negative) gate potential reaches its threshold value when Q'_s, as given by (4.2.51), reduces to zero. That, for approximate calculations, leads to the following equation:

$$V_T = Q_s/C_{GB0} = q \cdot Q'_s \cdot W \cdot L/C_{GB0} \tag{4.2.53}$$

which, after substitution of (4.2.51) and (4.2.52), leads to

$$V_T = \frac{q \cdot d(x)}{\varepsilon} \cdot \left(Q'_{ind} - \frac{\varepsilon}{q^2 d(x)} (q \cdot \Phi_B + E_F - \Delta E_C) \right). \tag{4.2.54}$$

Much more general expressions may be found in literature. Second order effects such as the influence of the channel width to the threshold voltage were also studied and experimentally verified.

Threshold voltage values of about −4 V were reported.

To better understand the properties of a HEMT and the behavior around threshold Fig. 4.2.85 depicts the static input and transfer characteristics of a HEMT obtained by simulation. Figure 4.2.85a represents the static input characteristic, i.e., the gate current as a function of the gate-to-source voltage. As can be seen a built-in positive voltage exists due to the spontaneous polarization which keeps the current low even for small positive voltages. That would correspond to the lower line in Fig. 4.2.83. For negative voltages negative gate current of approximately 0.5 μA was obtained.

Looking the Fig. 4.2.85b one may recognize that even though negative voltages are necessary to fully stop the drain current, the transistor is practically in off state even for small positive voltages. The complete static transfer characteristic of this transistor for $V_{DS} = 100$ V is depicted in Fig. 4.2.85c.

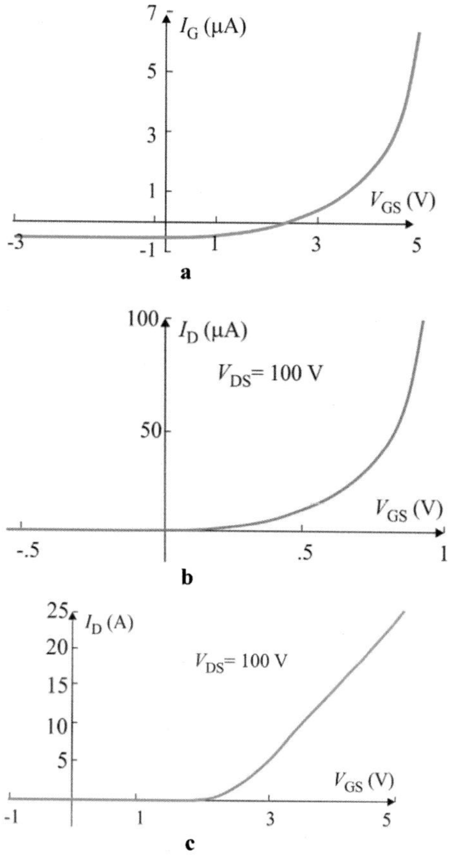

Fig. 4.2.85 Static characteristics of a HEMT. **a** Input characteristic, **b** transfer characteristic (inlet), and **c** transfer characteristic (full)

The temperature dependence of the threshold voltage was studied in literature. It was found that in the temperature range between 0 and 300 °C the threshold voltage is practically constant for samples grown on Si, Sapphire, or GaN.

4.2.3.3.17 The HEMT Model

For the ohmic region of the HEMT characteristics one usually applies the following

$$I_D = C_{GB0}\frac{W}{L} \cdot \frac{\mu_0}{1 + \frac{V_{DS}}{L \cdot K_s}} \cdot \left[(V_{GS} - V_T) \cdot V_{DS} - V_{DS}^2/2\right] \qquad (4.2.55)$$

where K_s is the critical field for GaN as given in Table 4.2.1. In this expression a formula for the mobility dependence on the electric field in the channel is built, $K = V_{DS}/L$ being the field. The same expression for approximation of the mobility may be inserted in the model of the SiMOS transistor (4.2.37).

If the intrinsic source and drain resistances R_S and R_D are to be considered the following substitutions must be included into (4.2.55):

$$V_{GS} = V_{GS0} - R_D \cdot I_D \tag{4.2.56a}$$

$$V_{DS} = V_{DS0} - R_D \cdot I_D \tag{4.2.56b}$$

where V_{GS0} and V_{DS0} are measured at the *outside* terminals of the device. For practical implementation of the model, however, it is better to create a series circuit constituted of two resistors [defined by (2.56)] and a current source defined by (4.2.55).

The drain-to-source saturation voltage is obtained from $dI_D/dV_{DS} = 0$ to be

$$V_{DSsat} = L \cdot K_s \cdot \left\{\sqrt{1 + 2 \cdot (V_{GS} - V_T)/(L \cdot K_s)} - 1\right\} \tag{4.2.57}$$

and the corresponding saturation current is

$$I_{Dsat} = \frac{1}{2}\mu_0 \cdot C_{GB0} \cdot K_s^2 \cdot W \cdot L \cdot \left(\sqrt{1 + \frac{2 \cdot (V_{GS} - V_T)}{L \cdot K_s}} - 1\right)^2 \tag{4.2.58a}$$

which is approximated by

$$I_{Dsat} = \frac{\varepsilon}{d}\mu_0 \cdot K_s \cdot W \cdot (V_{GS} - V_T). \tag{4.2.58b}$$

The last two expressions in fact represent the transfer characteristic of the HEMT which is illustrated in Fig. 4.2.85a, b.

The output characteristics of the same transistor are depicted in Fig. 4.2.86. In this diagram part of the characteristics for inverse ($V_{DS} < 0$) polarization of the transistor is given to illustrate that in this situation instead of V_{GS}, V_{DG} plays major role in definition of the drain current value.

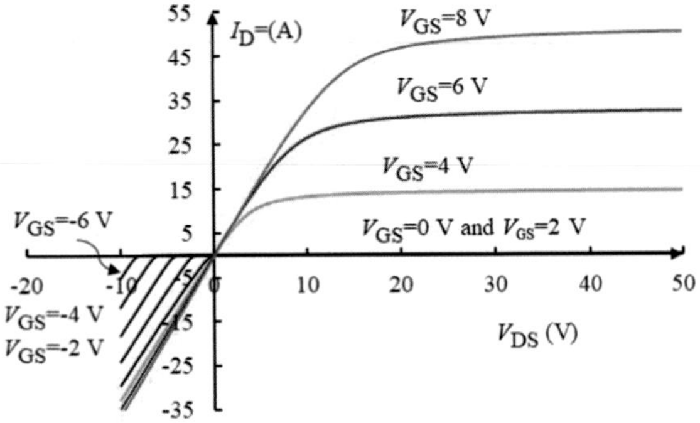

Fig. 4.2.86 Output characteristics of a HEMT at room temperature

4.2.3.3.18 Switching with HEMT

We will proceed first with the simulation of the circuit of Fig. 4.2.63a in order to get the dynamic voltage transfer characteristic of the HEMT. The changes made to the circuit of Fig. 4.2.63a were $V_{XX} = 5$ V and $L = 1$ pH (in the first simulation) and $L = 1$ nH (in the second). The switch was again turned on at $t = 10$ μs with his own transition time of 1 ps.

The simulation results are depicted in Figs. 4.2.87 and 4.2.88.

Figure 4.2.87 depicts the output voltage after transient for $L = 1$ pH. Extremely low delay time of 0.92 ns may be observed the overall transition time being of approximately 2 ns.

To check for the influence of the gate inductor one more simulation was performed with $L = 1$ nH. Significant changes may be observed in the output response as depicted in Fig. 4.2.88. Namely, oscillations may be seen in the response in the transition period redefining the overall transition time to about 8 ns. It is not shown here but our experiments with higher values of L showed long lasting oscillations making the definition of the transition time very difficult if not impossible. Considering these results and having in mind that inductance of 1 nH is not difficult to collect while mounting the transistor on the printed circuit board (PCB) and that L is not the only reactive element that will become influential after mounting, one may conclude that extreme care is to be implemented during the PCB design and mounting.

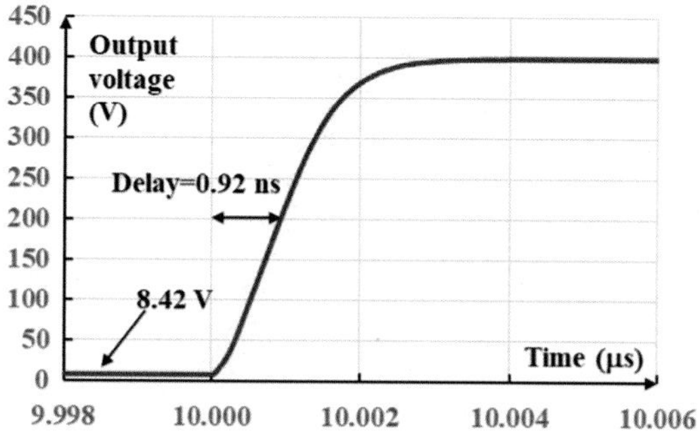

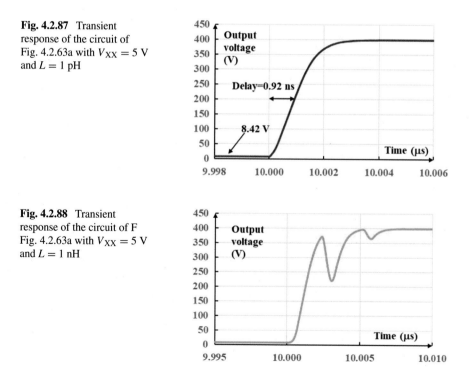

Fig. 4.2.87 Transient response of the circuit of Fig. 4.2.63a with $V_{XX} = 5$ V and $L = 1$ pH

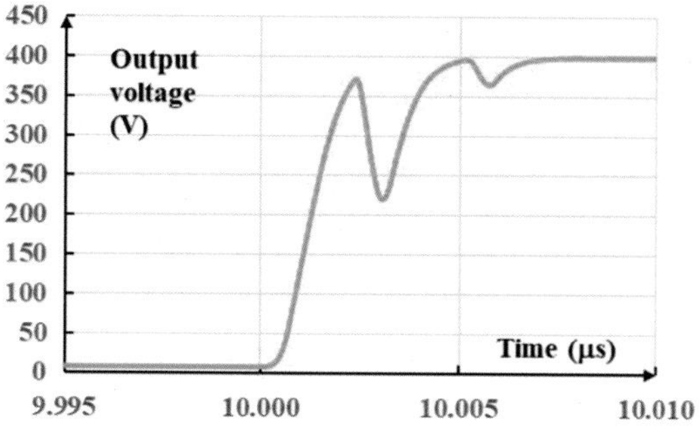

Fig. 4.2.88 Transient response of the circuit of F Fig. 4.2.63a with $V_{XX} = 5$ V and $L = 1$ nH

A relatively high value of $V_{DS} = 8.42$ V (when the transistor is in on-state) may be observed from Fig. 4.2.87 which could be deduced from the output characteristics since for the circuit of Fig. 4.2.63a $V_{DD}/R_L = 25$ A (see Fig. 4.2.53). Lower values of R_L would lead to even larger minimum value of V_{DS}.

4.2.3.3.19 SiC MOSFET

Here again we will get reminded the fundamental advantages and disadvantages of SiC as semiconductor material over silicon. That would be as follows:

- According to Table 4.2.1, SiC has smaller mobility of the electrons or, what is the same, a slab of SiC under the same electrical field will exhibit lower velocity of the carriers and, consequently, will have smaller current than a silicon one. In other words, in order to reduce R_{DSon} in components based on SiC the concentration of impurities has to be higher. That is important since the resistivity of parts of the component being on the path of the main current is reduced and leading to lower losses, i.e., R_{DSon}.
- Significantly larger value of the breakdown field K_C together with lower intrinsic concentration (Table 4.2.1) leads to higher breakdown voltages.
- Significantly smaller capacitances (Table 4.2.3) lead to faster switching.
- Larger maximal carrier velocity (Table 4.2.1) allows for larger currents at high fields.
- Higher operating temperature: Because of its high melting temperature, the SiC device can operate well over 400 °C—much higher than the maximum allowable junction temperature of standard silicon technology (150 °C). This property results in significant cost reduction of the cooling system since less expensive cooling materials and methods can be used. Even in the extremely high ambient temperature, enough temperature difference can be obtained to take the heat out of the semiconductor package. The ambient air temperature can be as high as 100 °C without any concern.
- Higher thermal conductivity: SiC has thermal conductivity about 3 times higher than that of silicon. Therefore, heat dissipation can be conducted from within the semiconductor with a much lower temperature drop across the semiconductor material. These properties make SiC ideal for high-power (>1200 V, >100 kW), high-temperature (200–400 °C) applications, but also suitable for less stringent usage as well.

Speaking on high voltages Fig. 4.2.89 depicts the span of maximum voltages for devices based on the three materials considered here.

To summarize, potential exists for a component based on SiC to exhibit many advantages over a silicon-based device when components intended to be used in power switching are considered.

We will consider here a SiC-based MOS transistor which, as does his silicon-based counterpart, may have several different structures. Here we will consider a

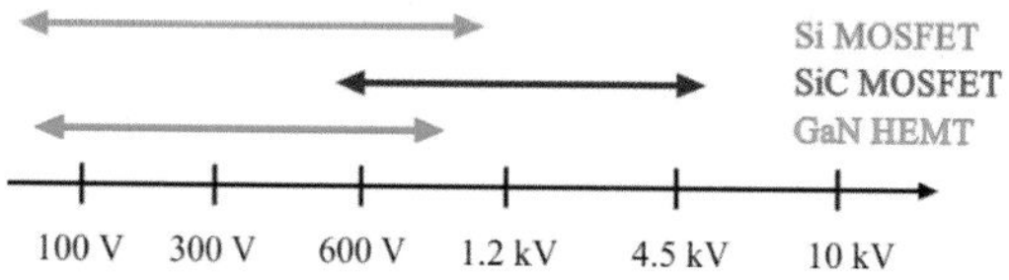

Fig. 4.2.89 Possible rated voltages for devices produced in the three materials

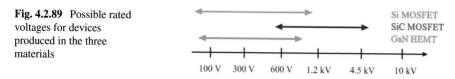

structure which is the same as in the case of the Si-based MOS which is depicted in Fig. 4.2.49.

Of, course, due to the structure, all considerations discussed for the Si-based device will be held here with the fact that due to the smaller intrinsic concentration, larger bulk concentration, and higher breakdown field, some dimensions will be accommodated (reduced). Note that the reduction of the distance between the channel end and the drain contact will lead to reduction of the resistance being connected in series with the channel, i.e., to reduction of R_{DSon}.

In the next the properties of a SiC transistor will be considered first.

Figure 4.2.90 depicts the output characteristics of this transistor.

As can be seen the salient property of SiC MOS transistor is relatively high threshold voltage. That may be observed also from the static transfer characteristic depicted in Fig. 4.2.91. Voltages between the gate and the source of order of 7 V are needed to achieve drain currents of order of 30 A. The ratings are $V_{DSmax} = 900$ V, $I_{Dmax} = 35$ A and $R_{DSon} = 65$ mΩ.

From the safe operating area (SOA) diagram depicted in Fig. 4.2.92 one may deduce that, for pulses shorter than 10 μs, maximum dissipation of 7 kW is allowed. That is a performance that may be rarely found if at all.

It is to be noted however that for low voltages the drain current is limited to much lower values due to the limiting value of the R_{DSon}. That means that for drain voltages less than 10 V the current rating is decreased too. Similar conclusions could be deduced for the MOS transistor's SOA depicted in Fig. 4.2.64.

To check for the effects of the Miller capacitance the dynamic behavior of the gate circuit was investigated by simulation of the circuit of Fig. 4.2.61a. The first results are depicted in Fig. 4.2.93 for $L = 0.1$ μH. Two values of the slope of the input signal were exercised: 50 and 500 V/ms. As can be seen for the slower signal

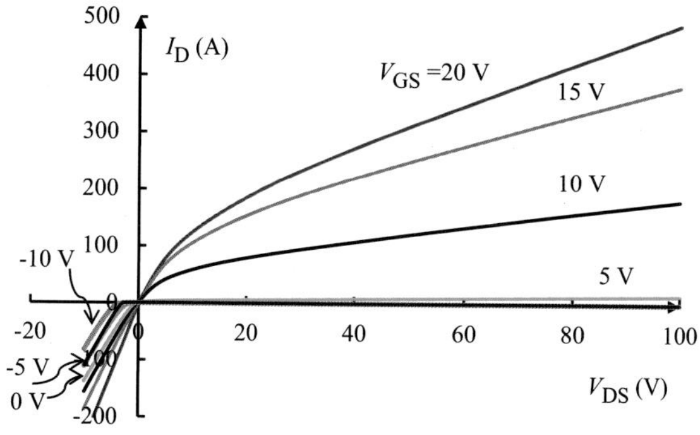

Fig. 4.2.90 Output characteristics of a SiC MOS transistor

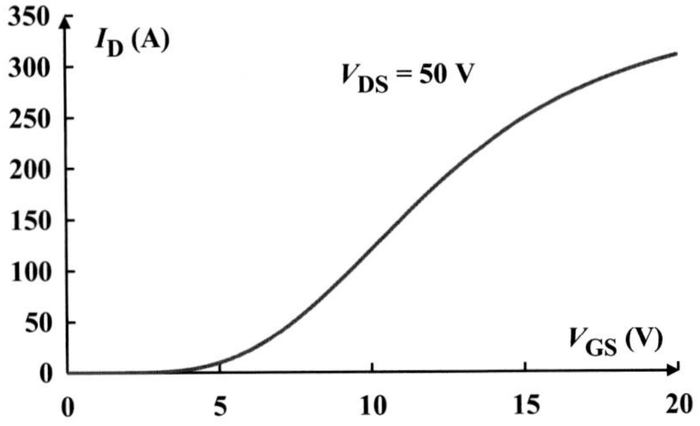

Fig. 4.2.91 Static transfer characteristic of a SiC MOS transistor

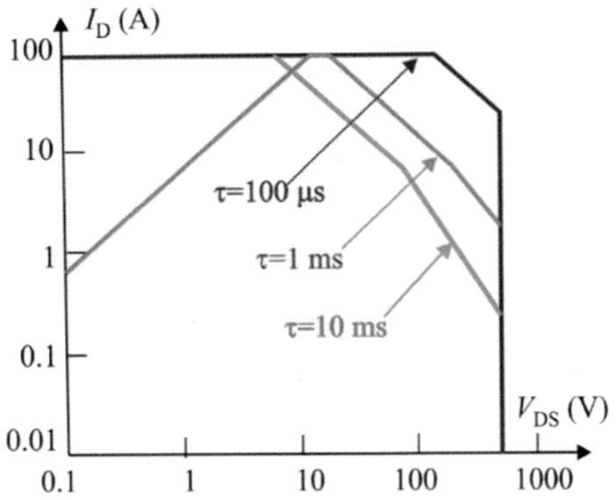

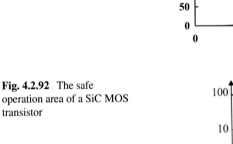

Fig. 4.2.92 The safe operation area of a SiC MOS transistor

the peak gate current is about 150 μA while the faster signal produces gate current of a value larger than 1.7 mA.

The same circuit but with $L = 1$ μH was simulated again, and no significant difference was recognized except for oscillations which arise at the moments of change of the slope ($t = 0$ and $t = 25$ μs) for the case of the slope of 500 V/ms. This response is depicted in Fig. 4.2.94. Larger values of L (not shown here) induce oscillations lasting all the time with much more noticeable amplitudes.

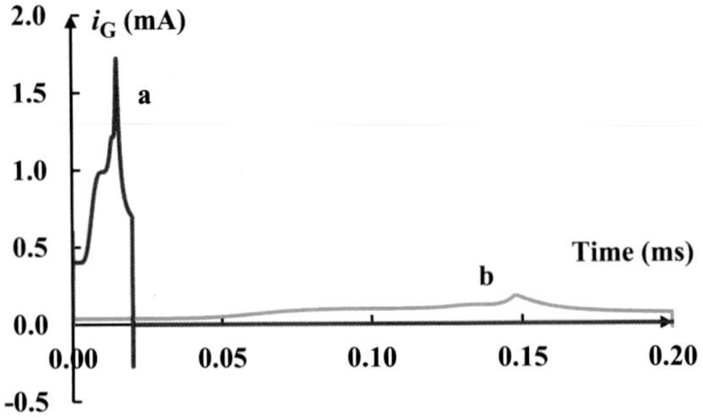

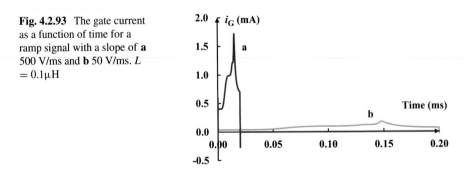

Fig. 4.2.93 The gate current as a function of time for a ramp signal with a slope of **a** 500 V/ms and **b** 50 V/ms. $L = 0.1$ μH

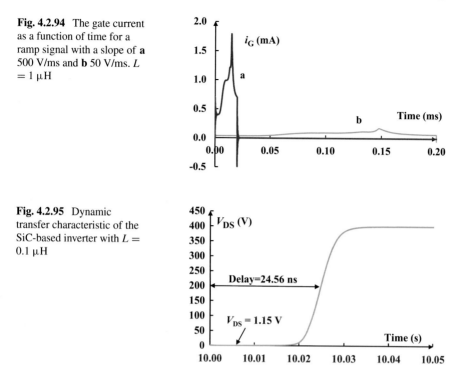

Fig. 4.2.94 The gate current as a function of time for a ramp signal with a slope of **a** 500 V/ms and **b** 50 V/ms. $L = 1\ \mu H$

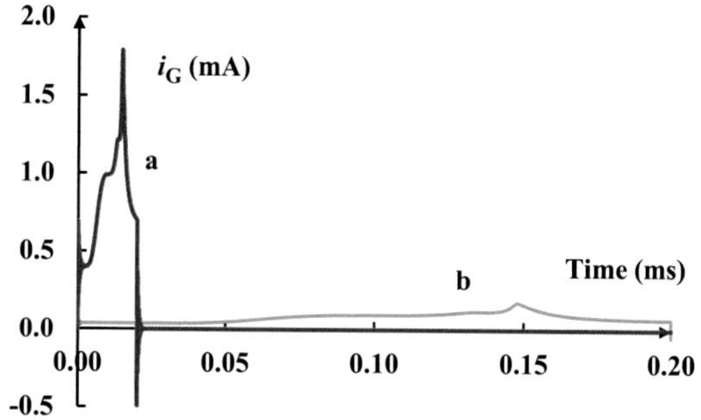

Fig. 4.2.95 Dynamic transfer characteristic of the SiC-based inverter with $L = 0.1\ \mu H$

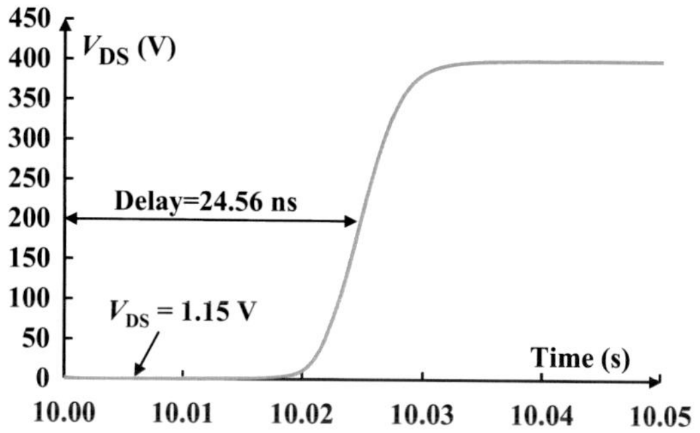

The dynamic transfer characteristic is obtained again by simulation of the circuit of Fig. 4.2.63a with the value of V_{XX} risen to 20 V. Here (Fig. 4.2.95) the result for $L = 0.1\ \mu H$ is depicted.

Delay of approximately 25 ns may be observed accompanied with a V_{DS} voltage for the on-state of 1.15 V only. This, from the delay point of view, puts the SiC switch in between the MOS and the HEMT (while not forgetting the difference in the inductance value). From the point of view of minimum voltage, the SiC transistors are preferable than both MOS and HEMT.

As expected, with risen value of L, oscillations may be observed in the output voltage as depicted in Fig. 4.2.96 for $L = 1\ \mu H$.

4.2.4 Thyristors

Historically, thyristors were the first solid-state device used as power switch. The name comes from the construct thyr(atron) + (trans)istor the thyratron being an old device, i.e., a gas-filled tube used as a high-power electrical switch and controlled rectifier. These were implemented in thyristors-based circuit breakers for high, medium, and low voltage implementations. A high variety of implementations are

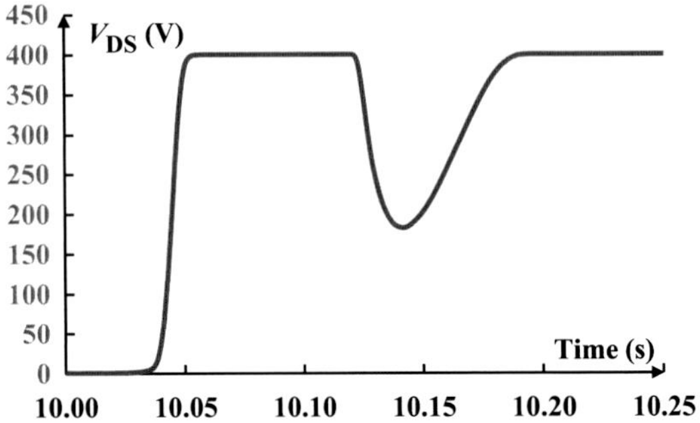

Fig. 4.2.96 Dynamic transfer characteristic of the SiC-based inverter with $L = 1\ \mu H$

reported related to DC circuit breakers, too. Here a short description of its functionality will be given although it is not used in power amplifiers. The reason for that was to have all power components described at a single place in the LNAE.

The thyristor or Silicon Controlled Rectifier (SCR) is a four-layer structure with three external terminals: anode, cathode, and gate. Its structure is depicted in Fig. 4.2.97a. The top P-layer is used as an anode, A (positive voltages for normal operation are expected) while the bottom N-layer is considered cathode, K. The internal (second from the top) N^- layer is not connected while the internal (third from the top) P-layer may be disconnected which makes the thyristor behave as a Shockley diode. If connected, however, it is referred to as gate, G, and the device is named SCR. The schematic symbol of the SCR is depicted in Fig. 4.2.97b.

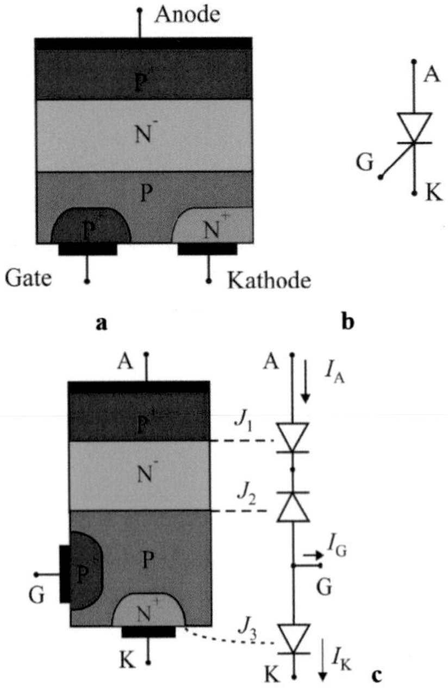

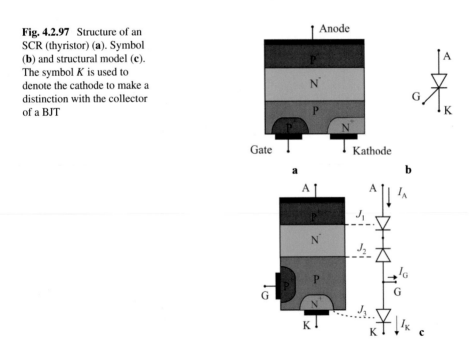

Fig. 4.2.97 Structure of an SCR (thyristor) (**a**). Symbol (**b**) and structural model (**c**). The symbol *K* is used to denote the cathode to make a distinction with the collector of a BJT

To facilitate the explanation of working mechanism of this device we will start with the Shockley diode first. To that end the model depicted in Fig. 4.2.97c will be used with the gate disconnected.

As may be seen there are three p-n junctions within this diode being enumerated from one to three going from the top toward the bottom of the figure. On the left-hand side the same structure as in Fig. 4.2.97a is depicted but properly rearranged to allow for the explanations.

The characteristic of the Shockley diode is depicted in Fig. 4.2.98a. One may recognize several regions in that characteristic. First is the inverse blocking region which will be examined with V_{AK} starting from the origin. With reference to Fig. 4.2.97c, 4.2.99a depicts the polarization of the junctions of the Shockley diode in the case of inverse blocking. The junctions J_1 and J_3 are inverse biased while the junction J_2 is forward biased. Since at least one junction in the series connection is inverse biased I_A is equal to its inverse saturation current incremented by the leakage current. Now, since $I_A \approx 0$ A, J_2 is practically not biased and consequently the whole of V_{AK} is distributed on J_1 and J_3 only. Large inverse voltages ($|V_{AK}|>|V_{IB}|$) will produce avalanche breakdown in the inversely biased diodes and consequently in the Shockley diode.

Next is the so-called forward blocking region in which the increase of V_{AK} produces only a very small current. With reference to Fig. 4.2.97c, Fig. 4.2.99b depicts the polarization of the junctions of the Shockley diode in the case of forward blocking. The junction J_2 is now backward biased while both junctions J_1 and J_3 are forward biased with no current and voltage since I_A is now equal to its inverse saturation current of J_2 incremented by its leakage current.

After V_{AK} reaches the firing voltage (V_{FB}) (after breakdown of J_2) the voltage decreases abruptly to a value approximately equal to a voltage of a forward biased diode. Further increase of the anode-to-cathode polarization results in fast rise of I_A while V_{AK} remains almost constant. This part of the characteristic is named forward conducting region.

The part of the characteristic between the forward blocking and the forward conducting regions being characterized by a negative resistance is referred to as the regenerative region. One says that "firing" happens in the device.

Figure 4.2.99c depicts the situation within the device in the case of forward conducting. Since large current is experienced while very low voltage is observed one may conclude that all three junctions are forward biased as depicted. For that to happen it is necessary for the potential of the inner P^- layer to be simultaneously higher than the potentials of both neighboring n-layers by a margin not smaller than the built-in potentials of the corresponding junctions. That in turn asks for a large amount of positive charge (holes) to be accumulated in the P^- layer during the regeneration. Now, the forward bias voltage of J_2 is subtracted from the forward bias voltages of J_1 and J_3 to produce a small value of V_{AK}.

There are two important points that are to be mentioned for this device. First, the device is not bilateral, i.e., there is no regenerative effect for negative values of V_{AK} as can be seen from Fig. 4.2.98a. Second, and more important, is the fact that even for forward biasing the characteristic is not reversible. Namely, after firing one cannot go back into the regenerative region. One can "extinguish" the device by reducing V_{AK} to zero or by reducing I_A below a threshold value named a holding (or latching) current, I_H.

The Shockley diode is seen as a trigger switch when firing SCR. Typical ratings are $V_{FB} = 40$ V, $I_H = 10$ mA, $V_{IB} = 20$ V, $C_{AKIVAK=0} = 40$ pF, and $I_{Amax} = 10$ A.

The firing process of the pnpn structure may be controlled by purposely injected holes (positive charge) into the P^- layer via a terminal added to it. That terminal is named gate, and the new device is the SCR. The firing with the help of the gate current is frequently explained using the two-transistor model as depicted in Fig. 4.2.100. Here Fig. 4.2.100a represents another rearrangement of the structure to allow for the two transistors to be recognized. On the top is a pnp BJT whose emitter is connected to the anode and the collector to the gate. Its base is connected to the internal N^- layer. On the bottom is a npn BJT whose collector is connected to the internal N^-

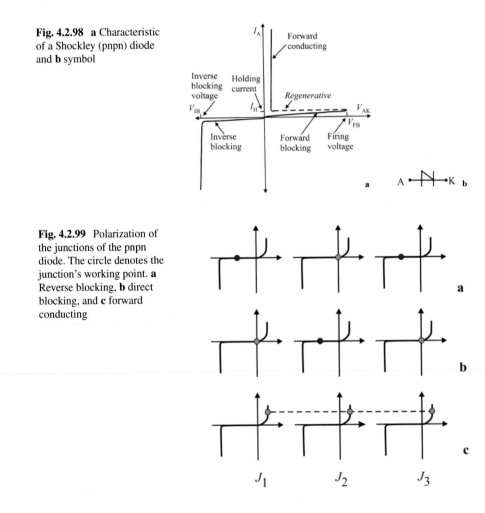

Fig. 4.2.98 **a** Characteristic of a Shockley (pnpn) diode and **b** symbol

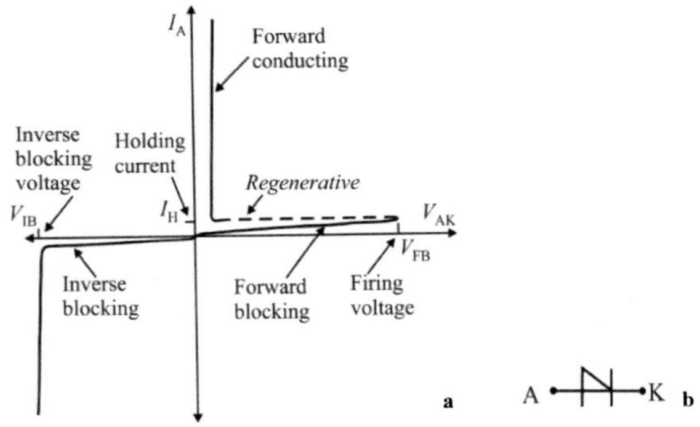

Fig. 4.2.99 Polarization of the junctions of the pnpn diode. The circle denotes the junction's working point. **a** Reverse blocking, **b** direct blocking, and **c** forward conducting

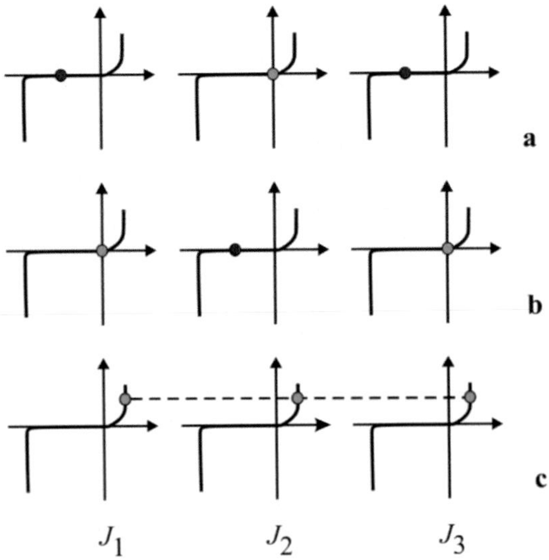

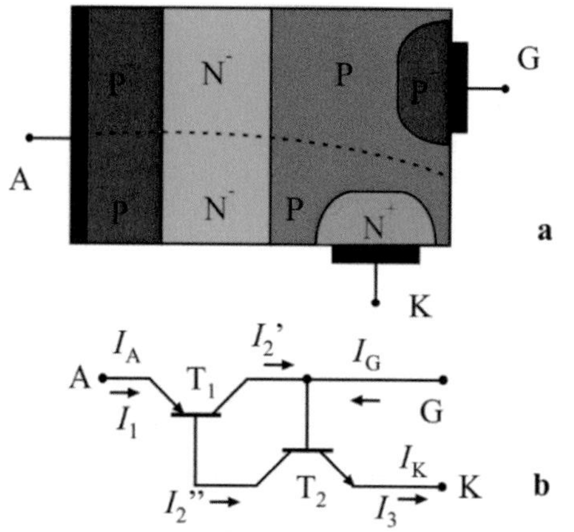

Fig. 4.2.100 Two-transistor functional model of the pnpn structure

layer while its emitter is connected to the cathode. Finally, its base is connected to the gate. One may easily see that a loop is formed: Base T_1-Collector T_1-Base T_2-Collector T_2-Base T_1. When the loop gain becomes larger than unity this loop will exhibit regenerative behavior due to the positive feedback.

To come to the condition for a regenerative process one may analyze the currents in the circuit of Fig. 4.2.100b. Assuming that the BJTs have equal inverse collector saturation currents incremented by the leakage current (here denoted as $I_{c0}/2$) based on the Ebers-Moll model, the following equations may be written for the collector currents of T_1 and T_2, respectively,

$$I_2' = \alpha_1 I_1 + I_{c0}/2 \tag{4.2.59}$$

$$I_2'' = \alpha_2 I_3 + I_{c0}/2. \tag{4.2.60}$$

where α_1 and α_2 are the corresponding forward current gain coefficients defined by (4.2.26).

For the whole circuit one may write

$$I_A = I_K = I_1 = I_3 = I_2' + I_2'', \tag{4.2.61}$$

which leads to

$$I_A = \frac{I_{c0}}{1 - (\alpha_1 + \alpha_2)}. \tag{4.2.62}$$

By substituting α with β and assuming (for simplicity) $\beta_1 = \beta_2$, one obtains

$$I_A = \frac{I_{c0}}{1 - 2 \cdot \beta/(1 + \beta)}. \tag{4.2.63}$$

Now, the value of α and β for very small currents is small so that $\alpha_1 + \alpha_2 \ll 1$ and I_A is small and constant. When V_{AK} rises from zero, the leakage current (added to I_{c0}) is following what gradually increases the value of α and β as depicted in Fig. 4.2.36 (looking from the origin to the right). At the moment when $\beta = 1$ condition for a regenerative process is established and the device "fires."

By bringing a gate current so that $I_{B2} = I_2' + I_G$ the conditions for regeneration are relaxed since $I_{B1} = \beta \cdot I_{B2} \approx \beta \cdot (I_G + \beta \cdot I_{B1})$ or $I_{B1} \approx \beta \cdot I_G/\left(1 - \beta^2\right)$.

This effect may be observed in Fig. 4.2.101 where characteristics of an SCR are depicted for various values of the gate current. Larger gate currents will lead to smaller forward blocking voltages (V_{FB}).

This phenomenon may be expressed in the opposite way. For fixed V_{AK}, even if it is small, the SCR may be fired by proper value of the gate current. Since, however, after firing, there is no need for the positive gate charge to be supported from outside of the device, I_G may be set back to zero, i.e., the firing may be realized by a short pulse of the gate current.

This property of the SCR puts it in front of the power BJT for switching applications since no current is needed to keep the device in on-state. That significantly reduces the design requirements for the gate (base) controlling circuit and the overall dissipation.

Still, however, both the power BJT and the SCR behave themselves as current controlled switches.

High input current versus negligible input current is not the only difference between the thyristor and the voltage-controlled devices discussed earlier here. Namely, there is a limitation on the speed of rise of the anode voltage of the switched-off thyristor. This is usually referred to as $(\mathrm{d}v/\mathrm{d}t)_{\max}$. Due to this phenomenon the switched-off thyristor may be unintentionally switched on at relatively *low anode voltages* while the gate terminal is disconnected (without a gate current).

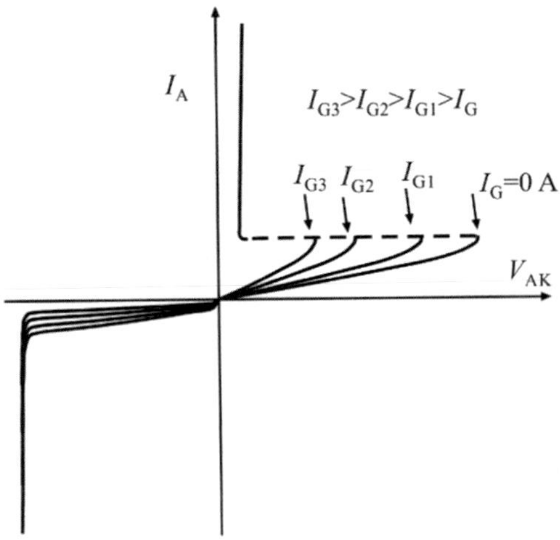

Fig. 4.2.101 Characteristics of an SCR for various gate currents

The phenomenon may be explained based on the depiction in Fig. 4.2.102a. Here the depletion capacitances of the p-n junction are shown as they appear in the device. When in off-state no anode current is flowing through the component so that $i_{p1} = i_{p2} = i_{p3} = C_{ek} \cdot (dv_{AK}/dt)$. Here C_{ek} is the series connection of C_{pn}, C_{np}, and C_{GK}. The gate voltage in this case may be expressed as

$$v_{GK} = \int_0^\tau \frac{1}{C_{GK}} \cdot C_{ek} \cdot \left(\frac{dv_{AK}}{dt}\right) \cdot dt. \tag{4.2.64a}$$

Here τ is the time instant of observation. To simplify, we will assume here that all three quantities under the integral are constant in time so that, after integration, one gets

$$v_{GK} = \Sigma \cdot \tau \tag{4.2.64b}$$

where

$$\Sigma = \frac{1}{C_{GK}} \cdot C_{ek} \cdot \left(\frac{dv_{AK}}{dt}\right) \tag{4.2.64c}$$

When Σ is large enough, i.e., when dv_{AK}/dt is large enough, v_{GK} eventually becomes larger than the built-in potential of the corresponding p-n junction and conditions are established for the thyristor to be fired.

No such phenomenon is possible in the case of a voltage-controlled device as depicted in Fig. 4.2.102b. As long as the resistance which is playing the role of the short circuit between the gate and the source is small enough, no gate voltage will be developed (the gate voltage will be smaller than the threshold voltage of the device)

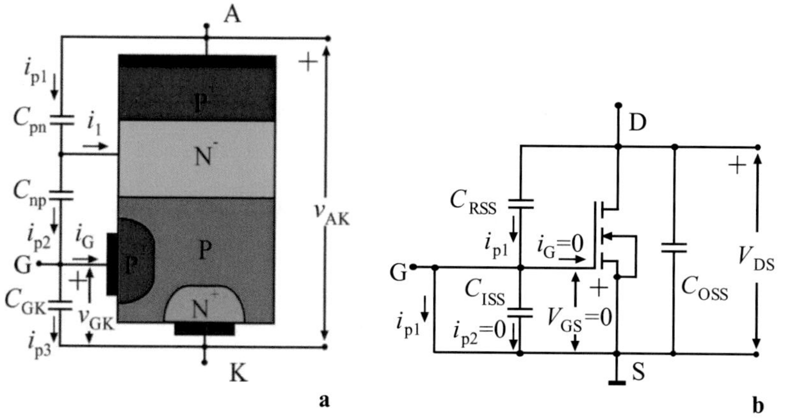

Fig. 4.2.102 **a** Explanation of the firing of a thyristor by fast change of the anode voltage and **b** reluctance of a voltage-controlled device to be switched on by fast change of the output voltage

and no anode (drain) current may be established. In the cases when very large values of the current i_{p1} are expected, one may use a Light Activated Silicon Controlled Rectifier (LASCR), also known as a Light Triggered Thyristors (LTT), as the device is acting as the short circuit.

In addition, since the SCR cannot be switched off by its input (gate) current, when implemented in AC circuit breakers, one is to wait until the next zero crossing of the power voltage.

Finally, let us consider the influence of an impedance connected between the gate and the cathode to the SCR's gate circuit as in Fig. 4.2.103. Let it, first, be a resistor R_{G}. This resistance is often built into SCR by applying the cathode contact to a part of the gate area during metallization. This results in a partial short circuit between the gate and the cathode which is manifested by connecting a small resistor of the order of a few ohms in parallel to the p-n junction. Regardless of whether this resistor is added externally or originated from within the SCR, it provides a path for the inverse leakage current of the J_3 junction. Therefore, part of the thermally generated current as well as part of the current of capacitor C_2 at high dv/dt is closed over this resistance. It is now clear that the presence of this resistance makes the SCR less sensitive to temperature as well as to dv/dt. Of course, the sensitivity for switching is now reduced (simply, a larger external gate current is needed to achieve switching), but with low-power thyristors, an increase in I_{GA} (the gate current needed for firing) is often desirable in order to eliminate the uncertainty in operation due to ignition using random small current pulses.

If a capacitor or inductance is used as Z_{G} in the gate circuit as in Fig. 4.2.103 the situation is somewhat more complex. The capacitor, first of all, reduces the sensitivity to high-frequency noise, which means that short-term (high-frequency) unwanted impulses will not be able to switch the thyristor. In addition, the capacitor reduces the sensitivity to dv/dt as it is connected in parallel to C_3 and thus increases the amount of charge required for direct polarization of the J_3 junction. Of course, the external capacitor has no effect on low-frequency (slow) excitation signals. The disadvantage of connecting this capacitor is reflected in the fact that the required rate of increase of the I_{G} (the gate current) increases and that it can cause a shutdown

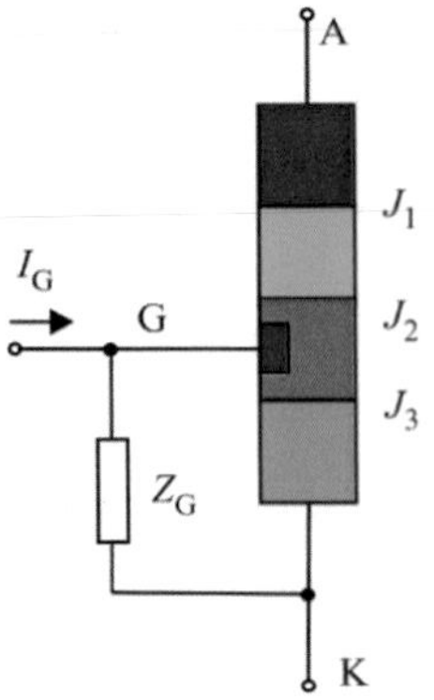

Fig. 4.2.103 Impedance in the gate of a thyristor

error since the capacitor maintains direct polarization of the gate even after the anode current falls below I_H. The inductance in the gate circuit causes a differentiating effect and therefore reduces the sensitivity to slow changes, while maintaining the sensitivity to sudden changes. The influence on the switching time at the switch-off is opposite to the influence of the capacitor, which means that the switch-off time can be significantly reduced due to the negative (opposite) current generated by the coil when the SCR is switched off.

The SCR was and is implemented in many solutions where current switching is sought. It is worth mentioning that in addition to the ubiquitous silicon-based SCRs, experimental silicon carbide devices have been produced. This should allow for either physically smaller or higher power capable devices.

Here is a short list of typical ratings of a medium power thyristor: $I_H = 10$ mA, $V_{AK\ (on\ state)} = 1.6$ V, $(dv/dt)_{max} = 20$ V/(μs), $I_{Amax} = 10$ A, $I_{G\ max} = 5$ mA, and $V_{IB} = 400$ V.

4.2.4.1 GTO

To overcome the inability of the thyristor to be switched off by a gate voltage, a new device was developed named gate turn-off thyristor (GTO). A cross section of this device is depicted in Fig. 4.2.104a. Interdigital structures are built on both sides of the die to allow for more large currents to flow (due to the positive temperature coefficient of a bipolar device) and to reduce the inverse breakdown voltage of the gate-to-cathode diode. In that way, for a device which is in on-state, after breakdown of the gate junction, electrons are injected into the P layer at the gate side which neutralizes the positive space charge of holes and reduces the anode current to zero.

As a price to be paid for the ability to fully control the switching from the gate terminal a device with larger on-state voltage is obtained. It is slower, too, since much larger gate voltage is necessary to establish the gate current when switching off. Nevertheless, the implementation of a GTO seriously reduces the complexity of the controlling circuit and the device had found broad applications especially in DC power switching sub-systems.

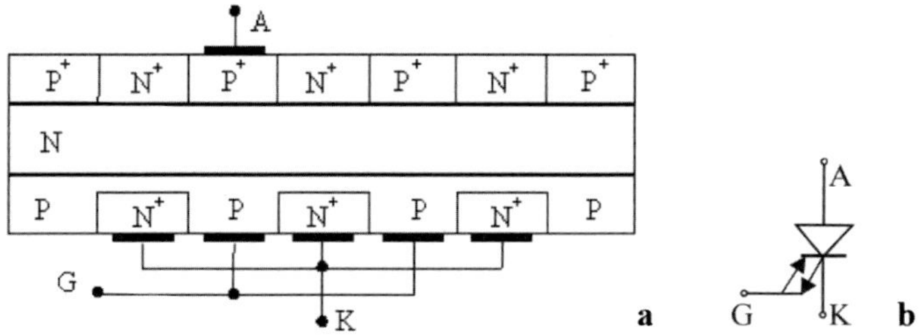

Fig. 4.2.104 **a** Structure of a GTO and **b** schematic symbol

Figure 4.2.104b depicts the schematic symbol for the GTO.

In the case when very high currents and fast switching are required, the parasitic components within the controlling circuitry become too influential. That was the main driving force behind the development of a new device named integrated gate commutated thyristor (IGCT). As the name implies in this case the complete controlling circuitry is integrated on the same die with the thyristor to enable more effective switching. It is still, however, a current controlled bipolar device.

4.2.4.2 Triac and Diac

The triac named from *Triode for Alternating Current* is a 3-terminal bidirectional device which can conduct current in both directions.

A triac is a component that can be considered as two SCRs connected in (anti) parallel, i.e., in such a way that the cathode of one is connected to the anode of the other and vice versa while the gate is common. As such, it can conduct current in both directions if the ignition conditions are met. The gate signal will be positive or negative during ignition. Due to its ability to control high power of DC signals, the triac is gaining increasing use in cars for regulating lighting, engine speed, or heating. Due to the similarity with SCR, the notions of triac limitations will not be redefined here, but the mode of operation and the given performance will be described.

Figure 4.2.105 shows the symbol, the general characteristics of the triac, and several cross sections. The triac symbol indicates a two-way flow of the gate current using two oppositely oriented diodes. The main terminals are marked with A_1 and A_2 since the terms cathode and anode lose their meaning. Based on Fig. 4.2.105c we see that the triac is a complicated component containing from A_2 to A_1 two oppositely oriented PNPN structures. First, we have the $P_1N_1P_2N_2$ structure where the area P_2 has the role of a gate, and then the structure $P_2N_1P_1N_0$ where the role of the gate is taken over by the area N. In this figure possible situations of triac polarization are shown. The upper half of Fig. 4.2.105c shows the possibility of firing by means of a negative or positive current pulse at the gate in the case when A_2 is at a more positive potential than A_1. It can be seen that when the voltage at the gate is positive the current flows from the gate to A_1 via the forward biased P_2N_2 junction. Therefore, conditions are created for the structure $P_1N_1P_2N_2$ to conduct.

When the signal at the gate is negative, the triac can also be fired with the difference that now the gate current flows from P_2 to N. Of course, the magnitude of the gate current required to ignite at negative anode voltage is not the same as at positive voltage at the gate due to asymmetry and the influence of the anode current on the distribution of charge in the area P_2. The rest of Fig. 4.2.105 refers to the firing process at positive and negative gate currents when A_1 is at a higher potential than A_2.

As shown in Fig. 4.2.105c the work in the first quadrant of the characteristic, with the exception of the negative gate current, is equivalent to the work of an SCR. This means that the component can be converted from non-conductive to conductive

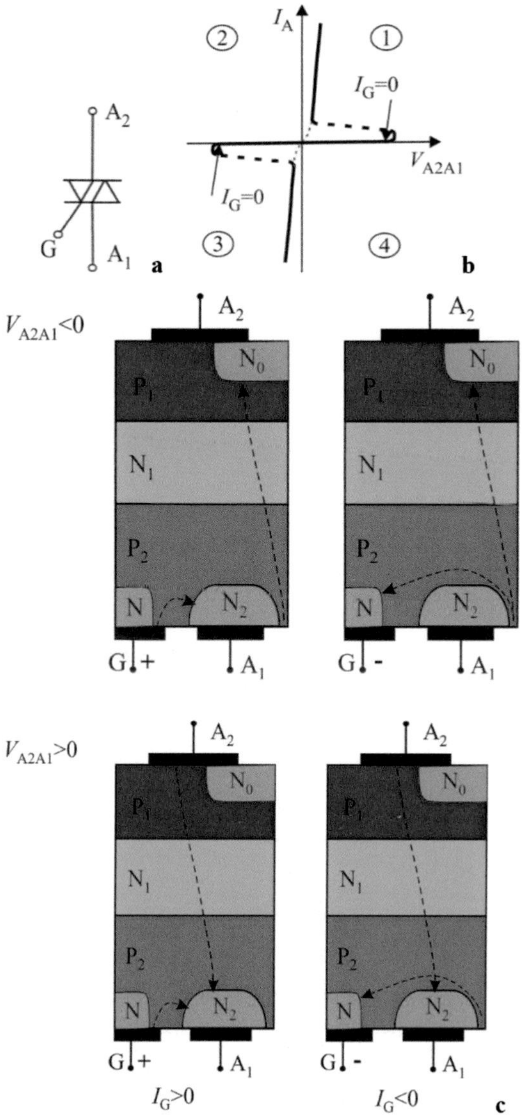

Fig. 4.2.105 Triac, **a** symbol, **b** generalized characteristic, and **c** process of firing

state by exceeding the maximum direct voltage or by supplying the appropriate gate current. When the triac is conducting, the gate current also loses control until the anode current is reduced to a value below the holding current. Firing requires similar values of current (tens of milliamperes) or voltage (1–2 V).

The triac parameters are defined in the same way as for SCR and the typical values for low and medium power thyristors coincide. However, the switching-off time is of greater importance for the triac since the SCR is conducting only in one half-cycle and has enough time to switch-off during the negative half-cycle. The triac conducts

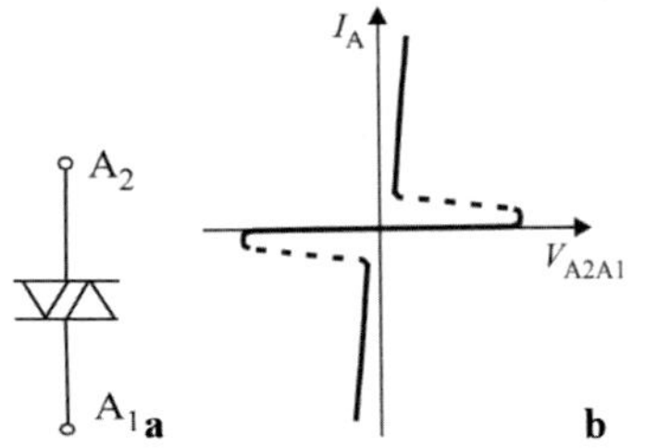

Fig. 4.2.106 Diac **a** symbol and **b** generalized characteristic

in both halves of the AC voltage between A_1 and A_2, so it needs to be switched off quickly in order to realize any concept of control (so as not to conduct all the time).

If from the structure of Fig. 4.2.105c one removes the area N and the corresponding terminal of the gate, a new component may be created named diac. The name comes from *Diode for Alternating Current*. It is a 2-terminal bidirectional semiconductor. It is used together with a triac to conduct current in both directions in AC phase-control, dimming, speed-control, and power-control applications (such as protection against high voltage -when connected in parallel to the component it protects).

The characteristic and symbol of the diac are shown in Fig. 4.2.106. It is easy to see that the diac represents two PNPN diodes connected in opposition.

4.2.4.3 UJT and CUJT

Unijunction Transistor (UJT) is one of the oldest and simplest semiconductor components. As its name suggests, the UJT (or double-base diode) is a three-terminal component containing only one p-n junction. It is mainly a switching component that is similar in characteristics to thyristors and is therefore considered here. UJT is most often used in oscillators, frequency dividers, and circuits for triggering (converting to the conductive state) of power components such as thyristors, generators of different waveforms, and the like. It is considered that UJT performs a number of functions in these circuits in a very efficient way.

Figure 4.2.107a shows the cross section of the UJT, which is reminiscent of the earlier way of making. The component consists of an N-type plate with a low concentration of impurities at the ends of which metal contacts marked B_1 and B_2 are made. These terminals are called bases.

The plate is in fact a resistor whose resistance is determined by concentration and dimensions. P diffusion with high concentration of impurities was formed on one side of the plate. We call this part of the UJT the emitter. Figure 4.2.107b shows the symbol for UJT. It resembles the structure of a component, with the arrow pointing in the direction of the p-n junction. Figure 4.2.107c shows the equivalent circuit of the UJT. It should be noted that the circuit resembles a simple voltage divider formed by the resistors r_{B1} and r_{B2}, with the emitter diode contacted in the middle. Therefore, the point X is at a fixed potential determined by the size of the connected battery

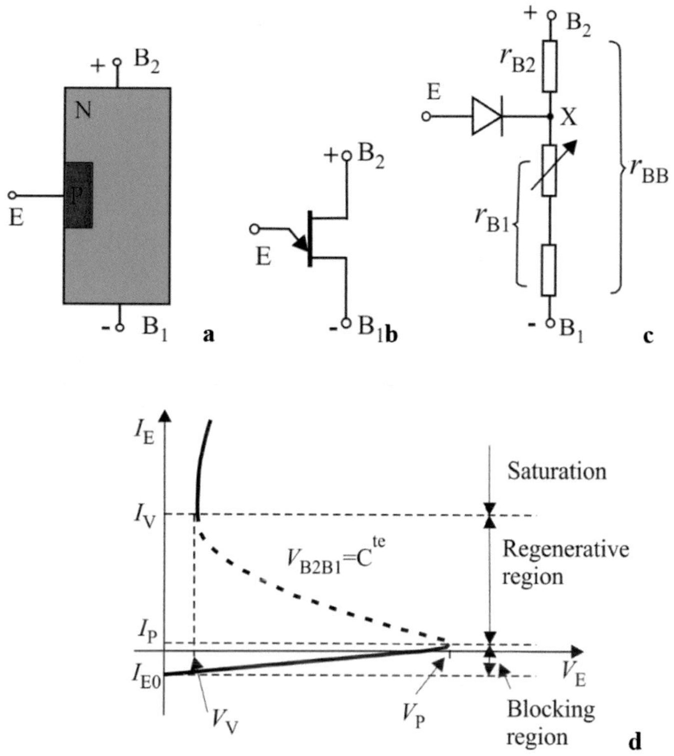

Fig. 4.2.107 UJT, **a** structure, **b** symbol, **c** electrical equivalent circuit, and **d** generalized characteristic for a single value of V_{B2B1}

between bases B2 and B1 (V_{B2B1}). For the potential of this point we have:

$$V_X = r_{B1} V_{B1B2}/(r_{B1} + r_{B2}) \tag{4.2.65}$$

As long as the emitter potential is at a lower level than the cathode of the diode (V_x), it is backward biased, and the emitter current consists of its inverse current as in Fig. 4.2.107d. When the emitter potential increases above V_X by the magnitude of the diode threshold, the diode conducts, and a characteristic regenerative process occurs in the UJT.

The resistance r_{B1}, in Fig. 4.2.107c, is divided into a constant (r_s) and variable (r_n) part. The latter refers to the body of the semiconductor between the p-n junction and the ohmic contact. When the diode conducts, the space charge in the area from the emitter to the base B_1 increases, which means that the concentration increases (injection of holes from the emitter, to maintain electrical neutrality, causes injection of electrons from B_1) and reduces the specific resistance of the semiconductor. This results in a decrease in r_{B1} and thus in V_X. A decrease in V_X at a given emitter potential leads the diode to the region of higher direct biasing or to the region of higher diode (emitter) current.

It is easy to see that the process is regenerative and soon the voltage between the emitter and B_1 drops to a small value (r_n drops to a very small value), and the emitter current rises sharply. The magnitude V_x is of decisive importance for the magnitude of the fining voltage V_P, and thus the magnitude of V_{B2B1}. By changing the battery voltage, V_P also changes and the braking part of the characteristic shifts (for higher V_{B2B1} to higher voltages V_P). The UJT parameter that tells the value of the firing voltage for a given battery voltage is:

$$\eta = r_{B1}/(r_{B1} + r_{B2}) \quad (4.2.66)$$

If we denote by V_D the voltage at the forward biased emitter junction, then also applies

$$\eta = (V_P - V_D)/V_{B1B2}. \quad (4.2.67)$$

Depending on the type of UJT, η can take values from 0.4 to 0.9. The following parameters of UJT are also of interest. The resistance between the bases r_{BB} determines the consumption of the component in the off-state. It is of the order of kilohms and exhibits a pronounced temperature dependence with a positive temperature coefficient (about 0.5%/K). The inverse leakage current of the emitter is less than microamperes and has the usual temperature dependence. The current at the time of firing (I_P) is of the order of microamperes, and the current immediately after firing (at the boundary of the negative resistance area) is a few milliamperes (I_V). The minimum voltage when UJT is in on-state (V_V) is of the order of a few volts.

Before we consider other components that have a similar structure or characteristics as UJT, some of the simplest circuits with UJT will be shown.

Figure 4.2.108a shows an oscillator circuit using UJT. It is a circuit that, thanks to the presence of the V_{BB} battery and the characteristics of the UJT, generates a periodic waveform at the output. It works as follows. When the power supply (V_{BB}) is switched on, the UJT is switched off and the capacitor C starts to be charged via R (point A in Fig. 4.2.108b). From B_2 to B_1 flows the current I_1 determined by r_{BB}. When the voltage on the emitter (on the capacitor) reaches V_P, the emitter becomes forward biased and r_{B1} drops to a very small value, allowing a large emitter current to flow (point B in Fig. 4.2.108b). Now a small resistance is connected in parallel to the capacitor and it gets discharged quickly via r_{B1} and R_1 (point C). The current through R_1 rises sharply and a short-term pulse is generated (as in Fig. 4.2.108b) since the capacitor is quickly discharged via a small resistor. The minimum value of the voltage on the emitter corresponds to the point on the characteristic with coordinates (V_V, I_V). Now the UJT gets switched off again (the emitter potential smaller than the base potential) and the process is repeated. By selecting R and C, the time constant of the capacitor charging, the slope of the leading edge of the emitter pulse, and thus the signal frequency can be controlled.

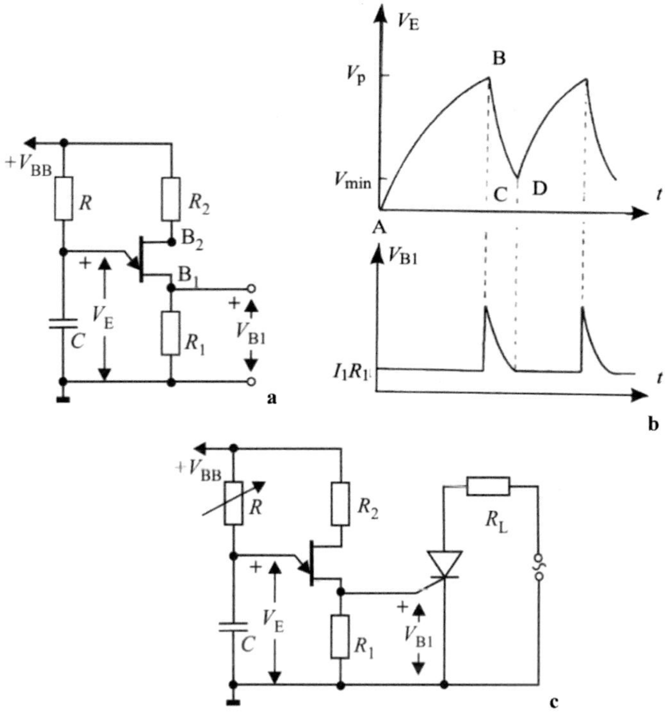

Fig. 4.2.108 Oscillator using UJT, **a** the circuit, **b** the waveforms, and **c** implementation in firing control of an SCR (switching alternating current)

Such an oscillator can be used to control the SCR gate as shown in Fig. 4.2.108c. At the time of pulse generation at B_1, if the anode voltage is positive, the SCR will get fired and remain conductive until the anode voltage becomes zero or negative. By changing *R*, the frequency (period) of firing is controlled, and thus the waveform of current (and power) through the consumer R_L.

Modern UJT is produced by planar technique (unlike Fig. 4.2.107a). Like other semiconductor components, it has its own complementary component, which is labeled CUJT (Complementary Unijunction Transistor). The most important advantages of CUJT are significantly less η as well as about three times less required voltage V_{BB} which makes this component compatible for integration. In addition, since electrons are the main carriers in the emitter, it can be used for higher frequencies. Finally, the inverse saturation current is significantly lower. The structure of an CUJT is depicted in Fig. 4.2.109a. Figure 4.2.109b shows its symbol. Its characteristic is identical to that of UJT, except that the signs on the coordinate axes are opposite. Figure 4.2.109c shows the corresponding oscillator with CUJT.

Although it has a characteristic similar to UJT, CUJT actually has a completely different structure. It acts as a four-layer component to which two resistors have been added as in Fig. 4.2.109a. The mode of operation of this component is easily

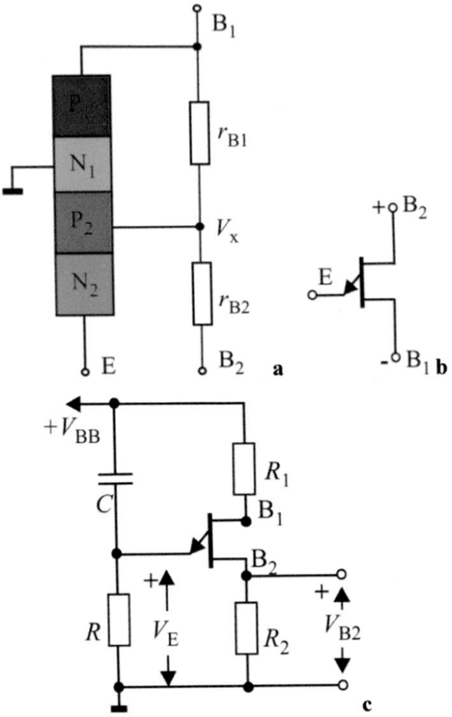

Fig. 4.2.109 CUJT **a** structure, **b** symbol, and **c** oscillator using CUJT

recognizable if we keep in mind that the area P_2 plays the role of the SCR's gate. If $V_E < (V_X - V_\gamma)$, all conditions for SCR firing occur.

Since resistors r_{B1} and r_{B2} have the same temperature coefficients, the temperature coefficient of voltage V_X is negligible, so CUJT has an increasing application in circuits with precise frequency control.

With a suitable combination of a four-layer structure and two resistors as in Fig. 4.2.110a creates a programmable single-junction transistor (PUT). The resistors R_1 and R_2, in this case, are usually located outside the housing (connected separately). The ability to choose r_{BB} is the basis of PUT's programmability. The symbol for PUT is shown in Fig. 4.2.110b together with the resistors.

Since the N_1 region now takes over the role of the gate, firing will not occur as long as the emitter potential is lower than V_X incremented by the threshold of the P_1N_1 diode. This means that by selecting the voltage at B_2 and by selecting the ratio of R_1 and R_2 we can program the firing of the PUT. It is important to note that the potential of the N_1 region determines the moment of ignition, and not the gate current, which was the case with SCR. This suggests that R_1 and R_2 can also be high resistance resistors.

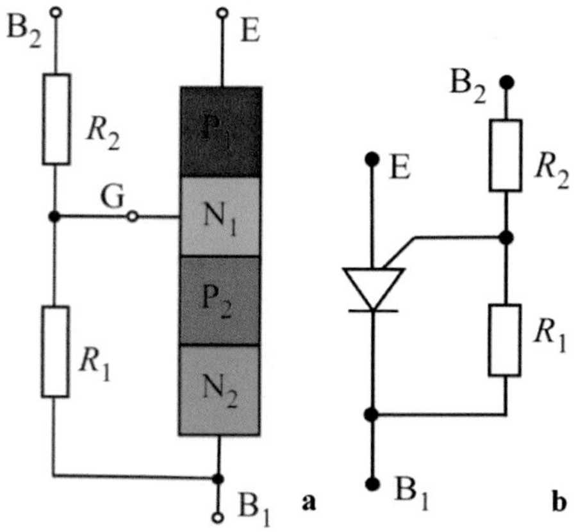

Fig. 4.2.110 PUT **a** structure and **b** symbol

4.3 Basic Theory of Large Signal Amplification

Abstract This is, according to the title of the book, its essential part. We will try here to introduce the reader into the concepts implemented for power amplifier design. Details will be given allowing understanding the main design suits. The reader will be ackwainted with the wide variety of possible solutions and the nishes where these solutions fit in. Of course, not all solutions deserve equal attention so that some will be defined only. In the text given, JFET, BJT, and MOS transistors are used which does not limit the applicability of the structures since IGBTs and HEMT may be easily implemented as substitutions.

4.3.1 Introduction

While the main requirement in electronic signal amplification usually is achieving a prescribed voltage gain; at the end, it comes to produce signals with large amplitudes of both the voltage and current. Of course, their product, the power delivered to the load, is frequently in focus as such. To achieve this one is to tackle problems related to braking the ratings in voltage, current, and power. In addition one is to confront the non-linearity of the transistor's characteristics and to avoid non-linear distortions of the signals. Finally, having in mind that the amplification in fact means conversion of the energy supplied by the power supply source(s) into the energy delivered to the load, one is to control the efficiency of this conversion.

For a given circuit the signal becomes large when it starts to exceed its ratings being that the maximum load voltage, the maximum load current, the maximum power dissipated at the active element or the maximum THD.

In any case, the amplifier handling large signals delivers increased power to the load. That is why large signal amplifiers are most frequently identified as power amplifiers (PAs).

To ensure efficient power transfer to the load one is to accommodate the output resistance of the power amplifier to the load resistance, usually being the resistance of a loudspeaker or of an antenna, which is expressed in several Ohms or several dozens of Ohms. That is achieved in two ways. Historically, when the only amplifying component was the electron valve, the only solution possible was to use a transformer

V. B. Litovski, *Lecture Notes in Analog Electronics*, Lecture Notes in Electrical Engineering 958, https://doi.org/10.1007/978-981-19-6528-9_3

to couple the load to the output of the amplifier. That is done nowadays in very high-power audio amplifiers due to the linearity that may be achieved. Of course, the transformer is in use in modern semiconductor-based circuits, too. Mostly, in the very same way. With the advent of BJT technology, however, a possibility was discovered to synthesize a power amplifier exhibiting low output resistance so that modern circuits, especially the integrated ones, use these ideas.

When it comes to the value of the power to be pronounced high, in addition to the limitations related to the power devices as such, one has to consider the signal frequency. Namely, it is not equally difficult to produce power at low and high frequency. For example, at audio frequencies one may easily purchase online a 1000 W amplifier, while at 1 GHz one can do that for a 10 W amplifier only. Probably a measure of the difficulty to develop a power amplifier may be expressed by the product of the power and the frequency (Pf). So for the audio amplifier, just mentioned, we would have $Pf = 1000 \cdot 10{,}000 = 10^6$ WHz, while for the high-frequency case it would be $Pf = 10 \cdot 10^9 = 10^{10}$ WHz. This does not mean that making a high-frequency PA is ten thousand time more difficult or more expensive, but still is a measure of the severe problems introduced as we go high.

Now we come to the main issue when designing PA: the choice of the location of the quiescent operating point of the active (amplifying) element. That will decisively influence all properties of the PA.

When looking the current of the active element one may categorize roughly the choices as continuous and as discontinuous.

Continuous would mean, here, that the DC current of the active element is not interrupted during time, and its value does not depend on the signal value. To achieve that the quiescent operating point should be at the center of the active area of the output transistor or what is the same, at the middle (most linear part) of the transfer characteristic. We say that in this case the amplifier is working in Class A. That is illustrated in Fig. 4.3.1. The quiescent point for this case is denoted Q_{A}. It is approximately at the half of the maximum allowable current of the output device. To avoid distortions the operating point should not migrate significantly which means that the input signal (sinusoid in brown in Fig. 4.3.1) should have a small amplitude.

When in Class A the AC component is continuous, too. In that way the conducting angle of the output transistor is $\alpha = 2\pi$.

Further improvement of the efficiency in continuous mode is achieved in Class G and Class H amplifiers. One may say that, practically, these are virtually continuous since, here, the amplifier is provided with two (or with a single but variable) power supply sources to which it switches depending on the amplitude of the input signal.

The discontinuous case may be seen as a twofold. In the first case the discontinuity of the DC current of the output active element comes from the location of the quiescent operating point only.

In that way, when the quiescent operating point is located at the edge of the cut-off, we say that the amplifier will operate in Class B (in magenta in Fig. 4.3.1). That is illustrated in Fig. 4.3.1. As one can see the conducting angle here is $\alpha = \pi$. The transistor is conducting in one half of the signal's period. To get full output sinusoid, an additional transistor is compulsory. It will create the other half of the load current.

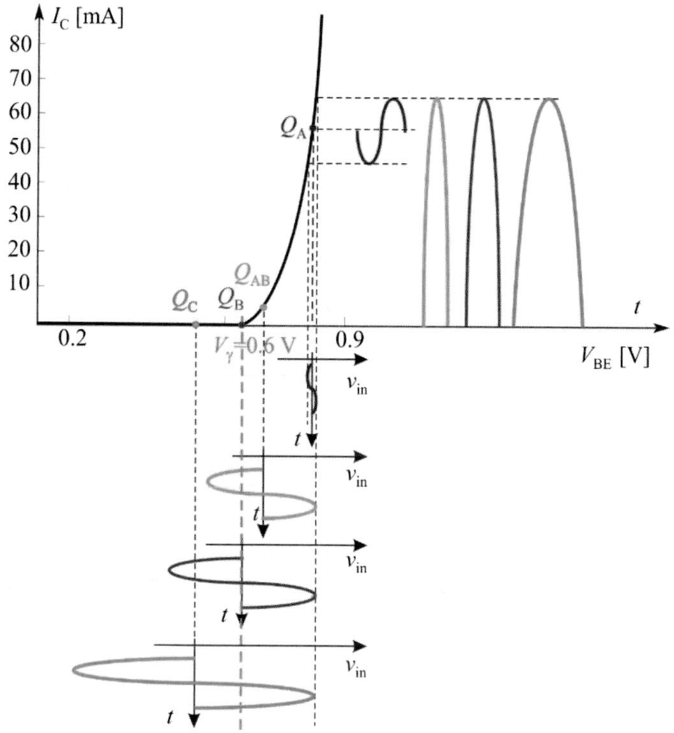

Fig. 4.3.1 Explanation of classes in power amplification nomenclature

A question arises as to why should one interrupt the device's current and spend one more transistor? It is related to the efficiency. Namely the DC component of a sinusoid that is running a half of the period is π times smaller than its amplitude. So that, in Class B two goals are achieved. First, the DC component of the output device current, which amounts for the larger part of the power supply current, is signal dependent. No signal, no power consumption. Second, even if the largest possible amplitude is present at the output, the DC component will be π times smaller which is less than half of it.

As can be seen from Fig. 4.3.1 the quiecent operating point of the amplifier working in Class B falls at the knee of the transfer characteristic. At this point, even at the conducting side of the transfer characteristic, non-linearity is present at most. To avoid that, if distortions are a crucial requirenemt, the quiescent point is partly shifted toward the conducting part of the characteristic. That, of course, stands for both devices the one vorking in the positive and the other working in the negative part of the signals period. In this way the Class AB amplifiers (in blue in Fig. 4.3.1) become much better from linearity point of view. The price is an idle current which continuously runs through the output transistors and consequently, reduces the efficiency.

Furher improvement in efficiency is obtained in the Class C amplifier (in green in Fig. 4.3.1). The quiescent operating point is now shifted deep in the non-conducting part of the transfer characteristic, and the conducting angle is less han π. Here, the restoration of the sinusoid at the output is not possible by simple regenerating the negative half-period since it too has conducting angle less than π. The solution is to filter the pulsed output current in order to eliminate its harmonic components. Due to the abundance of the harmonics one will have to use a highly selective filtering circuit with a narrow passband arround the frequency of the signal. That is achieved by passive (LC) resonant filtering circuits. So, the Class C amplifiers do have higher efficiency but are of no use for low-frequency signal amplification.

The second method implemented, applying dicontinuity of the current of the active element, is based on switching the output transistor. In all cases the useful signal is impregated into the pulse train controling the switch. In fact different schemes of modulation are impemented most of which are out of the scope of this book.

Still, even here, we may recognize two categoories. In Class D and Class G amplification a low-pass filter is used to restore the useful signal, and these may be effectively used in audio application. On the other side, in Class E and Class F amplification selective circuits are in use, and these are convenient for radio frequency (RF) power amplification.

In the sequel we will briefly go through all these variants of the power amplifier in order to give to the reader a basis to continue his specialization and to capacitate him for self-education.

4.3.2 Class A Amplifiers

In the case of power amplifiers in Class A, the quiescent operating point is located in the central part of the characteristics of the active element. This section will describe the procedures for the analysis and design of a single-stage amplifier. In the first case, it is assumed that the circuit elements are known, while in the second that they should be determined on the basis of the given performance of the circuit. In both cases, the starting point is the known characteristics of the active element.

Let us mention that under the influence of non-linear distortions, the DC component of the output current changes, so that the operating point shifts. Since the output current contains, in addition to the basic, also higher harmonics, the useful power should be defined only in relation to the basic harmonics. Distortions reduce the useful power and the efficiency. This can be easily seen by taking into account the increase in the DC component of the output current (dissipation power increases) and the decrease in the fundamental harmonic (useful power decreases).

The analysis that will be performed here will take into account that the distortions will not be greater than the acceptable ones and will not take into account their impact on the useful power and the efficiency. This small approximation will not affect the general conclusions, and the results obtained will be good enough for practical application.

Analysis and synthesis of power amplifiers consist in selecting the active element(s), determining the coordinates of the position of the operating point in the absence of a signal, determining the maximum amplitude of the input signal and the resistance of the load so that it receives the required useful power, determining distortions that should be less than prescribed, and determining dissipation on the active element that should be less than or equal to the maximum allowed.

Due to the different nature of the active elements, the analysis will differ somewhat. Therefore, amplifiers with JFET and BJTs will be considered separately.

4.3.2.1 Power Amplifier Stage Using a JFET

The "classical" basic amplification stage in Class A using a JFET is shown in Fig. 4.3.2a. It uses a transformer to enable impedance matching between the transistor's output and the load. Hence, we use the name "a transformer coupled" amplifier. It can be noticed that the primary of the transformer is connected between the power supply and the drain, and the biasing of the gate is performed via R_S. In the analysis that follows, the transformer will be considered ideal.

The DC potential of the drain is equal to V_{DD} so that the DC load line is determined by

$$I_D = V_{DD}/R_S - V_{DS}/R_S, \tag{4.3.1}$$

while the load line in the field of transfer characteristics is given by

$$I_D = -V_{GS}/R_S. \tag{4.3.2}$$

The DC load line is shown in Fig. 4.3.2 as a dashed green line. The slope of the line is $-1/R_S$, and as can be seen from the picture, it is relatively large. This is because in a PA the direct current at the quiecent operating point is high, so that the required values of V_{GS} are achieved with small values of the resistance R_S.

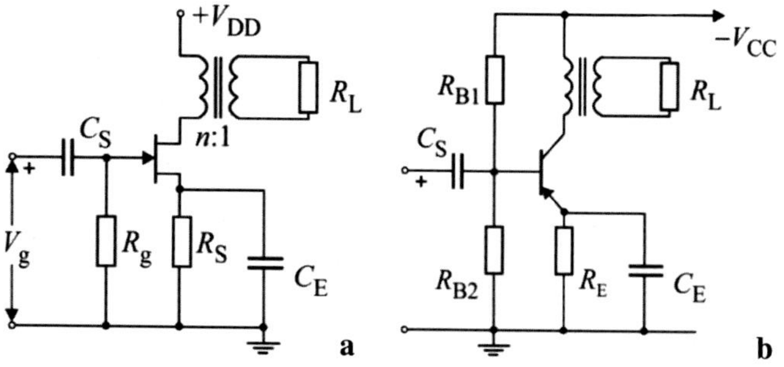

Fig. 4.3.2 Transformer coupled amplifier **a** with a JFET and **b** with a BJT

Based on this, the position of the quiecent operating point can be determined. For a given component, its coordinates are controlled by the values of V_{DD} and R_S.

Of course, the opposite problem is also of interest. Determine the circuit elements so that the power amplifier is optimal. In order to come to the procedure for determining the elements of the circuit, i.e., for the synthesis of amplifiers, we will introduce some additional considerations.

Temperature instability will not be taken into account when analyzing this amplifier. This was done in order to identify other important elements for the analysis. Temperature stability will be discussed in the analysis of amplifiers with bipolar transistors, and the conclusions can be applied here as well.

For the AC component the resistor R_S is short-circuited, while in the drain circuit we now find the mapped load at the primary of the transformer as

$$R_D = n^2 R_L \tag{4.3.3}$$

which determines the slope of the AC load line shown in blue. Figure 4.3.3 shows the output characteristics of JFET that will be used to determine the value of R_D or n. In order to obtain maximum useful power, the AC load line is set to touch the power hyperbole. This criterion was discussed during the graphical analysis of the amplifier in LNAE_Book 2, 2.2.1.14. The quiescent point is selected to be at the contact of the load line and the maximum power hyperbola. If so, the value of R_D can be obtained from the slope of the maximum power hyperbola at the point of contact as

$$R_D = \left.\frac{-1}{\partial I_D/\partial V_{DS}}\right|_{\substack{V_{DS}I_D = P_{Dmax} \\ I_D = I_{DQ}}} = P_{Dmax}/I_{DQ}^2. \tag{4.3.4}$$

The equation of the load line under these conditions is given by

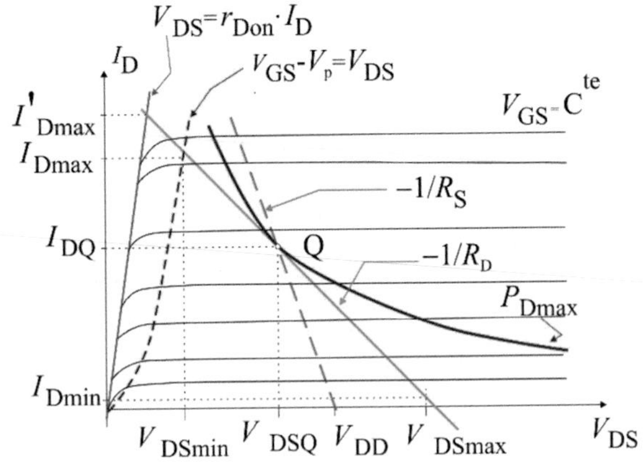

Fig. 4.3.3 Class A amplifier with JFET. Graphical analysis

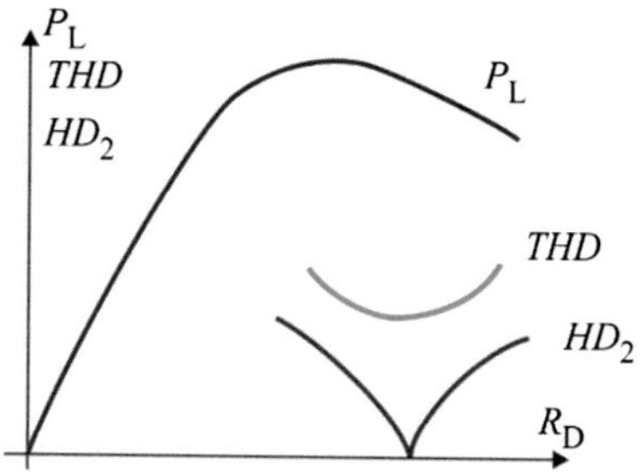

Fig. 4.3.4 Dependence of the useful power (P_L), *THD* and HD_2 on R_D

$$V_{DS} - V_{DSQ} = -R_D(I_D - I_{DQ}). \tag{4.3.5}$$

Depending on the value of R_D or n, or in other words, depending on the value of I_{DQ}, it can occupy different positions. It would be logical to assume that R_D is chosen so as to achieve power maching, i.e., that $R_D = r_D$ is valid. This condition, however, does not have to lead to an optimum in terms of both useful power and distortion, so the determination of R_D is slightly different. Figure 4.3.4 shows the dependences of the useful power (P_L) and the total harmonic distortion (*THD*) on the R_D value. The maximum useful power is close to the minimum THD, i.e., close to the value of the resistance R_D for which the value of the second harmonic (HD_2) is approximately zero. It is logical to choose this value of R_D as a load resistance. It is determined from the condition that $HD_2 = 0$. Bearing in mind that it is

$$HD_2 = \frac{I_{max} + I_{min} - 2I_Q}{2(I_{max} - I_{min})} \cdot 100\% \tag{4.3.6}$$

which leads to

$$I_Q = [I_{max} + I_{max}]/2. \tag{4.3.7}$$

This means that the quiescent point should be chosen to lie on the hyperbola so that it is in the middle between the currents I_{max} and I_{min}.

Based on the previous considerations, we came to two basic criteria for the design of power amplifiers with JFET in Class A using transformer coupling. First, it is assumed to be valid (4.3.7). In this expression we have three unknown quantities. One of them, however, we can easily choose. It is a minimum current I_{Dmin} that should be large enough so that the quiecent operating point (at low currents) does not go deep into the area of non-linear characteristics. Thus, two unknowns remain: I_{DQ} and I_{Dmax}, while one equation is available.

We will use the following reasoning to determine I_{Dmax}. At high currents, the operating point leaves the active area (linear operation) if it moves from the area of voltage saturation to the ohmic area. The boundary between these is given by

$$V_{GS} - V_p = V_{DS}. \tag{4.3.8}$$

I_{Dmax} is obtained in the intersection of this line and the AC load line. In addition, the values of V_{GSmax} and V_{DSmin} are obtained. In other words, in this point (4.3.2), (4.3.5) and (4.3.8) become

$$V_{\mathrm{GSmax}} = -R_{\mathrm{S}} I_{\mathrm{Dmax}}, \tag{4.3.9a}$$

$$V_{\mathrm{GSmax}} - V_{\mathrm{p}} = V_{\mathrm{DSmin}} \tag{4.3.9b}$$

and

$$V_{\mathrm{DSmin}} - V_{\mathrm{DSQ}} = -R_{\mathrm{D}}\left(I_{\mathrm{Dmax}} - I_{\mathrm{DQ}}\right). \tag{4.3.9c}$$

With the introduction of this system, another unknown was introduced into the account: V_{DSQ}. It is the voltage between the drain and the source at the quiescent operating point. We can eliminate it if we keep in mind that the operating point is on the maximum power hyperbole, i.e.,

$$V_{\mathrm{DSQ}} = P_{\mathrm{Dmax}}/I_{\mathrm{DQ}}. \tag{4.3.10}$$

By combination of (4.3.4), (4.3.7), (4.3.9a), (4.3.9b), and (4.3.10) the Eq. (4.3.9c) becomes transformed into a second order one:

$$I_{\mathrm{DQ}}^2 - \frac{2P_{\mathrm{Dmax}} + I_{\mathrm{Dmin}} V_{\mathrm{DD}}}{2V_{\mathrm{DD}} + V_{\mathrm{p}}} \cdot I_{\mathrm{DQ}} + \frac{2P_{\mathrm{Dmax}} I_{\mathrm{Dmin}}}{2V_{\mathrm{DD}} + V_{\mathrm{p}}} = 0, \tag{4.3.11a}$$

which has a solution as

$$I_{\mathrm{DQ}} = \frac{2P_{\mathrm{Dmax}} + I_{\mathrm{Dmin}} V_{\mathrm{DD}}}{2\left(2V_{\mathrm{DD}} + V_{\mathrm{p}}\right)}\left[1 + \sqrt{1 - \frac{8P_{\mathrm{Dmax}} I_{\mathrm{Dmin}}\left(2V_{\mathrm{DD}} + V_{\mathrm{p}}\right)}{(2P_{\mathrm{Dmax}} + I_{\mathrm{Dmin}} V_{\mathrm{DD}})^2}}\right]. \tag{4.3.11b}$$

All other quantities related to the design of the amplifier can now be easily determined. Before we present the results of a project, we notice that if we choose $I_{\mathrm{Dmin}} = 0$ we get a very simple expression for the current at a quiescent operating point:

$$I_{\mathrm{DQ}} = P_{\mathrm{Dmax}}/\left(2V_{\mathrm{DD}} + V_{\mathrm{p}}\right), \tag{4.3.11c}$$

where we should not forget that it is an N-channel transistor, i.e., that V_{p} is a negative number.

Here is an example. In a Class A power amplifier with JFET using a transformer the load is $R_{\mathrm{L}} = 4\Omega$. Given $V_{\mathrm{DD}} = 20$ V and the parameters of the transistor $V_{\mathrm{p}} =$ -4 V, $P_{\mathrm{Dmax}} = 4$ W and $I_{\mathrm{Dmin}} = 10$ mA find the coordinates of the quiescent operating point, the limitting values of the currents and voltages, and the element values of the circuit (R_{S} and n).

As a solution of this example we get $I_{DQ} = 0.22$ A, $V_{DQ} = 18.4$ V, $I_{Dmax} = 0.425$ A, $R_D = 84.5\ \Omega$, $n = 4.6$, $V_{DSmin} = 0.845$ V, $V_{Gsmax} = -3.15$ V, and $R_S = 7.4\ \Omega$.

The highest value of the voltage between the drain and the source is of interest because it must not exceed the breakdown voltage of the transistor. In this regard, amplifiers with inductive load have some specifics that need to be considered here. Namely, the inductive load in the drain, during the negative half-cycle of the AC component of the drain current, opposes the reduction of the current by delivering back to the circuit the energy it has accumulated during the positive half-cycle. The situation is equivalent to a circuit with two power sources. One is the V_{DD} source and the other is the transformer's inductance. Therefore, the instantaneous values of the drain voltage are significantly higher than the value of the supply voltage. As can be seen from Fig. 4.3.3, the maximum voltage reaches the value obtained as a solution of Eq. (4.3.5) under the condition $I_D = I_{Dmin}$. One gets

$$V_{DSmax} = V_{DSQ} - R_D\left(I_{Dmin} - I_{DQ}\right) \tag{4.3.12}$$

For the above example, the numerical value of the maximum voltage between the drain and the source is $V_{DSmax} = 32.7$ V. Roughly calculated, the value of the maximum voltage is close to twice the value of the supply voltage. This should be taken into account when choosing a transistor. The transistor used here must have a maximum allowable voltage greater than 32.7 V. If this is not the case, the amplifier must be redesigned by moving the AC load line translationally downwards so that its right end intersects the line $I_D = I_{Dmin}$ at a point whose abscissa is less than the maximum allowable component voltage. In the intersection of this new AC load line and the curve given by (4.3.8) a new value for I_{Dmax} is obtained, and in the middle between I_{Dmax} and I_{Dmin} a new quiescent operating point is determined, i.e., a new value for R_S. Of course, since the load line have been shifted translationally, the values of R_D and n had not changed.

Based on Fig. 4.3.3 for the largest amplitudes of alternating signals on the drain we have

$$V_{DSm} = \left|V_{DSQ} - V_{DSmin}\right| = \left|V_{DSmax} - V_{DSQ}\right| \tag{4.3.13}$$

and

$$J_{DSm} = \left|I_{DQ} - I_{Dmin}\right| = \left|I_{Dmax} - I_{DQ}\right| \tag{4.3.14}$$

So, for the maximum value of the useful power we get

$$\begin{aligned} P_L &= V_{DSm} J_{Dm}/2 = \left(V_{DSQ} - V_{DSmin}\right)\left(I_{DQ} - I_{Dmin}\right)/2 \\ &= \frac{V_{DSQ} I_{DQ}}{2}\left(1 - \frac{V_{DSmin}}{V_{DSQ}}\right)\left(1 - \frac{I_{Dmin}}{I_{DQ}}\right), \end{aligned} \tag{4.3.15}$$

while for the maximum value of the efficiency one may derive the following

$$\eta_{\max} = \frac{P_L}{P_B} = \frac{V_{DSm}J_{Dm}/2}{V_{DD}I_{DQ}} = \frac{1}{2}\frac{V_{DSm}J_{Dm}}{(V_{DSQ} + I_{DQ}R_S)I_{DQ}}$$
$$\approx \frac{1}{2}\frac{V_{DSm}J_{Dm}}{V_{DSQ}I_{DQ}} = \frac{1}{2}\left(1 - \frac{V_{DSmin}}{V_{DSQ}}\right)\left(1 - \frac{I_{Dmin}}{I_{DQ}}\right). \quad (4.3.16)$$

The ideal maximum possible value of the efficiency (for $V_{DSmin} = 0$ and $I_{Dmin} = 0$) is 50%. In practical cases, the efficiency is about 40%. This means that 40% of the battery power is delivered to the load, and the rest (most) is spent on the active element. In the example calculated above the values: $V_{DSQ} = 16.7$ V, $V_{DSmin} = 0.785$ V, $I_{DQ} = 0.24$ A, and $I_{Dmin} = 0.01$ A were valid, so $\eta = 0.457$ was obtained from (4.3.16). In this case the battery power is $P_B = V_{DD}I_{DQ} = 20{\cdot}0.24 = 4.8$ W, and the efficiency given by (4.3.16) is approximately calculated using the dissipation power on the transistor *in absence of a signal* whose value is $P_d = I_{DQ}V_{DSQ} = 0.24{\cdot}16.7 = 4.008$ W. The maximum value of useful power is $P_L = (16.7–0.785){\cdot}(0.24–0.01)/2 = 1.83$ W. Therefore, the exact value of the maximum value of the efficiency is $\eta = P_L/P_B = 0.38$, or $\eta = 38\%$.

By analyzing the expression (4.3.16), we conclude that the value of the selected current I_{Dmin} affects the value of the effiiency. However, having in mind that it is a small current, it can be said that its practical impact on the effiiency is negligible even when a relatively large I_{Dmin} is taken. The value of V_{DSmin}, however, is not so small and, if high efficiency is required, this feature of the JFET becomes its disadvantage. If a higher efficiency is insisted on, instead of (4.3.9b), the minimum value of the drain voltage can be taken as

$$V_{DSmin} = r_{Don}I_D, \quad (4.3.17)$$

which is shown in Fig. 4.3.3. Now we get a new value for I_{Dmax} (marked with I'_{Dmax}) and of course, a new quiescent point position. This requirement, however, abandones the criterion of minimal distortion and if adopted, a new criterion should be found for the selection of R_D.

It should be borne in mind that losses in the transformer should be included in practical calculations. Namely, in the transformer, due to ohmic losses in the windings and wasteful flux, some energy is consumed so that the total efficiency is obtained as a product

$$\eta = \eta_{transistor} \cdot \eta_{transforme} \quad (4.3.18)$$

The value of the efficiency of the transformer is higher than 90%.

Finally, let us mention that it happens that the required net power at the load is less than the maximum value given by (4.3.15). In such cases, however, the calculation should be performed as described here, and negative feedback should be used to reduce the gain to the desired value. In this way, reduction of non-linear distortions may be obtained.

4.3.2.2 Class A Amplifier with a BJT

A Class A power amplifier with a bipolar transistor is shown in Fig. 4.3.2b, while Fig. 4.3.5 shows the output and input characteristics of a typical transistor. It can be noticed that the output characteristic is very steep for low voltages on the collector so that $V_{CEmin} \approx 0$. Similarly, for a BJT we can assume that $I_{Cmin} \approx 0$, so in this case we get an efficiency very close to 50%.

Before we start the description of the procedure for the design such an amplifier, we will point out some of its specifics in relation to the circuit with JFET. Non-linear distortions will be considered first. In bipolar transistors, they originate not only from the non-linearity of the output characteristic but also from the non-linearity of the input (dynamic) characteristic shown in Fig. 4.3.5b. This statement will be illustrated by a practical example.

Let the operating point of an amplifier with a BJT whose characteristics are shown in Fig. 4.3.5 has the following coordinates: $I_{BQ} = -300\ \mu A$, $I_{CQ} = -31$ mA, and

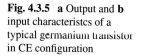

Fig. 4.3.5 **a** Output and **b** input characteristcs of a typical germanium transistor in CE configuration

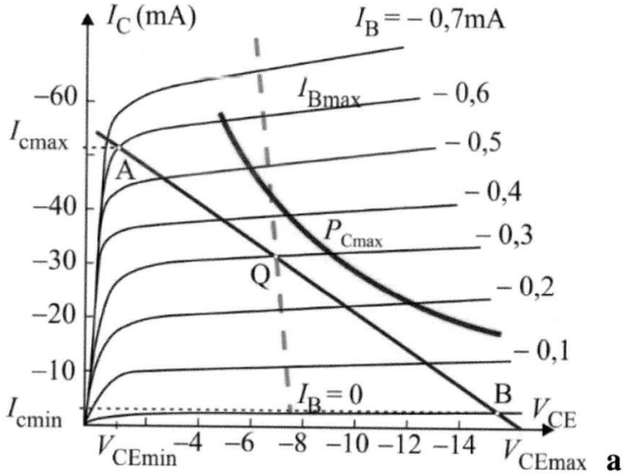

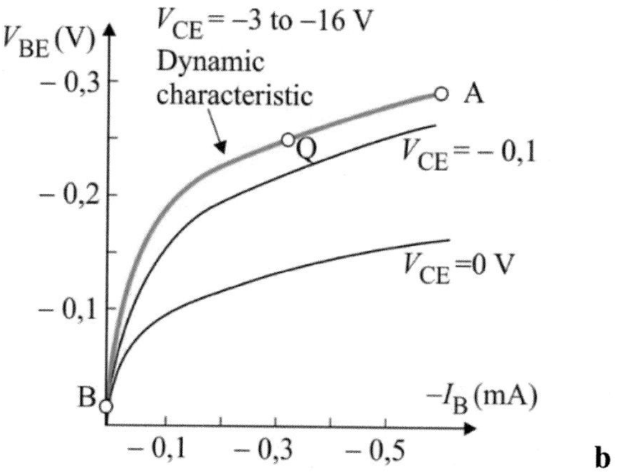

V_{CEQ} = -7.2 V. Let the transistor be excited by an ideal sinusoidal current source of amplitude $J_{Bm} = 300\ \mu A$. The total base current (DC and AC component) is shown in Fig. 4.3.6a. Due to this excitation, as can be seen from Fig. 4.3.5a, the operating point will move from point A with coordinates $I_{Bmax} = -600\ \mu A$, I_{Cmax} = -51 mA and $V_{CEmin} = -0.4$ V, to point B with coordinates $I_{Bmin} = 0$, $I_{Cmin} = -3$ mA and $V_{CEmax} = -16$ V. It is obvious that different amplitudes of the collector current are obtained in different half-cycles of the excitation signal. Thus, $|I_{Cmax}\text{-}I_{CQ}| = 20$ mA and $|I_{CQ}\text{-}I_{Cmin}| = 28$ mA. The waveform of the collector current is distorted as shown in Fig. 4.3.6b. Figure 4.3.5c shows the distorted collector voltage waveform where $|V_{CEQ}\text{-}V_{CEmin}| = 6.7$ V and $|V_{CEmax}\text{-}V_{CEQ}| = 8.8$ V.

From the dynamic input characteristic it can be seen that for $I_B = -300\ \mu A$, $V_{BEQ} = 0.24$ V. At point A it is for $I_B = -600\ \mu A$, $V_{BEmax} = -0.28$ V, and at point B for $I_B = 0\ \mu A$, V_{BEmin} = -0.05 V. Thus we have $|V_{BEmax}\text{-}V_{BEQ}| = 0.04$ V, and $|V_{BEQ}\text{-}V_{BEmin}| = 0.19$ V. The voltage between the base and the emitter is significantly distorted due to the non-linear input characteristic. The waveform of the input voltage is shown in Fig. 4.3.6d.

Let us now consider the waveforms of voltage and current when the amplifier is excited by an ideal voltage source. Now the V_{BE} voltage is sinusoidal. It is shown in Fig. 4.3.7a. The waveform of the base current, for the same operating point as in

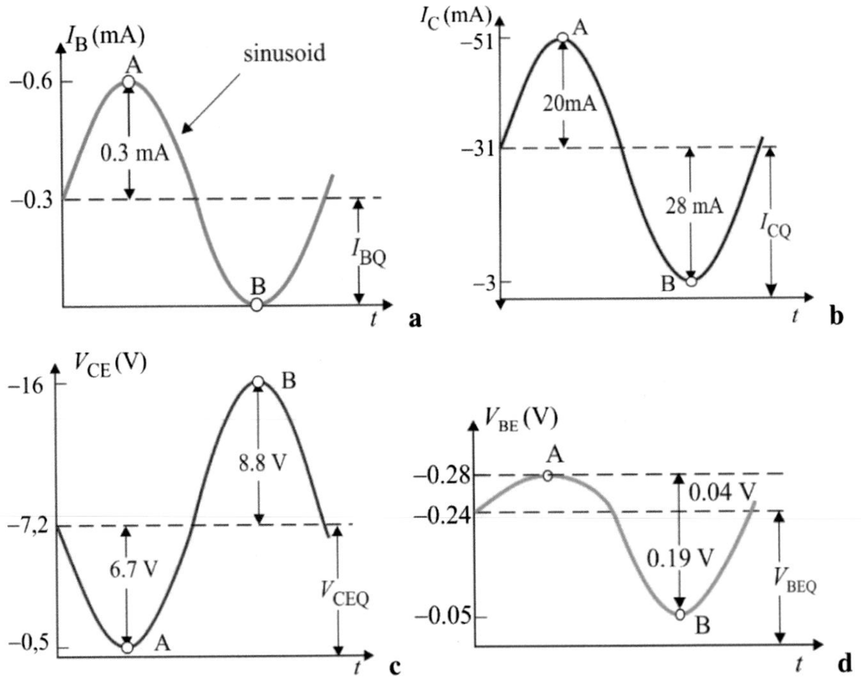

Fig. 4.3.6 Waveforms for the case of current excitation. **a** Input current, **b** output current, **c** output voltage, and **d** input voltage

the previous case, is shown in Fig. 4.3.7b. It is determined by the input characteristic depicted in Fig. 4.3.5b, where the amplitude of the excitation voltage is 0.04 V. It can be seen that the waveform is distorted. The amplitude of the positive half-cycle is 300 μA, and the negative half-cycle is 200 μA. Using Fig. 4.3.6a on the basis of the waveform of the base current, the waveforms of the collector current and the voltage between the collector and the emitter are determined. They are shown in Figs. 4.3.7c and d.

When considering non-linear distortions, by comparing the waveforms with Figs. 4.3.6 and 4.3.7 it can be concluded that the distortions are much smaller when the transistor is excited by an ideal voltage source. Since ideal source will not be used in an actual amplifier, let us consider the influence of the (finite) internal resistance of the voltage excitation source on non-linear distortions. For simplicity, the review will be performed as follows. As the internal resistance of the source grows from zero, the voltage source tends to become a current source. This means that the waveforms from Fig. 4.3.7 tend toward the waveforms of Fig. 4.3.6. If we observe the waveforms of the V_{CE} voltage from Fig. 4.3.6c and d, we can see that when the resistance of the source changes from zero to infinity, there is one value of this resistance for which the amplitudes of both halves will be equal.

This value allows for minimal distortion. Its numerical value is usually determined experimentally due to the non-linearity of the equations describing the circuit.

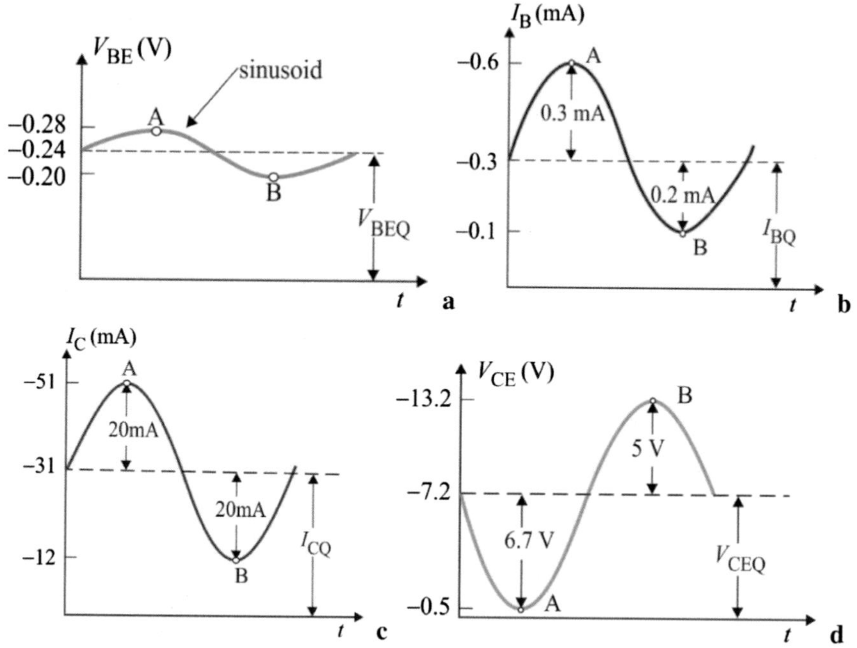

Fig. 4.3.7 Waveforms for voltage excitation. **a** Input voltage, **b** input current, **c** output current, and **d** output voltage

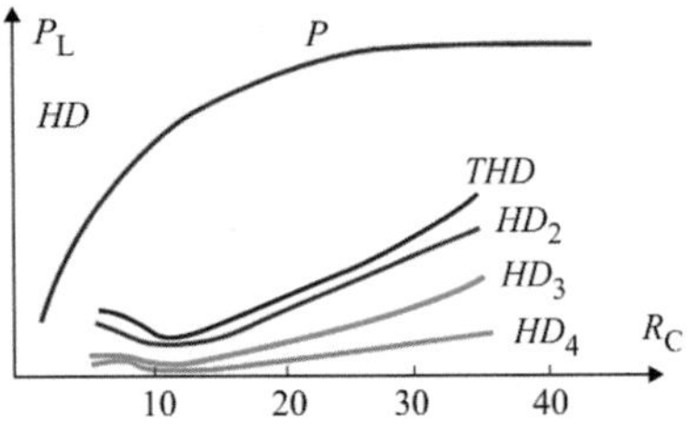

Fig. 4.3.8 Dependence of the useful power and the harmonic distortions on R_C

Since the non-linear distortions are smaller when an ideal voltage source is used, we conclude that the obtained value of the internal resistance of the source will be relatively small. Of course, this conclusion leads to the optimum from the point of view of distortion. From the point of view of power matching, however, it is necessary that the output resistance of the previous stage be equal to the input resistance of the power amplifier. In this case, at the input of the amplifier, since the signal strength is still low, the criterion of optimality from the point of view of power matching is less important.

Power matching and non-linear distortions, however, also depend on the value of load resistance. Figure 4.3.8 shows the dependences of useful power and various *HD* quantities on the value of load resistance. The value of the resistance R_C at which the distortions are the smallest is significantly lower than the one for which the maximum net power is obtained. In the area of small distortions, the useful power decreases rapidly. Therefore, the optimum is between the R_C resistance value for which the minimum distortions and the maximum useful power are obtained. These properties of the amplifier will be discussed when choosing the procedure for its design.

Let us consider the influence of temperature and dissipation on the properties of amplifiers. Figure 4.3.5a depicts a plot of the maximum collector dissipation curve P_{cmax}. This quantity, however, is not constant and depends on temperature. It is understood that the temperature of the collector junction is usually significantly higher than the ambient temperature. If good transistor cooling is provided, the junction temperature is slightly higher than the ambient temperature. Thus, by using the dependence $P_{cmax} = f(T)$, the minimum value of the maximum collector dissipation can be determined for a given maximum expected ambient temperature. The following relation is used for this purpose

$$T_J - T_0 = R_{th} \cdot P_C, \tag{4.3.19}$$

where as before, T_J is the temperature of the collector junction, T_0 is the ambient temperature, R_{th} is the thermal resistance, and P_C is the collector dissipation power. It should be taken into account that the upper limit of the maximum junction temperature for silicon transistor is $T_{maxSi} = 150$ °C, and for germanium transistor $T_{maxGe} = 60$ °C. The thermal resistance is usually listed in the component's datasheet together with the conditions under which it applies. Based on such judgments, the curve of maximum dissipation is plotted in Fig. 4.3.5a.

In Fig. 4.3.5a the load line for direct current is drawn. With the change of temperature, in the absence of a signal, the operating point will migrate while staying on this line. The new positions of the AC load line will also correspond to the new temperatures, since this is shifted in parallel, and its slope is determined by the resistance in the collector: $R_C = n^2 R_L$.

With the rise of temperature, if no special measures are taken, the operating point will be moved, along to the DC load line, to the area above the maximum allowable dissipation, so that the AC load line will partially lie above the hyperbole. That is unacceptable. It should be noted here that the condition for preventing self-heating: $V_{CE} \leq V_{CC}/2$, cannot be met due to low static resistance in the collector (resistance of the transformer's primary winding). Therefore, maximum attention should be paid to temperature stabilization.

When we plot the power hyperbola for the worst case (the highest ambient temperature), the AC load line is set as follows. From the known (set) or calculated values of the instability factor ($S = \Delta I_C/\Delta T$), the maximum change of the collector current ΔI_{CT} that will occur in the worst temperature case is calculated. The load line is set so that the minimum distance from the hyperbola power in the y-direction is larger than or equal to ΔI_{CT}. This distance should be taken to be slightly larger than ΔI_{CT} is, considering that the operating point also moves due to non-linear distortions.

When designing this amplifier, one should start from the diagram shown in Fig. 4.3.8. This diagram shows the dependence of useful power and harmonic distortions on the value of load resistance. There is a significant difference from the JFET case. The abscissa of the minimum *THD* and the maximum useful power differs significantly. In addition, the value of the useful power at the place of the minimum *THD* is significantly lower than the maximum, and at the same time, the value of the *THD* at the maximum useful power is significantly higher than its minimum. Therefore, it is necessary to introduce an additional criterion on the basis of which the value of the load resistance will be adopted. For the transistor of Fig. 4.3.8 one compromise solution would be to choose $R_C = 22$ K. The only limitation in relation to the selected value for R_C refers to the abscissa of the point at which the AC load line will touch the power hyperbola. Since the selection of R_C also automatically determines the V_{CEQ} as

$$V_{CEQ} = \sqrt{P_{Cmax} \cdot R_C}, \tag{4.3.20a}$$

we can immediately determine whether the R_C value is acceptable. Namely, if a value is obtained for V_{CEQ} that is larger than or equal to the value of the supply voltage V_{CC}, we conclude that the choice of R_C is bad and that a lower value should be chosen. However, if this value of V_{CEQ} or R_C is insisted on, the V_{CC} can be declared variable and adjusted to this choice. After applying the procedure described above, in our further calculations it can be considered that R_C, and thus n is a known quantity. It remains to determine the position of the quiecent operating point, and from it the values of other resistances.

First of all, based on the maximum useful power, we determine the value of the collector current under the condition that the AC load line touches the quiecent operating point:

$$\left(I'_{\mathrm{CQ}}\right)^2 = P_{\mathrm{Cmax}}/R_{\mathrm{C}}. \tag{4.3.20b}$$

The actual position is somewhat lower, i.e.,

$$I_{\mathrm{CQ}} = I'_{\mathrm{CQ}} - \Delta I_{\mathrm{CQ}}, \tag{4.3.20c}$$

which means that the load line has been translationally shifted. The value of the base current is determined from $I_{\mathrm{BQ}} = I_{\mathrm{CQ}}/\beta$, and the value of R_{E} from the DC load line:

$$V_{\mathrm{CC}} \approx V_{\mathrm{CE}} + R_{\mathrm{E}} I_{\mathrm{C}}, \tag{4.3.21a}$$

where the expression is approximate since $I_{\mathrm{E}} \approx I_{\mathrm{C}}$ was taken. Based on these results and the procedure shown in LNAE_Book 3, the resistances R_{B1} and R_{B2} can be determined.

The AC load line for this circuit now is

$$I_{\mathrm{C}} - I_{\mathrm{CQ}} = -\left(V_{\mathrm{CE}} - V_{\mathrm{CEQ}}\right)/R_{\mathrm{C}}. \tag{4.3.21b}$$

The minimum value of the voltage on the collector is obtained at the intersection of the AC load line and the line

$$I_{\mathrm{C}} = V_{\mathrm{CE}}/r_{\mathrm{Don}}, \tag{4.3.22}$$

which approximates the characteristics of the transistor in the field of current saturation. The coordinates of the intersection are

$$\begin{aligned} I_{\mathrm{Cmax}} &= \frac{\left(V_{\mathrm{CEQ}} + R_{\mathrm{C}} I_{\mathrm{CQ}}\right)}{R_{\mathrm{C}} + r_{\mathrm{Don}}} \\ V_{\mathrm{Cemin}} &= r_{\mathrm{Don}} I_{\mathrm{Cmax}} \end{aligned} \tag{4.3.23}$$

The maximum value of the collector voltage is obtained at the intersection of the AC load line and the horizontal line $I_{\mathrm{C}} = I_{\mathrm{Cmin}}$. One gets

$$V_{\mathrm{CEmax}} = V_{\mathrm{CEQ}} + R_{\mathrm{C}}\left(I_{\mathrm{CQ}} - I_{\mathrm{Cmin}}\right). \tag{4.3.24}$$

Since this was not taken into account, with the load line determined in this way, it may happen that the maximum instantaneous value of the voltage on the collector becomes higher than the maximum allowed value. This is also true when talking about the maximum allowable collector current. In doing so, it should be checked whether these two limiting values are exceeded in the case when the operating point is migrating (this means that in (4.3.23) and (4.3.24) should be put I'_{CQ} instead of I_{CQ}.). In this case, I_{Cmax} and V_{CEmax} are the largest and should still be less than the maximum allowable values. If this is not the case, an even lower value for the I_{CQ} should be adopted, and the procedure is repeated. This activity is justified also because the values of the circuit elements are now known, so the component ΔI_{CQ} related to the temperature increase can be determined. New load line is set, and new values for I_{Cmax} and V_{CEmax} are determined. This is to be iterated until temperature stable operation is ensured without exceeding voltage and current ratings.

4.3.2.3 Push–pull in Class A Power Amplifiers

It is known that from the point of view of application, both for voltage or power amplifiers, at low frequencies, bipolar transistors are a more suitable component than JFET due to incomparably higher gain. In Class A power amplifiers with bipolar transistors, however, as we saw in the previous section, the issue of non-linear distortions is not resolved. Although not equally problematic, in JFET amplifiers, it can be observed that if the maximum useful power was insisted on, the problem of non-linear distortions would remain unsolved. Therefore, attempts were made to satisfy both the requirement for minimum non-linear distortions and the for maximum useful power. One of the existing solutions is the use of symmetric coupling or a coupling better known as push–pull.

The special combination of two active elements with identical characteristics enables obtaining twice the useful power with a significant reduction of non-linear distortions as compared to the single transistor stage. This coupling is called push–pull and is shown in Fig. 4.3.9 for amplifiers with JFETs and with bipolar transistors. It is easy to see that a push–pull coupling is achieved by coupling two identical stages so that the power supply circuit is common. The operation of these circuits can be analyzed using a generalized push–pull amplifier, which is shown in Fig. 4.3.10. In its implementation, it was considered that the voltage drop across the source (emitter) resistor is much smaller than the voltage drop across the active element v_{DS} (v_{CE}) and that the base current is much smaller than the collector current.

At the input of a "classical" push–pull amplifier there is an input transformer T_{r1}. The secondary of this transformer has three terminals. The ends of the secondary winding are connected to the input terminals of the active elements while the middle terminal is short-circuited for the AC signals to the common terminals of the active elements (source or emitter) or the ground. In this way, the input signals of the active elements are of the same amplitude and of opposite phase. With the voltage changes on the primary winding of the transformer T_{r1}, both signals at the inputs of the active elements will change in the same amount, but in the opposite direction.

The load is also connected via a symmetrical transformer. The primary winding has three terminals. The end terminals are connected to the output terminals of the active elements, and the middle terminal is connected to the common terminals or to ground (for AC signals). A current of one active element flows through each half of

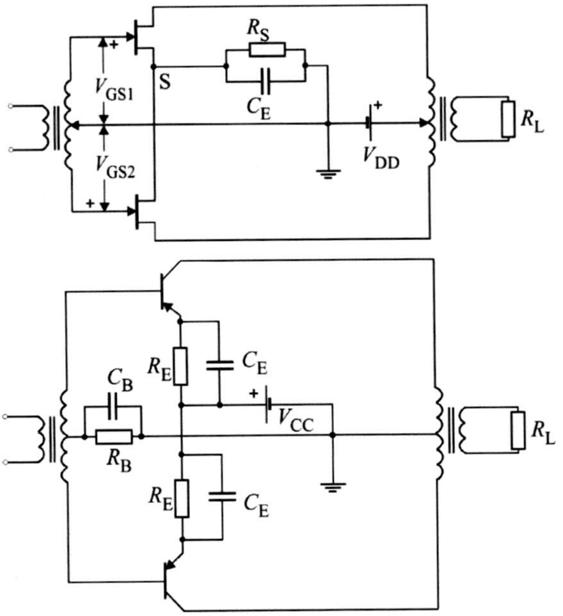

Fig. 4.3.9 Push–pull amplifier in Class A. **a** Using JFETs and **b** using BJTs

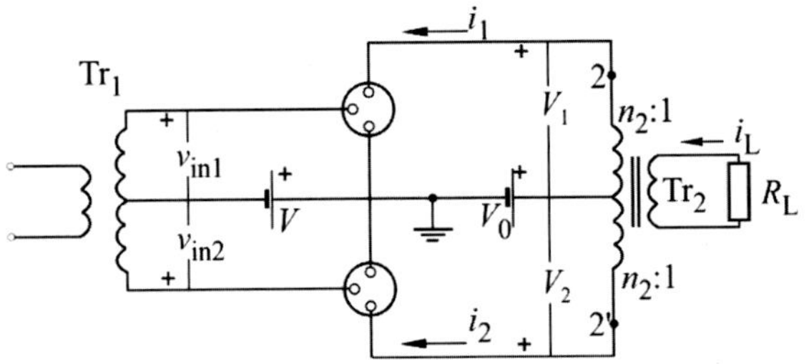

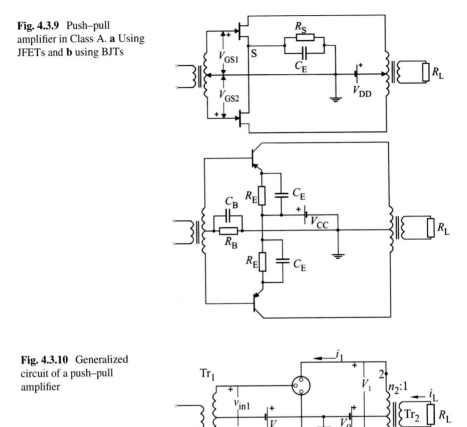

Fig. 4.3.10 Generalized circuit of a push–pull amplifier

the primary. Since the input signals are of the same amplitudes and opposite phase, if the active elements are identical, the output currents are also of the same amplitudes and opposite phase. With this in mind, based on notation from Fig. 4.3.10, for a simple periodic excitation we can write the following:

$$i_1 = I + J_{1m}\cos(\omega t) + J_{2m}\cos(2\omega t) + J_{3m}\cos(3\omega t) + \ldots \tag{4.3.25}$$

and

$$\begin{aligned} i_2 &= I + J_{1m}\cos(\omega t + \pi) + J_{2m}\cos[2(\omega t + \pi)] + J_{3m}\cos[3(\omega t + \pi) + \ldots \\ &= I - J_{1m}\cos(\omega t) + J_{2m}\cos(2\omega t) - J_{3m}\cos(3\omega t) + \ldots \end{aligned} \tag{4.3.26}$$

The current at the secondary winding of T_{r2} may be obtained as

$$n_2 i_1 - n_2 i_2 = 1 \cdot i_L, \quad (4.3.27)$$

or

$$i_L = n_2(i_1 - i_2) = n_2[2J_{1m}\cos(\omega t) + 2J_{3m}\cos(3\omega t) + \ldots] \quad (4.3.28)$$

The current through the load does not contain the second and in general, even harmonics. They were canceled by the use of symmetrical coupling. As the amplitude of the second harmonic is larger than all other harmonic components, in a push–pull amplifier, the *THD* is significantly reduced as compared to power amplifiers with single active element.

Both transistors operate in Class A. An increase in the current of one follows a decrease in the current of the other in the same amount. That is why the name push–pull has been adopted for the symmetrical coupling.

The total resistance that is mapped to the terminals of the transformer T_{r2} between nodes 2–2' is given by

$$R'_{22} = (2n_2)^2 R_L = 4n_2^2 R_L. \quad (4.3.29)$$

Half of this resistance is seen in the output circuit of each active element:

$$R_C = \frac{R'_{22}}{2} = 2n_2^2 R_L. \quad (4.3.30)$$

In the circuit with Fig. 4.3.9a the input bias voltage is obtained automatically by means of the R_S resistance. The capacitor C_E is a short circuit for the AC component of the current. Since R_S is common to both active elements, for the total current flowing through it we can write

$$i = i_1 + i_2 = 2I + 2J_{2m}\cos(2\omega t) + 2J_{4m}\cos(4\omega t) + \ldots \quad (4.3.31)$$

This current does not contain the fundamental harmonic, so the C_E capacitor can be omitted. Bearing in mind, however, that the transistors cannot have completely identical characteristics, which means that there will still be a small part of the fundamental harmonic frequency, it is usually left in the circuit. It is more important, however, that the direct current through R_S is twice as high as the direct current of the transistor 's drain, so it is valid:

$$V_{GS} = -R_S(2I). \quad (4.3.32)$$

Based on this, we conclude that half of the value for R_S is needed than is the case of an ordinary CS amplifier.

In the circuit of Fig. 4.3.9b the required direct current of the base is provided through the resistor R_B. It is short-circuited for AC signals by the C_B capacitor so as not to affect the input current. The R_E resistor in the emitter circuit is used for temperature stabilization of the operating point. It should be noted that even in the case of amplifiers with JFETs, the polarization can be realized using two R_S resistors.

Each active element delivers the same useful power to the load R_L. The total useful power on the common load $R_{22'}$ is twice as high. From the point of view of choosing the most favorable operating conditions, the same conditions apply to each active element as previously described for a power amplifier with a single active element. However, in the graphical analysis of the push–pull amplifier, the static characteristics of each element cannot be used individually. The reason is the mutual influence of the active elements, since both conduct current at the same time and are not separated from each other. Therefore, the graphical analysis is performed by deriving an equivalent push–pull characteristic that applies to both active elements at the same time.

If we introduce a new variable $i_{eq} = i_1 - i_2$, the Eq. (4.3.27) may be rewritten as

$$i_L = n_2 i_{eq}. \tag{4.3.33}$$

Now both active elements from Fig. 4.3.10 can be replaced by one whose output current is i_{eq} and whose turn ratio is n_2:1. This equivalent circuit is shown in Fig. 4.3.11. If the input signal (voltage or current) is denoted by x_{in}, then the output characteristic of each element individually is represented by the functions

$$i_1 = f(x_{in1}, v_{out1}) \tag{4.3.34}$$

$$i_2 = f(x_{in2}, v_{out2}) \tag{4.3.35}$$

where the transistors are assumed to have identical characteristics. Now, for the output voltage we have

$$v_{out1} = V_0 - v_1 \tag{4.3.36}$$

$$v_{out2} = V_0 - v_2 \tag{4.3.37}$$

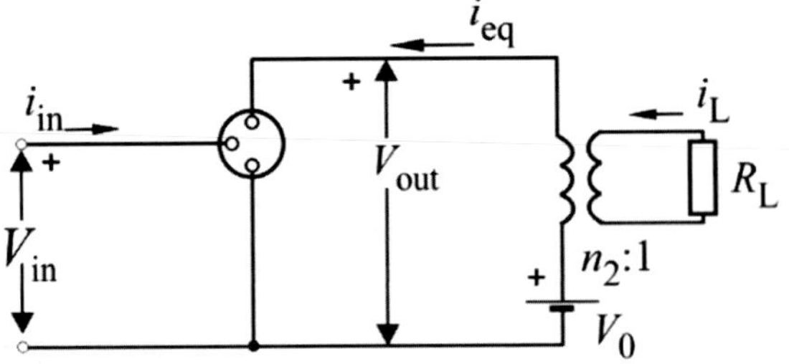

Fig. 4.3.11 Equivalent circuit of a push–pull amplifier

where v_1 and v_2 denote the AC components of the output signals, i.e., voltages. In this case, $v_2 = -v_1$. The input signal can be represented in a similar way

$$x_{\text{in1}} = X_{\text{in0}} + x_{\text{int1}} \tag{4.3.38}$$

$$x_{\text{in2}} = X_{\text{in0}} + x_{\text{int2}} \tag{4.3.39}$$

where x_{int1} and x_{int2} are AC components of the input signal. Similarly, $x_{\text{int2}} = -x_{\text{int1}}$.
With this notation one may write the following:

$$i_{\text{eq}} = i_1 - i_2 = f(X_{\text{in0}} + x_{\text{int1}}, V_0 + v_1) - f(X_{\text{in0}} - x_{\text{int1}}, V_0 - v_1). \tag{4.3.40}$$

The characteristics of the equivalent active element can be obtained by summing the output characteristics of both active elements in such a way that the characteristics of the other are taken with the opposite sign. This conclusion is easily reached by analyzing the expression (4.3.40). Therefore, in the analysis procedure, which is shown in Fig. 4.3.12, the characteristics of both active elements are opposite, so that the axis of increase of the output voltage for one transistor is at the same time the axis of decrease for the other. For $v_1 = 0$ V both characteristics have a common point Q for which $v_{\text{out1}} = v_{\text{out2}} = V_0$, where V_0 is the voltage of the power supply. The direct current component at this point is equal to zero since i_{eq} is given by (4.3.40).
For the sake of clarity, in Figs. 4.3.12t he operating points of the individual active elements are also shown. They are marked with Q_1 and Q_2. Since these characteristics of the active elements correspond to the absence of a signal, the equivalent operating point corresponds also to the case of the absence of a signal, i.e., $x_{\text{eq}} = 0$. So, one point on the equivalent characteristic is determined (point Q). We get the rest by varying v_1 at constant x_{eq} (in this case $x_{\text{eq}} = 0$). An increase in v_1 causes the point to move to the right on both characteristics by the same amount, and a decrease causes it to move to the left. By subtracting the currents at these points, we obtain

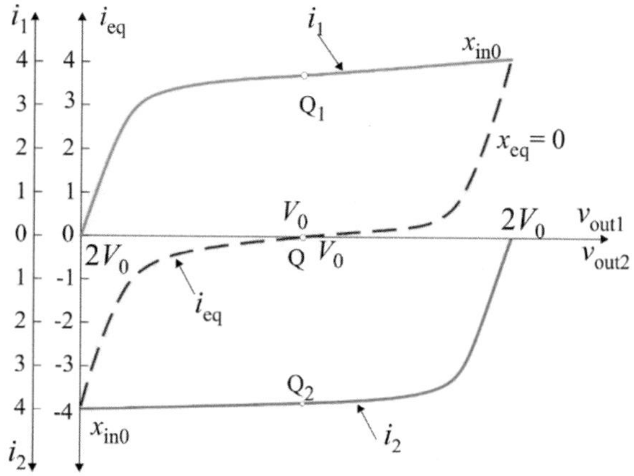

Fig. 4.3.12 Creating the equivalent characteristics of a push–pull amplifier

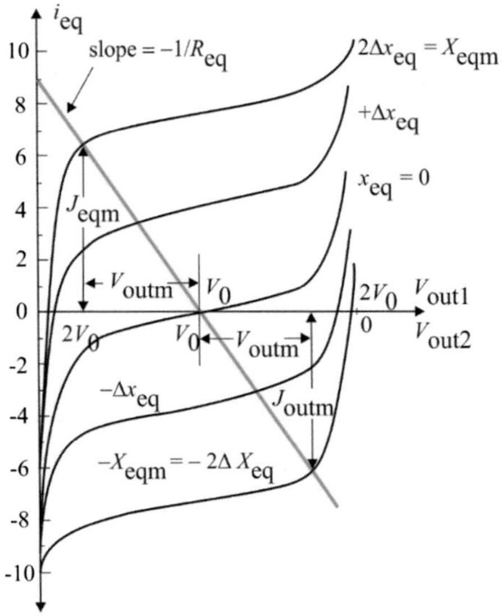

Fig. 4.3.13 The field of output characteristics of the equivalent element

the equivalent characteristic that is depicted in Fig. 4.3.11 shown by the dashed red line.

The rest of the characteristics of the equivalent active element are obtained in the following way. Let the input signal of the first active element increases by Δx_{in}. The static output characteristic for $X_{in0} + \Delta x_{in}$ corresponds to this value. In the case of the second active element, there will be a decrease in the signal, so the corresponding static characteristic will be the one marked with X_{in0}-Δx_{in}. Subtracting these two lines gives the equivalent characteristic for $x_{eq} = \Delta x_{in}$. By repeating this procedure, a field of equivalent output characteristics is obtained, which is shown in Fig. 4.3.13.

The equivalent load is given by

$$R_{eq} = n_2^2 R_L \tag{4.3.41}$$

where Eq. (4.3.33) and the equivalent voltage relation are taken into account. The value of R_{eq} determines the slope of the AC load line, which is shown in Fig. 4.3.13 in blue. With the help of this load line, the useful power and non-linear distortions are further calculated in the same way as with an amplifier with single active element.

Summarizing the achieved by the push–pull power amplifiers in Class A, we conclude that a very small *HD* can be achieved, which is of course favorable, and at the same time the dissipation on the active element is reduced to about (maximum) 60% of total power delivered by the power supply. This means that at slightly higher power levels, due to the low efficiency, expensive components must be installed, i.e., complicated and expensive cooling system. The construction of high-power amplifiers in Class A is unthinkable.

The first idea toward achieving higher efficiency is to eliminate the direct current of the active component, i.e., dissipation on it, in the absence of a signal. This significantly increases the efficiency. An amplifier operating in this mode is called a Class B amplifier, and this type of amplifier will be discussed in the following sections.

4.3.2.4 Transformerless Single Transistor Class A Power Amplifier with BJT

The role of the transformer in the power amplifier is to enable impedance matching (mainly at the output) or what is the same, maximum power transfer. Unfortunately the transformer has several drawbacks among which are weight, volume, power losses (reduces efficiency), limited frequency response, and simular. To avoid the transformer, however, one needs to find a compromise solution that would eliminate these drawbacks and in the same time will enable proper power transfer from the amplifier to the load.

The only practical answer to this request is the use of a common collector (CE) stage or as it is frequently referred to the emitter follower. The reason for that is its very low output resistance which may match the load resistance to which one may add its high-current gain. In addition this stage has large input resistance which allows for the previous stage to have large voltage gain not degraded by th CE amplifier. Finally, the CE stage has a very wide frequency response which is an important issue since power transistors have a reduced gain-bandwidth product.

To introduce this stage here we will start with its properties in processing sinusoidal signals as described in LNAE_Book 2.

The usual schematic of the CE amplifier intended to be used for power applications is depicted in Fig. 4.3.14a. The problem with this circuit is the value of R_E which limits the quiescent current of the amplifiying transistor. The improvement is depicted in Fig. 4.3.14b where it is substituted by a current source.

For the analysis of this amplifier stage we will suppose that the circuit element values are $V_{CC} = 12$ V, $I_E = 5$ mA, $I_B \approx 100$ μA, $R_B = 120$ kΩ, $R_g = 50$ Ω, $C_{s1} =$

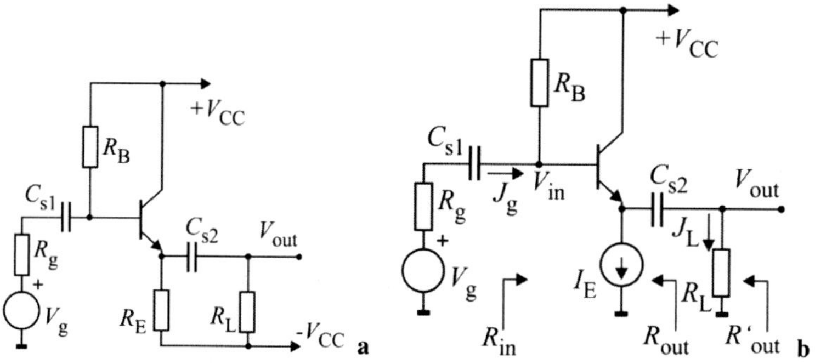

Fig. 4.3.14 Common collector stage. **a** Original and **b** improved

$C_{s2} = 100\ \mu F$. The load resistance is chosen to be $R_L = 8\Omega$. The transistor model parameters used in the analysis are $g_m = 50$ mA/V, $r_\pi = 1091\ \Omega$, $r_B = 576\ \Omega$, $r_C = 120$ kΩ, and $r_\mu = 3.3$ MΩ. We will consider that the output resistance of the current source is r_C.

The following expressions are taken from LNAE_Book 2 and accomodated to the proper application (having in mind the numerical values given above)

$$R_{in} \approx \left[r_b + r_\pi + (1 + g_m r_\pi) r'_e\right] R_B, \tag{4.3.42a}$$

$$r'_e = R'_E r_c (R'_E + r_c)^{-1} \approx R_L. \tag{4.3.42b}$$

$$A_v \approx \frac{g_m R'_E}{1 + g_m R'_E} \tag{4.3.42c}$$

$$A_c = \frac{J_L}{J_g} = \frac{V_{out}/R_L}{V_g/(R_g + R_{in})} \approx A_v \frac{R_{in}}{R_L} \approx \frac{R_{in}}{R_L} \tag{4.3.42d}$$

$$R_{out} \approx \frac{r_\pi + r_b + R'_B}{1 + g_m r_\pi} R_E. \tag{4.3.42e}$$

$$R'_{out} \approx \frac{r_\pi + r_b + R'_B}{1 + g_m r_\pi} R'_E. \tag{4.3.42f}$$

$$R'_B = R_B || R_g \approx R_g \tag{4.3.42g}$$

$$R'_E = R_E || R_L \approx R_L \tag{4.3.42h}$$

After substitution the numerical values we get $A_v \approx 1$, $A_c \approx 264$. This means that if the previous stage is a voltage amplifier with low output resistance, the overall power gain will be large and distributed, the CE staige being the one which provides for the current gain.

Note, the maximum amplitude of the output signal is only limitted by the maximum amplitude of the input signal so that large power can be delivered to the load R_L.

4.3.3 Class B Amplifiers

Class B power amplifiers are also realizad as push–pull structure. Here the operating point of each active element is chosen to be at the point where the output current stops flowing. Therefore, when there is no input signal, the output current of the active element is zero. The current through one of the active elements will flow only when the signal (excitation) is of such polarity that it brings the operating point of the active element into the active area. If thc excitation signal is sinusoidal, obviously, one active element will conduct only in one half-cycle of excitation, and in the other half-cycle it will be blocked. Therefore, the use of single active element only would lead to very large distortions, since the output signal would consist of a train of positive or negative pulses of sinusoidal shape.

By using a push–pull structure, this shortcoming is eliminated. The circuit schematic of Class B push–pull amplifier with BJTs is depicted in Fig. 4.3.15. The difference as compared to the push–pull Class A amplifier refers to the biasing of the input terminals. The V_{CC}-R-D branch is now used for polarization. The value of R is chosen so that a voltage equal to the conduction threshold of the base-emitter junction of the transistor is formed on the diode D. (Instead of a diode in this circuit, a resistor can be used with which we get poorer temperature stability. This will be discussed later.) So, the transistors will not conduct but there will be a voltage at the inputs that is at the conduction threshold and with the arrival of the input excitation signal, which is superimposed on the diode voltage, one will immediately start conducting and the other will be immediately cut-off.

For the analysis of the operation of this amplifier, the general circuit of the push–pull amplifier given in Fig. 4.3.10 will be used. The voltage of the battery V is chosen so that both active elements are in Class B, i.e., it is equal to the conduction threshold of the transistors.

In the absence of a signal $i_1 = i_2 = 0$. With the appearance of an alternating signal, the input terminal of one active element becomes more positive, and the input terminal of the other becomes more negative. In the first active element, current begins to flow, and the second is blocked. In the next half-period, the situation is reversed. Each active element conducts current in only one half-cycle. However, the current through the load contains a complete waveform. One half-period is given to it by one active element, and the other by the other.

Let us now turn to the analysis of the amplifier. Since while one active element is running, the other is cut-off and vice versa, there is no more interaction between the active elements, so the push–pull amplifier in Class B can be analyzed using the characteristics of single active element. Figure 4.3.16 shows the output characteristics of the generalized active element. The input signal is denoted by x_{in}. The output current and voltage are denoted by i_1 and v_1, respectively. The DC biasing of the input terminals (X_{in0}) is chosen so that the operating point is in the characteristic's

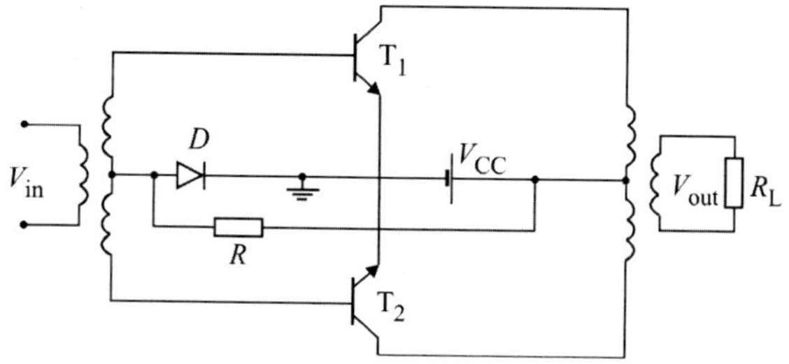

Fig. 4.3.15 One version of the realization of a push–pull Class B amplifier with BJTs

knee so that $i_1 = 0$ stands in the absence of a signal. When a positive half-cycle appears, the operating point moves along the load line whose slope is $1/R$ where

$$R = n_2^2 R_L, \tag{4.3.43}$$

since the load resistance is mapped only in one half of the winding of the whole primary (on the side of the active element that conducts).

According to Fig. 4.3.16 the total (maximum) useful power developed on a single active element is

$$P_{L_1} = \frac{1}{2}\left(\frac{1}{2} J_{1m} V_{1m}\right) = \frac{1}{4} J_{1m} V_{1m} = \frac{1}{4} J_{1m} (V_0 - V_{min}). \tag{4.3.44}$$

In this expression one half is used since the power is developed during half of the period while P_{L_1} is representing a full period. The useful power delivered by both elements (for the whole period) is

$$P_L = 2P_{L_1} = \frac{1}{2} J_{1m} (V_0 - V_{min}). \tag{4.3.45}$$

The power of the power supply source delivered to single active element is

$$P_1 = V_0 I_0 \tag{4.3.46}$$

where I_0 is the DC component of the output current. By developing the waveform of the output current from Fig. 4.3.16 into a Fourier series, the following value for the DC component is obtained

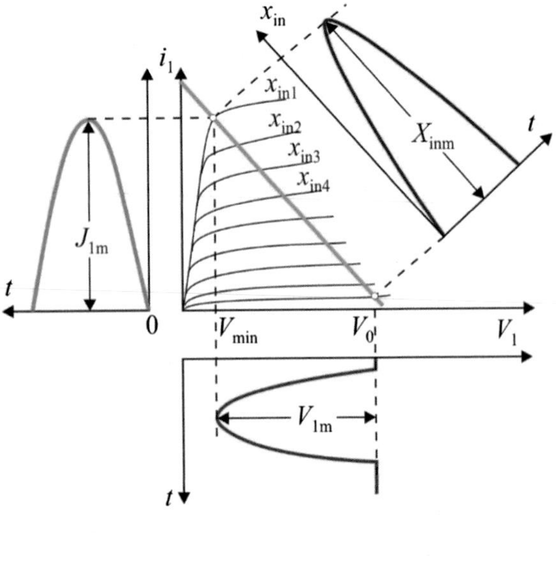

Fig. 4.3.16 Graphic analysis of the push–pull structure in Class B

$$I_0 = J_{1m}/\pi. \tag{4.3.47}$$

After substitution we get

$$P_1 = V_0 J_{1m}/\pi \tag{4.3.48}$$

The power given by the power source to the other active element is the same so that the total battery power is given by

$$P = 2V_0 J_{1m}/\pi \tag{4.3.49}$$

The efficiency of the push–pull structure per one active element is equal to the efficiency of the whole amplifier, so

$$\eta = \frac{P_{k1}}{P_1} = \frac{\pi}{4} \cdot \frac{V_0 - V_{min}}{V_0} = 0.785 \cdot \left(1 - \frac{V_{min}}{V_0}\right) \tag{4.3.50}$$

We were not talking here about the maximum efficiency since the above expression applies to any signal. When there is no signal, the efficiency is equal to zero, but that is not a problem, considering that the dissipated power is also equal to zero. At maximum excitation and with an ideal transistor, it can be considered that $V_{min} = 0$ V, so the efficiency would be 78.5%. An amplifier with bipolar transistors is closer to this condition than an amplifier with JFET, since V_{min} (or V_{DSon}) is smaller in bipolar transistors. Here we can remind that the efficiency of a Class A power amplifier is at most 50%, which indicates the advantages of Class B.

The dissipated power on one active element is

$$P_{d1} = P_1 - P_{L_1} = V_0 \frac{J_{1m}}{\pi} - \frac{V_{1m} J_{1m}}{2} = V_0 \frac{J_{1m}}{\pi} - \frac{1}{4} R \cdot J_{1m}^2. \tag{4.3.51}$$

From this expression it can be seen that the dissipation is equal to zero in the absence of a signal. As the input signal increases, so does J_{1m} and the dissipation. It reaches its maximum for

$$J_{1m} = \frac{2}{\pi} \frac{V_0}{R} \tag{4.3.52}$$

and takes the value of

$$P_{dt_{max}} = \frac{1}{\pi^2} \cdot \frac{V_0^2}{R} \tag{4.3.53}$$

If $V_{min} = 0$ is taken, an interdependence can be easily found between the maximum value of the dissipation power (4.3.53) and the useful power (4.3.44). It reads

$$P_{L_1} = (\pi^2/4) P_{d1_{max}} \approx 2.5 \cdot P_{d1_{max}}. \tag{4.3.54}$$

Hence the conclusion that in a push–pull Class B amplifier, the useful power per active element is two and a half times higher than the dissipated power. With Class A amplifiers, the maximum useful power of the circuit can reach up to half the battery power, which means that it is at best equal to the dissipated power. This comparison of Class A and Class B becomes more significant if the following example is considered. If an amplifier is needed delivering power to the load of $P_L = 20$ W, each element should deliver 10 W. In Class B, 4 W will be dissipated on each transistor while in Class A, 10 W. In the absence of signal in the Class B amplifier no power will be dissipated, while a Class A amplifier will dissipate as much as 20 W (in this case $P_d = P = V_0 I_o = 2P_L$). From this we conclude that the components installed in a Class B amplifier can have two and a half times less maximum dissipation power rating than those used in Class A, while both amplifiers provide for the same useful power to the load.

Continuing to compare the properties of amplifiers in Class A and Class B, distortions will be briefly mentioned. Since the distortions are defined in relation to the complete output signal, in Class B, under the condition of ideal symmetry, the even harmonics are also canceled. However, a Class B amplifier gives larger distortion than a Class A amplifier because the quiescent operating point is in the non-linear part of the characteristic. The reduction of non-linear distortions in Class B will be discussed later.

Only one active element can be used in Class B, but then the load impedance must be selective in order to eliminate unwanted harmonic components of the output signal. This is used in narrowband high-frequency power amplifiers.

Before proceeding to improvements that eliminate the shortcomings of the basic configuration of the power amplifier in Class B, let us first mention that the DC current component of the active element is not a constant, which, on the internal resistance of the power supply forms a voltage drop which is a function of the amplitude J_{1m}. Therefore, it is required that the internal resistance of the power supply be so small that the variable voltage drop across it is negligible. This is not so easy to achieve given that the DC current in power amplifiers is of the order of amperes. If this condition is not met, a variable voltage drop is created on the battery, which excites the amplifier stages that precede Class B, which results in unwanted feedback. On the other hand, if we want to increase the useful power in these circuits, at a given maximum component current, we need to increase the voltage dynamics. This is achieved by increasing the supply voltage. Thus, the power amplifier uses a relatively large V_{CC}, too. However, such a high supply voltage is not necessary to supply the previous amplifier stages. Moreover, the alternating voltage component of the power supply is more influential to the small-signal amplifiers that preced the power stage. Therefore, there is a need to use two power sources in such circuits. Such a solution, however, would be expensive, so instead, the large DC voltage is reduced and another smaller is formed. At the same time, the alternating component of the power supply is eliminated. The realization of such a power supply circuit is shown in Fig. 4.3.17. A voltage drop equal to the difference in the supply voltage values of the two sub-systems is created on the resistor R, and its value is selected based on the value of the DC supply current of the voltage (preceding) amplifier.

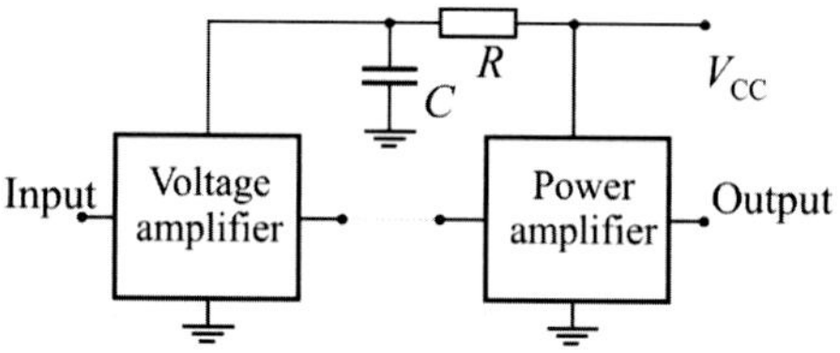

Fig. 4.3.17 Realization of the power supply circuit of a system containing cascade of voltage and power amplifiers

The capacitor C acts as a filter for the alternating component of power amplifier not allowing it to reach the preceding voltage amplification stages.

4.3.4 Phase Splitters

The first improvement that we will introduce in the circuit of Fig. 4.3.15 is realted to the elimination of the input transformer. We will repeat here that the transformer, as such, is a circuit that contributes more to amplitude and phase distortions at both low and high frequencies than active elements. Therefore, the transformer coupling can be used in a relatively narrow frequency range. In addition, the power transformer is an expensive component and requires a large PCB area, i.e., it cannot be integrated. In the following, electronic circuits that allow transformers to be eliminated will be presented, and in this section we will talk about circuits that can be used to eliminate the input transformer.

The input transformer serves to generate two signals whose amplitudes are equal, and the phases are opposite. That is why the electronic circuit, which will replace the transformer, is called phase splitter.

Two circuits schematics that can be used as phase splitters are shown in Fig. 4.3.18. The circuit of Fig. 4.3.18a is a differential amplifier with an asymmetric input in which, since it is used in an RC coupled amplifier, the biasing of the base terminals is specially provided. Supposing the circuit is symmetrical, we have (here we use the analogies: $h_{\text{inE}} = h_{11\text{E}}$; $h_{\text{rE}} = h_{12\text{E}}$; $h_{\text{fE}} = h_{21\text{E}}$; and $h_{\text{oE}} = h_{22\text{E}}$)

$$V_2 = A_{11} V_1 = -\frac{h_{\text{fE}} R_{\text{C}} (1 + h_{\text{fE}} + h_{\text{inE}}/R_{\text{E}})}{2h_{\text{inE}}[1 + h_{\text{fE}} + h_{\text{inE}}/(2R_{\text{E}})]} \cdot V_1 \tag{4.3.55}$$

and

$$V'_2 = A_{12} V_1 = \frac{h_{\text{fE}} R_{\text{C}} (1 + h_{\text{fE}})}{2h_{\text{inE}}[1 + h_{\text{fE}} + h_{\text{inE}}/(2R_{\text{E}})]} \cdot V_1 \tag{4.3.56}$$

If $R_{\text{E}} > > h_{\text{inE}}$, it becomes

$$V'_2 = -V_2. \tag{4.3.57}$$

That means two voltages of equal amplitudes and opposite phases are obtained.

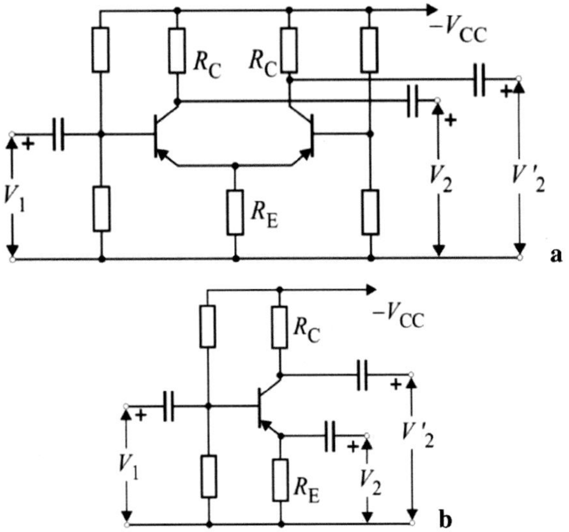

Fig. 4.3.18 Phase splitters. **a** Differential amplifier and **b** basic amplifier

For the circuit of Fig. 4.3.18b we have

$$V_2 = \frac{(1 + h_{fE})R_E}{h_{inE} + (1 + h_{fE})R_E} \cdot V_1 \approx V_1 \tag{4.3.58}$$

and

$$V_2' = -\frac{h_{fE}R_C}{h_{inE} + (1 + h_{inE})R_E} \cdot V_1 \approx -\frac{h_{fE}R_C}{(1 + h_{fE})R_E} \cdot V_1 \approx -\frac{R_C}{R_E} \cdot V_1. \tag{4.3.59}$$

Here we used $(1 + h_{fE})R_E >> h_{inE}$ which is usually the case. If we impose $R_C = R_E$, and having in mind that $h_{fE} >> 1$, one gets

$$V_2 = -V_2' \approx V_1. \tag{4.3.60}$$

Of course, in these development it is assumed that $h_{rE} = 0$ and $h_{oE} = 0$. If these quantities are also taken into account, somewhat more complex expressions are obtained, and the ratio of R_C and R_E should be determined from the conditions $V_2 = -V_2$. In doing so, one of these two resistances will be determined from other considerations.

Let us mention, finally, that instead of resistors R_C and R_E, appropriate dynamic resistances can be installed and that both of these circuits can also be implemented with JFETs.

For amplifiers realized as discrete circuits, a differential amplifier is more suitable for use, since the output impedances on both collectors are the same. The circuit of Fig. 4.3.18b has much lower output resistance on the emitter than on the collector terminal. When integrated, however, the transistor of Fig. 4.3.18b can be designed

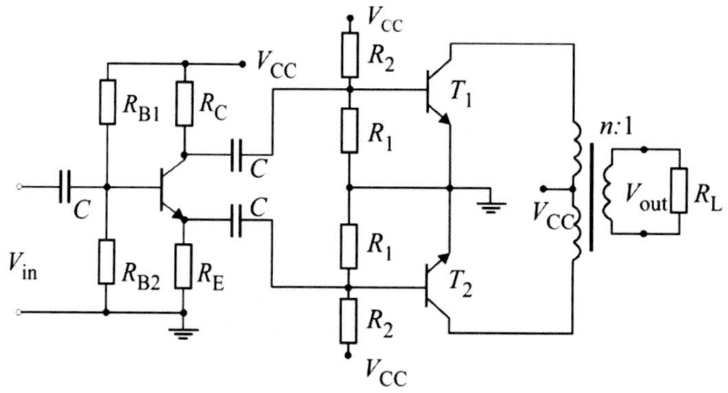

Fig. 4.3.19 Phase splitter implemented in a push–pull structure

to be symmetrical so that the values of the output impedances at both terminals are close to each other.

An example of the use of a phase splitter as a replacement to an input transformer is shown in Fig. 4.3.19.

The power amplifier in this circuit can operate in both Class A and Class B, and even in Class C. This will depend on the value of the voltage drop across resistor R_1. If the DC voltage on the resistor R_1 is smaller than the threshold of conduction of the base-emitter junction of the output transistor, the power amplifier will operate in Class C and this is not the mode intended for this circuit schematics. If the voltage drop across the resistor R_1 is equal to the threshold of the output transistor (V_γ), the output stage will operate in Class B. This mode of operation is possible and favorable. It is also possible to normally polarize the output transistors by selecting a sufficiently large value for R_1, so that Class A is obtained. Finally, if the resistance value of R_1 is chosen so that the V_{BE} voltage is only slightly higher of value V_γ, the output stage will work in the transitional Class AB. For now, it will be considered that the most favorable work is in Class B. For such work it is necessary the following condition to be satisfied

$$\frac{R_1}{R_1 + R_2} V_{CC} = V_\gamma. \tag{4.3.61}$$

This determines one of the resistances. To determine the other, a requirement on the value of the current gain of the amplifier may be used, which is also a function of the biasing resistances:

$$A_C \approx \frac{R_B h_{fE}}{R_B + h_{inE}}. \tag{4.3.62}$$

When considering the output circuit, it is obvious that the DC load line is defined by

$$V_{CE} = V_{CC} \tag{4.3.63}$$

It is, in fact, a vertical straight line (since there is no resistor in the circuit). The AC load line will be

$$V_{CE} - V_{CC} = -R_C I_C. \tag{4.3.64}$$

where $R_C = n^2 R_L$. When setting (4.3.62) it was supposed that $I_{CQ} = 0$. We now conclude that the value of n is determined from the condition that the AC load line has the largest possible slope, without exceeding the maximum dissipation or the maximum current of the transistor.

The basic disadvantage of the circuit we observed is in that it is still using a transformer. Since the amplitudes of the signals in the output circuit are significantly larger, the dimensions and the price of this transformer will be higher. Along with that are his other shortcomings. In addition to the mentioned disadvantage, the upper circuit can be considered extremely temperature unstable, which will be discussed later.

Therefore, instead of this circuit, circuits with a pair of complementary transistors that do not require an output transformer are most often used. The next section is dedicated to that.

4.3.5 Push–pull Amplifiers with a Pair of Complementary Transistors in Class A

By using two transistors of different conductivity types (one PNP and the other NPN), it is possible to obtain a push–pull power amplifier without the use of a transformer at the output as shown in Fig. 4.3.20. The transistors are complementary, and it is assumed that their input and output characteristics are the same, with opposite orientation of the voltages and currents. For the AC signal, the bases of the transistors are short-circuited, and for the DC signal they are separated. Biasing is performed via R_B so that both transistors operate in Class A.

Since the transistors are of the opposite type of conductivity, an increase in the AC signal at the bases causes an increase in the base current in the upper one, and a decrease in the lower transistor, which also applies to collector currents. Thus, the alternating components of the output currents are shifted by 180°. Therefore, an input transformer is not required.

Since two DC power sources were used and the currents of both transistors flow through the load, and since these are equal and of opposite sign, the total DC current through the load is zero. As with the push–pull amplifier using a transformer, the AC component of the load current does not contain even harmonics. When it comes to the fundamental harmonic, however, since the transistor currents are opposite in phase and are subtracted on the load, the total current of the fundamental harmonic

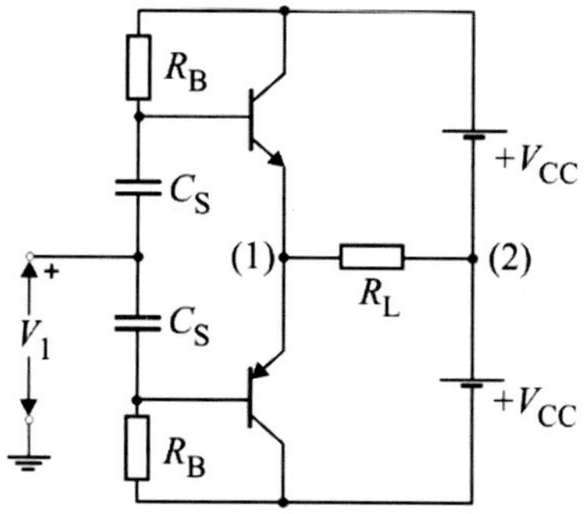

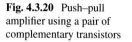

Fig. 4.3.20 Push–pull amplifier using a pair of complementary transistors

of the load is twice as high. Of course, no output transformer was required to perform addition of the currents.

The realization of the circuit with Fig. 4.3.20 poses one problem. Namely, usually one end of the load resistor is connected to the ground. Thus, one of the points, marked with (1) and (2), should be grounded.

If point (1) is grounded, the transistors will operate as a common emitter amplifier. Through R_B, in this case, negative feedback is also established for the AC signal. In order to remove it, the resistor R_B is realized as two series-connected resistors. The point in between these resistors is connected to ground via an additional capacitor. This realizes an RC filter that passes direct current to the base, and short circuits the AC signal to ground via the capacitor. An example of such a filter was shown in basic CMOS amplifier in LNAE Book 3.

Grounding point (1) has another drawback. In this case, neither terminal of the battery is grounded. This may be unacceptable from the point of view of the realization of the power supply.

These deficiencies can be remedied if point (2) is attached to the ground. In this case, the transistors perform as common collector amplifiers. The output impedance of the CC amplifier is small, which also becomes an advantage when the load resistance is low (which is usually the case) since the issue of power matching is solved per se. However, in this case, a higher input voltage is required to obtain the appropriate power at the load, since the voltage gain of the CC amplifier is much smaller than the gain of the CE amplifier.

In other words, since the CC amplifier has a voltage gain approximately equal to unity, two consequences should be taken into account. First, the previous stage should be designed to provide the maximum voltage amplitude at maximum excitation, and then, the output stage should be designed so that it has the maximum current gain in order to obtain the maximum power gain at the same time.

The amplifier of Fig. 4.3.20 can be set to work in Class A and in Class AB, but in the latter case it is necessary that R_B has a very large value.

4.3.5.1 Push–pull Amplifiers with a Pair of Complementary Transistors in Class B

Due to the advantages offered by Class B in terms of efficiency and useful power, the use of power amplifiers in Class B is much more frequent than those in Class A, and therefore much more attention will be paid to them. The basic schematic of this circuit is given in Fig. 4.3.21a.

When the input signal is positive transistor T_1 which is NPN type conducts and its output current flows through resistor R_L. Then the T_2 transistor which is of PNP type is switched off. Otherwise, when the input voltage is negative, transistor T_1 is switched off, and transistor T_2 conducts, providing current through the load. The amplifier works in Class B. For example, if the input voltage is a sinusoidal transistor T_1 is conducting during the positive half-cycle and transistor T_2 during the negative half-cycle of the input voltage. In this way, if distortions are overlooked, the voltage on the load will also be sinusoidal.

Precisely speaking, the circuit from Fig. 4.3.21a operates in Class C, however. Namely, the transistors start to conduct only when the voltage between the base and the emitter is greater than $|V_{BE}| = |V_\gamma| \approx 0.5$ V, which can be seen from the simplified transfer characteristic shown in Fig. 4.3.21b. To avoid this, it is necessary to provide a DC voltage between the bases of the transistor $V = 2 \cdot V_\gamma > 1.0$ V. This is achieved by connecting two diodes between the bases of the transistor that are constantly forward biased as shown in Fig. 4.3.22.

For this circuit, it should first be noted that it works with a single power supply which is shown here as an alternative to all other circuits in Class B discussed later. Since the coupling capacitors are shown, it is assumed that this circuit will be realized as discrete one. A pair of diodes were used to bias the p-n junctions of the transistors. The advantage of using series-connected pair of diodes as opposed to a resistor (on which the same DC voltage drop would be achieved) is reflected in the improved temperature stability. Namely, it is known that at constant current, the voltage at the p–n junction decreases with increasing temperature. This means that an increase in temperature in this circuit will reduce the potential difference between the bases.

In the circuit of Fig. 4.3.22 diodes having equal characteristics as the transistors are used, and they are mounted on the same cooler so that their temperatures are the same. The increase in temperature will accordingly reduce the potential difference

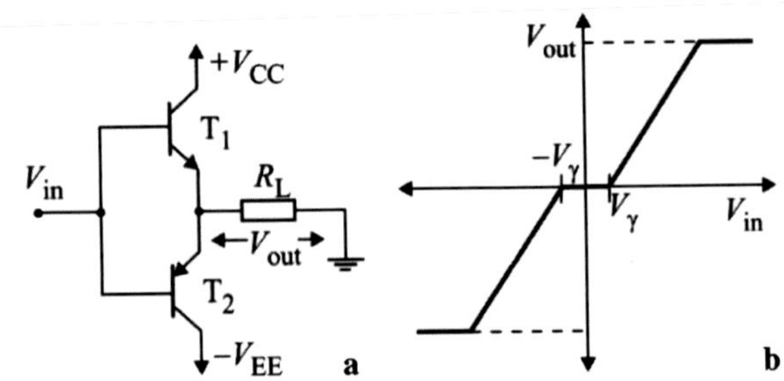

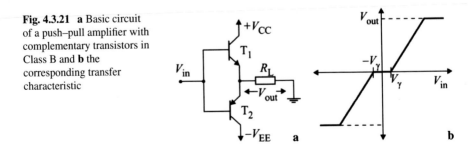

Fig. 4.3.21 **a** Basic circuit of a push–pull amplifier with complementary transistors in Class B and **b** the corresponding transfer characteristic

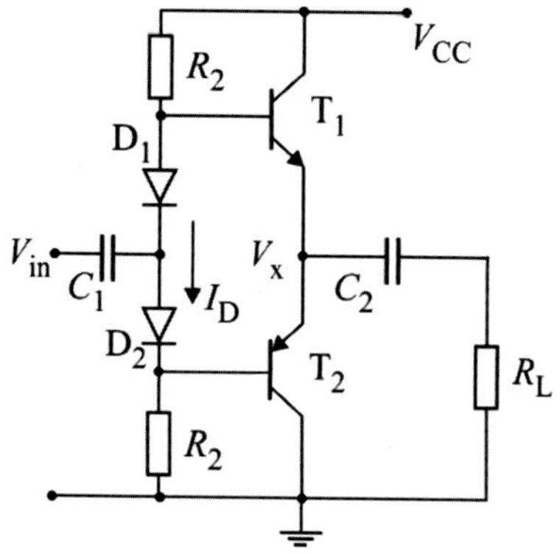

Fig. 4.3.22 Push–pull circuit in Class B with complementary transistors using single power supply

between the bases so that there will be no self-heating. Of course, the temperature stability can be further improved if a small resistor (usually 0.47Ω) is connected to the emitter of each of the transistors in turn, which has a stabilizing effect like any R_E (the so-called degenerated emitter).

Analysis and design of such an amplifier require knowledge of the internal resistance of the diode, which we can assume is typically about $r_D \approx 10\ \Omega$. It is important for the operation of the amplifier that the diode current is large enough for the diode to remain in the conductive mode for all values of the input voltage. In other words, the direct current of the diode must be larger than the maximum negative amplitude of the same current. In this way, the total current through the diode will not become negative, and the diode will remain forward biased. This limitation can be expressed as

$$I_D \geq |J_{Dm\,max}||| \tag{4.3.65}$$

where $J_{Dm\,max}$ is the largest amplitude of the AC diode current.

Since in absence of excitation the base currents and collector currents are equal to zero, we conclude that the potentials of the nodes in between the diodes and in between the emitters of the transistors are equal to half the supply voltage, i.e., in the absence of signal $V_x = V_{CC}/2$, V_x is depicted in Fig. 4.3.22. From there

$$I_D = (V_{CC}/2 - 0.7)/R_2, \tag{4.3.66}$$

where it was assumed that the diode voltage is 0.7 V.

The value of the base current at alternating excitation can be determined on the basis of the AC circuit shown in Fig. 4.3.23a. Since in each half-cycle only one transistor is active, the circuit contains a model of single transistor with $h_{rE} = 0$ and $h_{oE} = 0$ S. Except when specifically emphasized, the signal frequencies will be considered high enough that the capacitor impedance can be neglected.

For the maximum amplitude of the diode current at the negative half-cycle it is possible to write

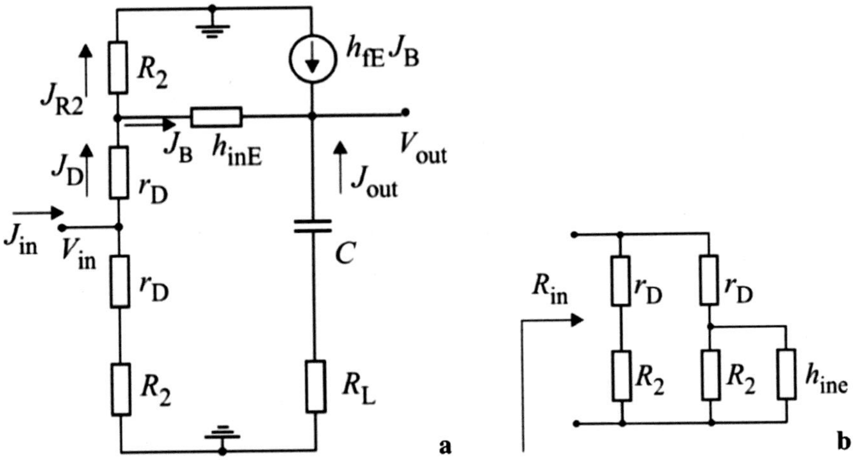

Fig. 4.3.23 Push–pull circuit in Class B with complementary transistors using a single power supply. **a** Equivalent circuit for calculation the gain and **b** equivalent circuit for calculation the input resistance

$$J_{\mathrm{Dm\,max}} = J_{\mathrm{Bm\,max}} + V_{\mathrm{in\,max}}/R_2. \tag{4.3.67}$$

On the other side, for the output node the following is valid

$$V_{\mathrm{in\,max}} = R_{\mathrm{L}}(1 + h_{\mathrm{fE}})J_{\mathrm{Bm\,max}} \approx R_{\mathrm{L}}h_{\mathrm{fE}}J_{\mathrm{Bm\,max}}. \tag{4.3.68}$$

By combining the last three expression one gets a relation between the maximum amplitude of the output voltage and the resistance R_2:

$$R_2 = h_{\mathrm{fE}}R_{\mathrm{L}}\frac{V_{\mathrm{CC}}/2 - 0.7 - V_{\mathrm{in\,max}}}{V_{\mathrm{in\,max}}}. \tag{4.3.69}$$

For the current gain of this circuit one gets

$$A_{\mathrm{C}} = \frac{J_{\mathrm{out}}}{J_{\mathrm{in}}} = -\frac{R_2}{2}\cdot\frac{1 + h_{\mathrm{fE}}}{R_0 + h_{\mathrm{fE}} + (1 + h_{\mathrm{fE}})R_{\mathrm{L}}} \approx -\frac{R_2}{2R_{\mathrm{L}}}. \tag{4.3.70a}$$

where $R_0 = R_2(r_D + R_2/2)/(R_2 + r_D)$ while in the approximation it was considered that $h_{fE}R_L$ is a much larger resistance than all the rest resistances in the circuit which, in general, may not be thrue. By analysis of the same circuit for the voltage gain one gets:

$$A = \frac{V_{\mathrm{out}}}{V_{\mathrm{in}}} = \frac{R_2R_{\mathrm{L}}(1 + h_{\mathrm{fE}})}{R_2r_{\mathrm{D}} + [h_{\mathrm{inE}} + (1 + h_{\mathrm{fE}})R_{\mathrm{L}}](R_2 + r_{\mathrm{D}})} \approx 1. \tag{4.3.70b}$$

In the circuit of Fig. 4.3.23b we used $h_{\text{ine}} = h_{\text{inE}} + (1 + h_{\text{fE}})R_{\text{L}}$. Accordingly, for the input resistance one may use the following expression

$$R_{\text{in}} = (r_{\text{D}} + R_2)||[r_{\text{D}} + R_2 h_{\text{ine}}] \tag{4.3.70c}$$

The efficiency of this circuit is given earlier while Class B analysis was performed by the expression (4.3.50) but the distribution of power between the elements of the circuit in the case when only one battery is used is slightly different, so the power balance will be reconsidered here.

The power delivered by the battery is

$$P_0 = V_{\text{CC}} I_0 = V_{\text{CC}} J_{\text{Cm}}/\pi. \tag{4.3.71}$$

where I_0 is the direct current of the battery which is related to the maximum amplitude of the collector current and as before, is given by $I_0 = J_{\text{Cm}}/\pi$. The following should be taken into account here. During the positive half-cycle of the input signal, current is established from the battery via the upper transistor, the capacitor C_2, and the load. The lower transistor is cut-off. Battery energy is distributed to the transistor (dissipation), load (useful power) and accumulates in the capacitor. Now, during the negative half-cycle of the input signal, the upper transistor gets cut-off, and the battery does not deliver current except for the voltage divider 2·(R_2-D) which refers to the input circuit. Instead, the accumulated energy in C_2 is returned to the circuit and the load through the load and the lower transistor. On the other hand, the maximum amplitude of the collector current is

$$J_{\text{Cm max}} = V_{\text{CC}}/(2R_{\text{L}}), \tag{4.3.72}$$

where the minimum voltage on the transistor is neglected, so the maximum useful power is

$$P_{\text{Lmax}} = \frac{V_{\text{in max}} J_{\text{Cm max}}}{2} = \frac{1}{2}\frac{V_{\text{CC}}}{2}\frac{V_{\text{CC}}}{2R_{\text{L}}} = \frac{V_{\text{CC}}^2}{8R_{\text{L}}}, \tag{4.3.73}$$

while the total dissipation on the transistors is

$$P_{\text{d}} = \frac{V_{\text{CC}} J_{\text{Cm max}}}{\pi} - \frac{J_{\text{Cm max}}^2 R_{\text{L}}}{2}, \tag{4.3.74}$$

We find its maximum if we differentiate this expression with respect to $J_{\text{Cm max}}$ and equate the derivative with zero. The corresponding current is $J_{\text{m}} = V_{\text{CC}}/(\pi R_{\text{L}})$, and the maximum dissipated power per transistor is

$$P_{\text{dmax}} = \frac{V_{\text{CC}}^2}{2\pi^2 R_{\text{L}}}. \tag{4.3.75}$$

As an example, if one uses a transistor with $\beta = 60$, if the needed useful power is 1.2 W, and if $V_{CC} = 12$ V, for an amplifier loaded with an 8 Ω loudspeaker, one gets $J_{Cm\,max} = \sqrt{2P_L/R_L} = 0.345$ A, $V_{x\,max} = R_L J_{Cm\,max} = 2.83$ V, and $R_2 = 60{\cdot}8{\cdot}(6–0.7–2.83)/2.83 = 419$ Ω. The maximum power dissipation per transistor is $P_{dmax}/2 = 12^2/(4{\cdot}3.14^2{\cdot}8) = 0.456$ W.

In order to take a closer look at the distortions that occur in this Class B amplifier, we will consider the waveform of the output current more carefully. To this end, Fig. 4.3.24 shows the equivalent transfer characteristic of a push–pull amplifier with JFETs (in red). The circuit is excited by a sinusoidal signal x_{in} (in green), and the waveform of the output current (i_{eq}) is also shown (in magenta). It is easy to notice that in the region of small currents, the output current deviates significantly from the sinusoid. This means that the distortions will be significant, i.e., significantly higher than those occurring in Class A. A qualitatively identical waveform of the output current corresponds to an amplifier with bipolar transistors.

If we want to reduce distortions, we can do it in two ways. First, negative feedback can be applied, which was partially discussed (by inserting small resistors in series with the transistor emitters) and will be also discussed later, and second, the quiescent operating point can be moved from the edge of the cut-off (or pinch-off) area to the area of low currents, which produces an amplifier operating in Class AB.

The transfer characteristic of the equivalent element in Class AB is shown in Fig. 4.3.25 The individual transfer characteristics are shown in blue and green, while the magenta line is the resulting one. It is more linear than the one in Fig. 4.3.24 so the distortions are smaller. Of course, in turn, the amplifier in Class AB will have less useful power and efficiency than that in Class B. The maximum useful power will be lower because the dynamic range of the input, and thus the output signal is reduced and the efficiency is smaller because for an amplifier operating in Class AB we always have direct current (denote I_L in Fig. 4.3.25) which means we always

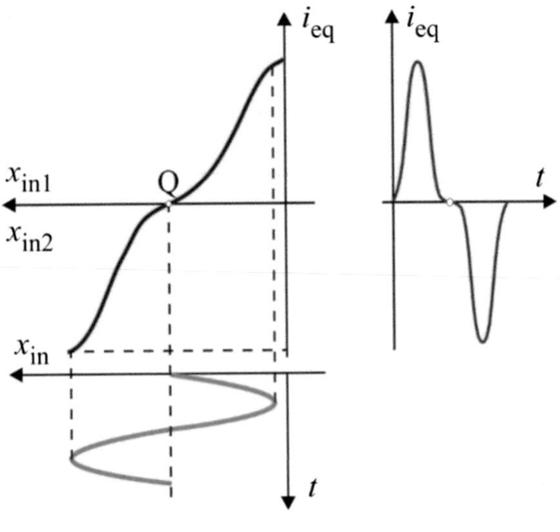

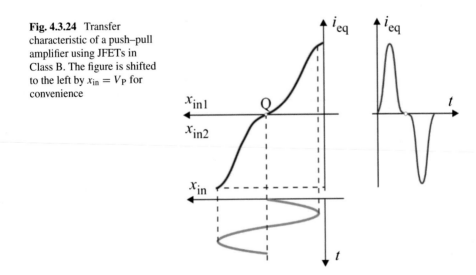

Fig. 4.3.24 Transfer characteristic of a push–pull amplifier using JFETs in Class B. The figure is shifted to the left by $x_{in} = V_P$ for convenience

have dissipation on the transistor. Since the Class AB is a compromise between high efficiency and low distortion, it has found the widest application in audio power amplifiers.

From the point of view of the physical realization of the power amplifier in Class AB, it is necessary that in the basic circuit with Fig. 4.3.21a instead of two diodes, an element is placed that provides a voltage that is slightly higher than the threshold of the two diodes in series. The first example of such a circuit is given in Fig. 4.3.26. Although two diodes for biasing of the input terminals will still be shown in some circuits later, the circuit with Fig. 4.3.26 should always be kept in mind. The essential difference with respect to a pair of diodes is that the point between the diodes is now inaccessible.

The circuit in principle of the newly formed amplifier is shown in Fig. 4.3.27a, and the corresponding transfer characteristic is given in Fig. 4.3.27b.

The transfer characteristic of the circuit of Fig. 4.3.27a does not pass through zero, i.e., even though the transistors have identical characteristics, when $V_{in} = 0$ V, $V_{out} \neq 0$. In order to eliminate this, it is necessary to ensure that the input voltage also has a DC component $V_{in} = V_{BE2}$. Therefore, a push–pull amplifier with a complementary pair is excited via a common collector amplifier where the voltage drop between the collector and the emitter provides this DC component. In the schematic of Fig. 4.3.28 this role is playing the transistor T_5.

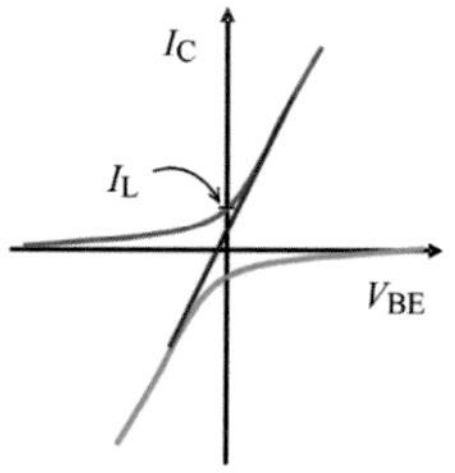

Fig. 4.3.25 Transfer characteristic of a Class AB push–pull amplifier. I_L is the idle current

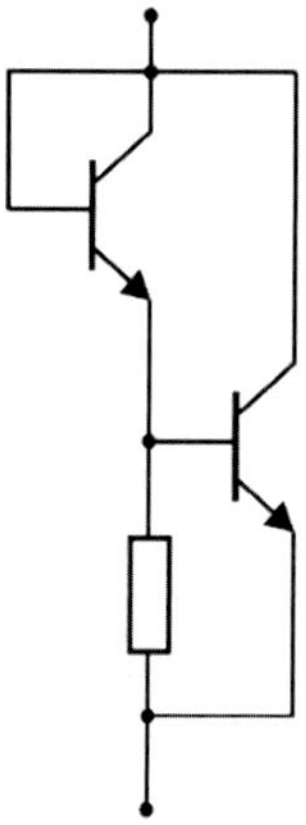

Fig. 4.3.26 A circuit convenient for biasing the input terminals of a Class AB push–pull amplifier

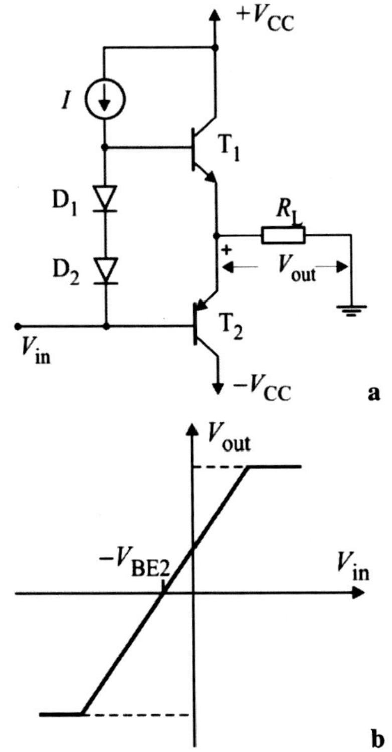

Fig. 4.3.27 a Complementary pair with diodes used to reduce distortions and **b** corresponding transfer characteristic

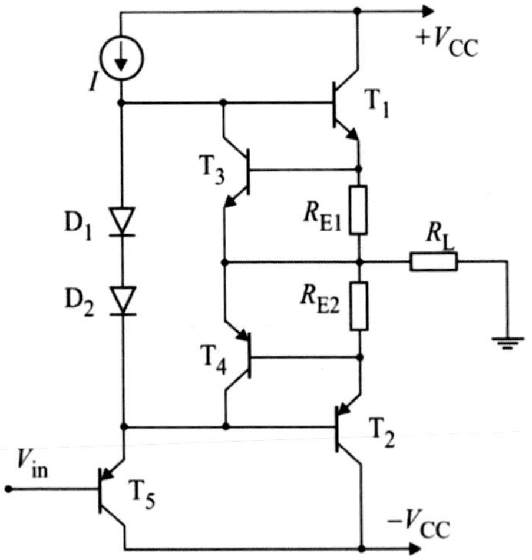

Fig. 4.3.28 Push–pull amplifier with a complemetary pair and short-circuit protection

There is another problem related to the practical implementation of push–pull power amplifiers using complementary pair. If the resistance of the load R_L for any reason becomes too small, the current through the transistors becomes too large. In the event that the common emitters of the transistors (the output of the amplifier) were shorted to ground while the excitation at the input is still present, the current through transistors T_1 and T_2 would increase so much that it would destroy them. This problem is solved by adding two new transistors to the basic circuit, as shown in Fig. 4.3.28.

Under normal operating conditions, when the resistance of the load is not too small, i.e., when the output is not short-circuited to ground, transistors T_3 and T_4 are switched off and do not conduct. This is provided by a voltage drop across resistors R_{E1} and R_{E2}. These resistances are small, on the order of hundreds of millions, so that the voltage drop across them is insufficient to bring the transistors T_3 and T_4 into the conducting state.

In the event of a short circuit at the output, the currents through transistors T_1 and T_2 begin to increase. Therefore, at the same time, the voltage drops across resistors R_{E1} and R_{E2} increase. Transistors T_3 and T_4 begin to conduct. Consequently, the base currents of transistors T_1 and T_2 are reduced and limited, regardless of the fact that the output is short circuited.

The resistances R_{E1} and R_{E2}, due to their small value, have little effect on the value of the gain of the complementary pair under normal conditions. In addition, they stabilize the complementary pair in terms of temperature.

4.3.5.2 Class B Push–pull Amplifier with a Darlington Pair

The production of high-power complementary NPN and PNP transistors (with identical characteristics) is a problem, even in integrated circuits. Therefore, it is simpler to replace the transistor T_2 of the basic circuit in Fig. 4.3.28a by a complementary Darlington pair equivalent to a PNP transistor. Such a solution is shown in Fig. 4.3.29

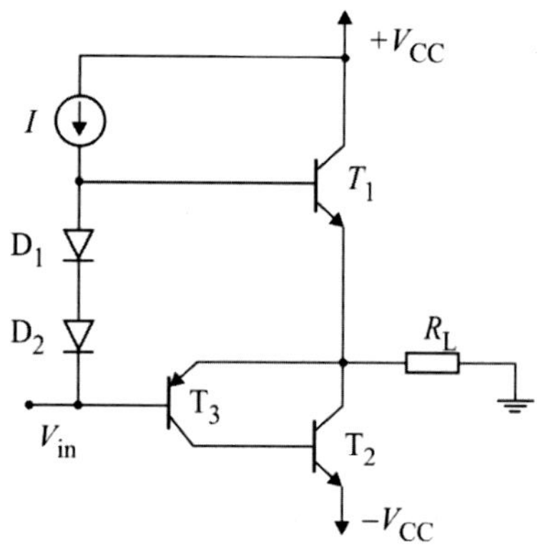

Fig. 4.3.29 Class B push–pull amplifier with a Darlington pair

Both transistors T_1 and T_2 are of NPN type, designed for power amplification and have identical characteristics. The transistor T_3 which is of PNP type is easier to manufacture because it is not a power transistor.

4.3.5.3 Push–pull Amplifiers with Power MOS Transistors

Push–pull amplifier with MOS transistors is realized equally as the bipolar ones. The solution that is often used in CMOS integrated circuits is shown in Fig. 4.3.30. The output complementary pair consists of transistors T_1 and T_2. Transistors T_5 and T_6 are connected like diodes and enable biasing of the gates of the output stage. The value of the total voltage drop on the pair T_5-T_6 depends on the current flowing through them which is determined by the constant current source T_6 or the reference voltage V_r. Transistor T_3 is an excitation or a CS amplifier transistor.

It is not necessary to prove much that in this case, too, the output transistors work as common drain amplifiers, and at will, they will be in Class AB or Class B.

When working with high-power amplifiers, DMOS, VDMOS, and VVMOS transistors are used in discrete circuits. However, the formation of a push–pull amplifier is difficult because it is difficult to produce N-channel and P-channel MOSFETs of the same characteristics when it comes to high-power transistors. Therefore, frequently, in a push–pull solution two N-channel MOSFETs are used as shown in Fig. 4.3.31. Such a circuit is called a push–pull amplifier with quasi-complementary symmetry.

Since both MOSFETs have the same type of conductivity, the excitation voltages at their gates must be of the same amplitude and opposite phases. This means that the circuit of Fig. 4.3.31 is excited by a phase splitter. Transistor T_1 acts as a common drain amplifier, and transistor T_2 acts as a common source amplifier. The excitation voltages V_{in} have also a DC component that serves to polarize the gates. This circuit will be discussed in more detail later.

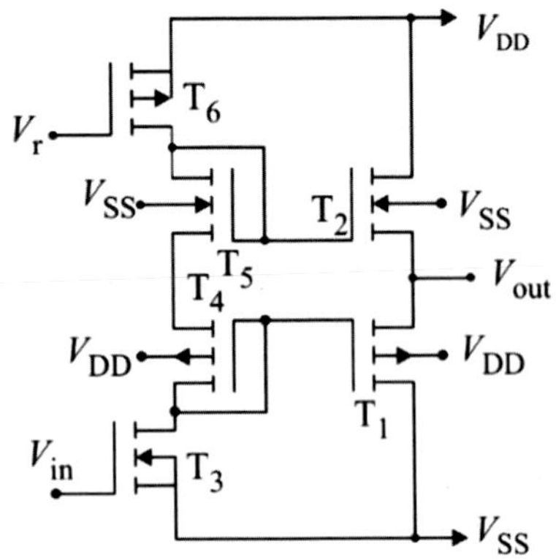

Fig. 4.3.30 Class B or Class AB push–pull amplifier using complementary MOS transistors

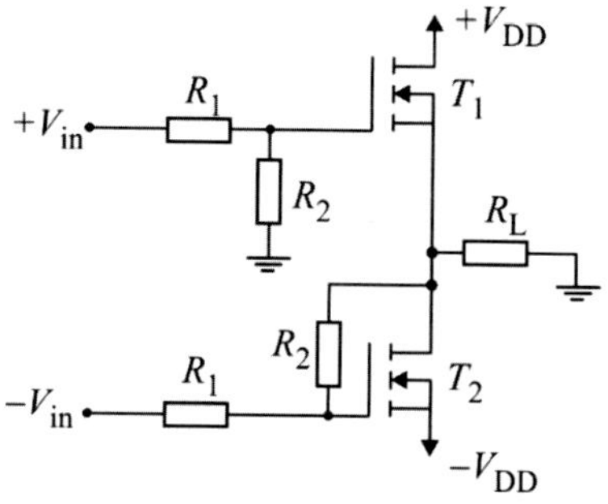

Fig. 4.3.31 Class A or Class AB push–pull MOSFET amplifier using quasi-complementary symmetry

4.3.6 Class C Power Amplifiers

Figure 4.3.32 depicts simplified Class C power amplifier circuits with a JFET and a bipolar transistor. The excitation source and the load are coupled with the amplifying circuit via DC isolating capacitors. C_s is a short circuit at the frequencies of the excitation signal. The coupling may be realized with coupled inductances (or transformers), which can be also exploited for power matching at the input or output. At the input and output of the active element there are parallel resonant circuits tuned to the same resonant frequency. We call this configuration tuned in tuned out (TITO) as opposed to the one when the Class C amplifier is used as a separate amplifying circuit, out of the RF analog signal processing chain, where a parallel resonant circuit is used at the output only. When calculating the value of the resonant frequencies of the resonant circuits, the values of the capacitances of the capacitors C_1 and C_2, respectively, include the input or output capacitance of the active element.

Using a separate power supply at the input, the operating conditions of the active element are set to correspond to Class C. That means the following. The biasing of the input terminals is such that no current flows through the active element in absence of input signal. When an input signal appears, the output current will not flow until the input diode (of the bipolar case) are forward biased. Therefore, the output current flows only during one part of one half-cycle of the input signal. During the entire second half, the active element is cut-off. In order to avoid the biasing battery of the input circuit of Class C power amplifiers, RC bias is most commonly used. The input electrode (gate or base) is forward biased in a small part of one half-cycle of the input signal so that a significant input current flows. In other words, in that part of the half-cycle, considerable energy is consumed at the input, so it is often necessary for the previous stage to be a power amplifier, too.

To ensure that the AC component of the output current does not flow through the power supply, the battery is isolated with a high-inductance coil. In this way, an amplifier in Class C is created, which is shown in Fig. 4.3.33 for the case of amplifiers with JFET.

A special problem in the design of Class C power amplifiers is the transmission of signals in the inverse direction via the Miller capacitance. Since the output current is

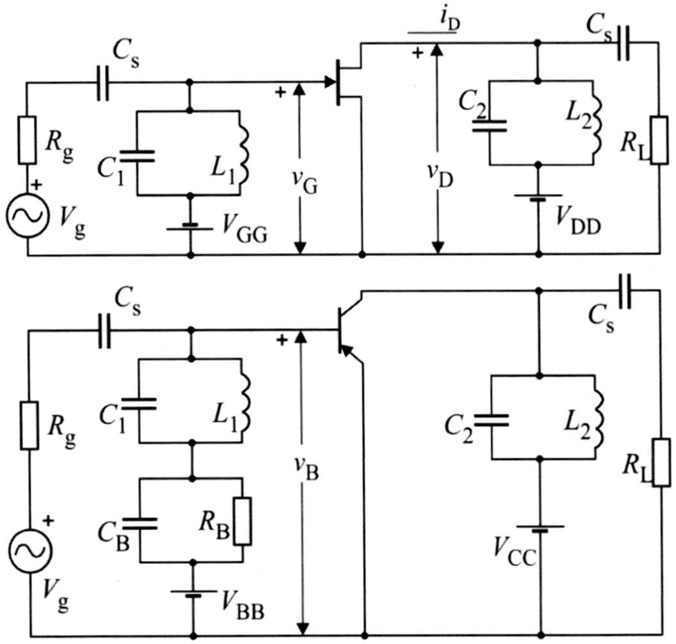

Fig. 4.3.32 Simplified schematics of Class C power amplifier with **a** JFET and **b** BJT

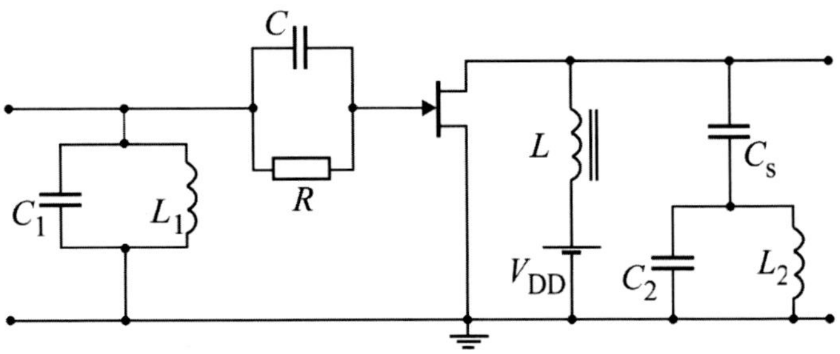

Fig. 4.3.33 Class C amplifier using RC biasing of the input and isolated power supply

rich in harmonics, and their amplitude is large, the amplifier can oscillate at frequencies where the impedance of the load impedance is inductive. This phenomenon is prevented by the so-called neutralization circuit. Through this circuit, a signal opposite in phase to that coming from Miller's capacitance is fed to the input, so that they are canceled.

In the following, some general conclusions will be drawn regarding the Class C amplifier. The analysis will be performed using the generalized schematic depicted in Fig. 4.3.34. In the case of a circuit with JFET, the exciting input voltage is sinusoidal while in the case of a circuit with a BJT, the exciting input current is sinusoidal.

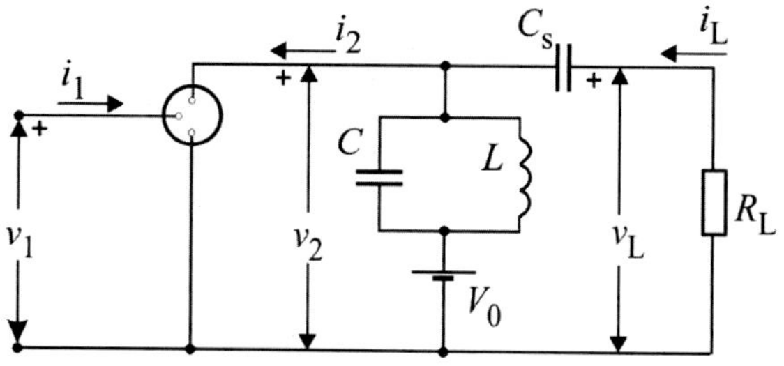

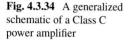

Fig. 4.3.34 A generalized schematic of a Class C power amplifier

Waveforms of currents and voltages in the circuit with Fig. 4.3.34 are shown in Fig. 4.3.35. The input voltage and current refer to the case when JFET was used. In the part of the positive half-cycle of the input signal, when the voltage is larger than the pinch-off V_{1P}, between the phase angles $-\alpha_1$ and $+\alpha_2$, the output current flows. If the excitation signal is large enough, an input current flows between the phase angles $-\beta_1$ and $+\beta_2$. It should be noted that the source (emitter) current is divided into gate (base) current and drain (collector) current so that at high gate (base) currents there may be a decrease in drain (collector) current, which is shown by a dashed line in the waveform i_2.

The output voltage is simple periodic despite the fact that the output current (i_2) is pulsed, thanks to the fact that the resonant circuit is tuned to the frequency of the fundamental harmonic. For the fundamental harmonic component of the current, the impedance of the resonant circuit is very large, and thus a voltage is formed whose frequency is equal to the fundamental one. For other harmonic components of the current, the impedance of the resonant circuit is small so that the voltage amplitudes of the higher harmonics are small. Note that the resonant circuit is narrowband, i.e., the width of its selectivity curve is significantly smaller than the resonant frequency (narrow band band-pass filter). Since the second harmonic is at a frequency twice as high as the resonant one, it is understandable that the impedance of the resonant circuit will be very small.

Therefore, it can be said that the Class C amplifier is a narrowband or selective amplifier. It should be mentioned that sometimes the resonant circuit at the output may be tuned to a frequency of another but higher harmonic. In such a case, we get a selective power amplifier that multiplies the frequency of the signal at the same time. This means that a signal of frequency f that is lower than the cut-off frequency of the transistor is fed to the input, and an amplified signal whose frequency is $2f$ (or more) is taken away, which now, since we already used the transistor, can be above its cut-off frequency. This trick can be used in radio transmitters where the antenna efficiency is higher if the signal transmitted is of higher frequency while the transistor's cut-off frequency is limited.

Let us now turn to some quantitative indicators. The instantaneous value of the dissipation power on the active element is

$$p_d = i_2 v_2. \tag{4.3.76}$$

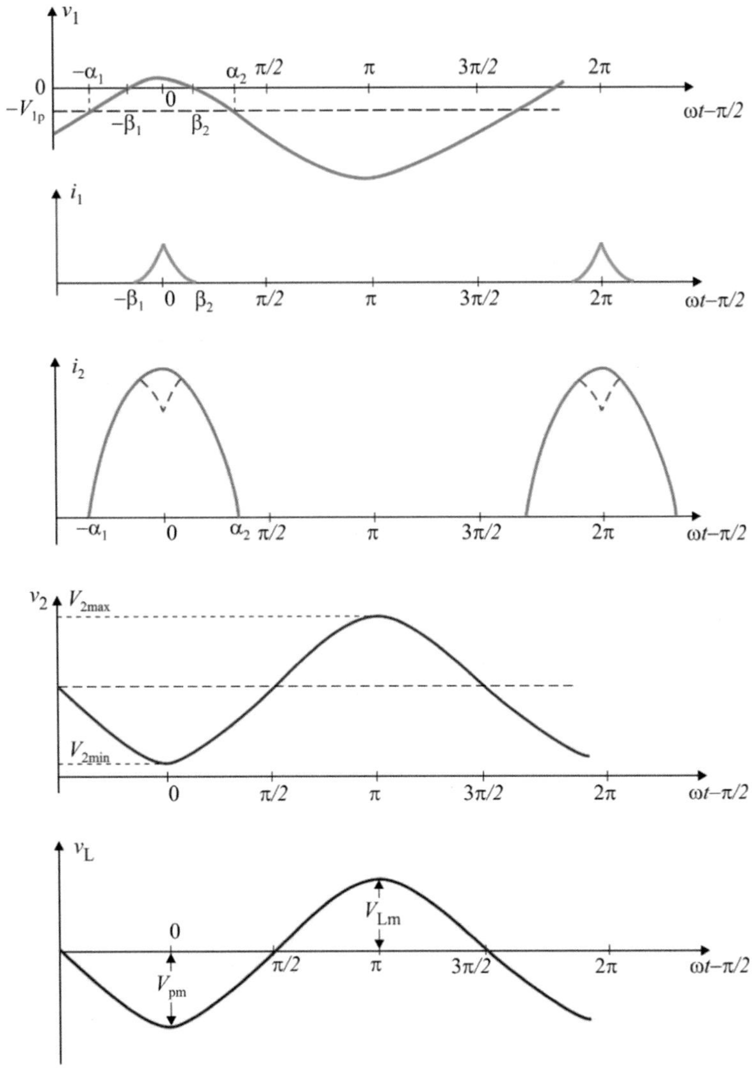

Fig. 4.3.35 Waveforms of the signals in a Class C power amplifier

This product has a low value not only because for most of the period $i_2 = 0$ but also because when i_2 takes its highest value, v_2 has its lowest value. The total dissipation is

$$P_d = \frac{1}{2\pi} \int_0^{2\pi} i_2 v_2 d(\omega t) = \frac{1}{2\pi} \int_{-\alpha_1}^{+\alpha_2} i_2 v_2 d(\omega t). \tag{4.3.77}$$

The power delivered by the power supply is

$$P = \frac{1}{2\pi}\int_{0}^{2\pi} i_2 V_0 \mathrm{d}(\omega t) = \frac{1}{2\pi} V_0 \int_{-\alpha_1}^{+\alpha_2} i_2 \mathrm{d}(\omega t). \tag{4.3.78}$$

while the useful power is

$$P_\mathrm{L} = \frac{1}{2} J_\mathrm{pm} V_\mathrm{pm} = \frac{1}{2}\frac{V_\mathrm{pm}^2}{R_\mathrm{L}}. \tag{4.3.79}$$

For the efficiency, according to the definition we have

$$\eta = \frac{P_\mathrm{L}}{P} = 1 - \frac{P_\mathrm{d}}{P}. \tag{4.3.80}$$

The exact values of these quantities can be determined if the values of voltages and currents are known at all times. This means that a non-linear circuit needs to be analyzed, which is usually done graphically or computer aided. We will not perform this analysis here, but we will estimate the value of the efficiency on the basis of approximate calculations. For this purpose, we assume that the angle of flow of the output current

$$\alpha = \alpha_1 + \alpha_2 \tag{4.3.81}$$

is small, so that the output voltage v_2 is practically constant and equal to $V_{2\mathrm{min}}$. Then, for the dissipation power we have

$$P_\mathrm{d} \approx \frac{1}{2\pi}\int_{-\alpha_1}^{+\alpha_2} i_2 V_{2\mathrm{min}} \mathrm{d}(\omega t) = \frac{V_{2\mathrm{min}}}{2\pi}\int_{-\alpha_1}^{+\alpha_2} i_2 \mathrm{d}(\omega t) \tag{4.3.82}$$

By combining (4.3.81), (4.3.79) and (4.3.77) one gets

$$\eta = 1 - V_{2\mathrm{min}}/\mathrm{V}_0. \tag{4.3.83}$$

Since it can be made that $V_{2\mathrm{min}} \approx 0$, for the case when the flow angle is small, the maximum value of the efficiency reaches 100%.

In practical circuits, $V_{2\mathrm{min}} > 0$ and is not negligible, so the maximum efficiency is less than 100%. Namely, the flow angle must not tend to a zero value, considering that, in such a case, the current flows in a short interval. Then, in order to provide sufficient useful power to the load, it is necessary for $J_{2\mathrm{m}}$ to tend to a very large value. $J_{2\mathrm{m}}$, however, is limited by the maximum allowable collector current (current ratings).

Based on all the above, we can conclude that the Class C power amplifier can be used very successfully for selective power amplification at RF (radio frequencies). It is most often used in the output stages of analog radio transmitters.

Let us also mention that the Class C power amplifier can also be realized as a push–pull circuit.

4.3.7 Class D Amplifiers

When it comes to audio frequency power amplifiers, the improvements obtained in Class C amplification are of no use since the frequency range is not compatible. We are due to use Class AB amplification in order to satisfy both the efficiency and linearity requirements. However, for high-power loads, the dissipated power becomes prohibitive since it asks for expensive cooling of the power transistors be it by large housing, large radiators, fans, or even water cooling. Here we think on both the price of the cooling gears and, of course, the power spent for cooling.

The efficiency is one of the most important aspect of power amplification. That, however, is signal dependent. This means that when the signal amplitude is in the middle of the normal range (or lower) the idle dissipated power of Class A or Class AB amplifiers becomes more dominant as compared to the output signal power.

What we need is as small the dissipation as possible. The idea for the solution comes from the digital domain, in fact CMOS, where in steady state of the pulse there is no power consumption. The only consumption happens during state transistion which lasts much shorter than the steady state. Hense, smaller dissipation is produced.

To exploit this idea in analogue signal amplification one needs first to create a pulse train in which the input signal is somehow impregated. Then, by use of this signal as a control between the power supply and the load, to allow for high power to be deliverd with low dissipation. What is needed as a final touch is to restore (reconstruct) the analogue waveform at the load side. This kind of power amplification is referred to as Class D. Since the active elements behave as switches we refer to this type of power amplifiers as the switched mode ones.

A system that may perform all these activities (a Class D power amplifier) is depicted in Fig. 4.3.36. It consists of the following parts. First, the pulse train is created using a pulse width modulator (PWM) circuit. In this case (there are altentives) the amplitude of the input signal is modulating the width of the pulses at the output so that for higher values of the input voltage wider pulses are obtained and vice versa.

The PWM principle of operation is depicted in Fig. 4.3.37. Here the modulating input signal (of frequency f_1) is compared with a given triangular (carrier) waveform of a fixed frequency (f_c) to produce a pulse train the width of the pulses being defined by the interception points of the two input signals. The pulse train so obtained is later used to control the switches.

One is to note that if the amplitude of the signal approaches the amplitude of the carrier, i.e., if

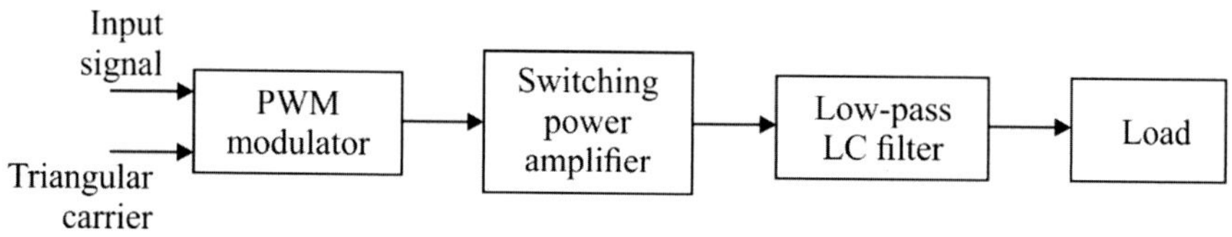

Fig. 4.3.36 Structure of a Class D power amplifier

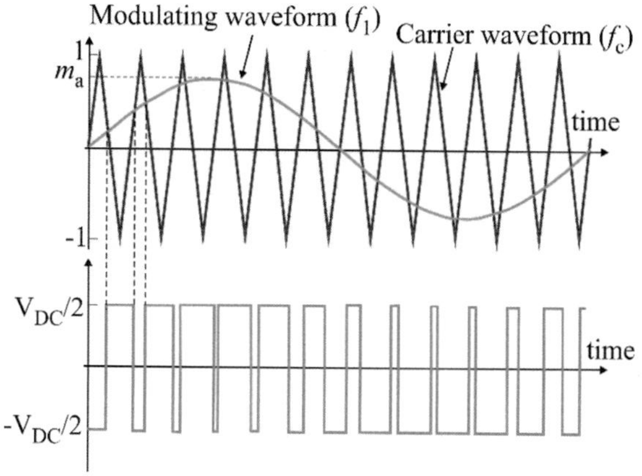

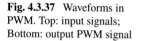

Fig. 4.3.37 Waveforms in PWM. Top: input signals; Bottom: output PWM signal

$$m_a = \text{Amplitude of the signal/Amplitude of the carrier} \tag{4.3.84}$$

approaches unity, the width of the pulses becomes small which means that such narrow pulses will be swallowed by the circuit parasitics, and no switching will be possible. This is of importance since, in theory, Class D, if $m_a = 1$, would have maximum efficiency of 100%. It is realistic to expect $\eta_{max} \leq 90\%$ which is much above Class A and Class AB.

Having the PWM pulse train we may control the power switches to deliver power to the load. The circuitry of that part will be given later.

In attempts to restore the input signal we face with the complex spectrum of the PWM pulse train. To this end we use a low-pass filter tasking it with the elimination of all harmonics. To get smooth waveforms at the output the frequency f_c is supposed to be much higher than f_1. The ratio

$$m_f = f_c/f_1 \tag{4.3.85}$$

together with the ratio of the amplitudes m_a are defining the location, the harmonic content, and the harmonic power in the output spectrum.

Based on the literature we produce Fig. 4.3.38 which depicts the structure of the spectrum of the output signal for $m_f = 15$ and $m_a = 0.8$. As can be seen the harmonic components are grouped around frequencies which are equal to $k \cdot m_f \cdot f_1, k = 1,2,\ldots$

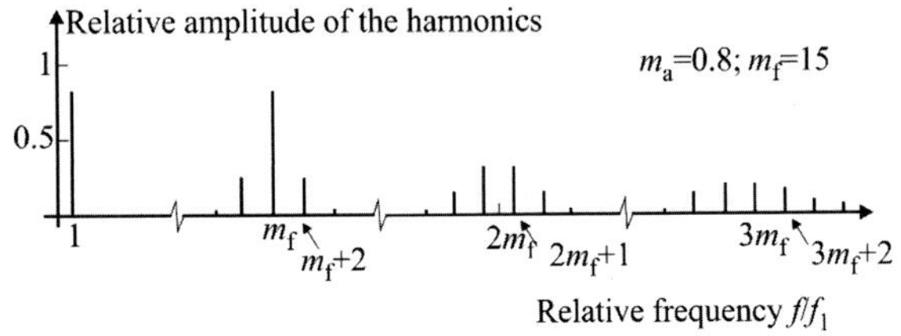

Fig. 4.3.38 Spectrum of the PWM's output voltage

We see from this diagram that the neerest harmonic to the input signal frequency is arround $m_f \cdot f_1$. If we do intend to amplify audio signals up to $f_{1\max} = 20$ kHz, since, to allow for proper attenuation, the cut-off frequency of the filter must be considerably lower than $m_f \cdot f_{1\max}$, we have to choose m_f of order of say 16 (four octaves) and above (i.e., $f_c > 360$ kHz). The larger the better. That, however imposes a need for fast switching high-power transistors (which was the topic of the previous chapter).

Figure 4.3.39 depicts the simplest version of the amplifier/filter/load connection. It is known as the half-bridge structure. Due to the structure which resembles the Buck DC-to-DC converter one frequently use the analogy for analysis of the waveforms in this circuit. On the other side, the active part resembles a CMOS inverter with the important difference that the N-channel transistor is now on the top of the P-channel one. That is becose fast charging of the coil L_1 is needed when switchin on to compensate for the remaining energy in the coil and to neutralize the space charge in the P-channel transistor's output capacitance. The N-channel transistor is of course faster per se. Note that there are improvements of the switching part of the circuit so that both transistors are of N-type conductivity. That, of course, will need a separate biasing circuit for the (new) lower N-channel transistor.

Diodes are included as limmiters.

Second order filter is used here which in most cases is not satisfactory since, in order to reduce the ripple on the load, will ask for large f_c.

Figure 4.3.40 depicts the differential or H-bridge architecture of the switching part. It is, in facts, twice as complex as the previous one and is loaded with a floating load. Accordingly, better performance is expected.

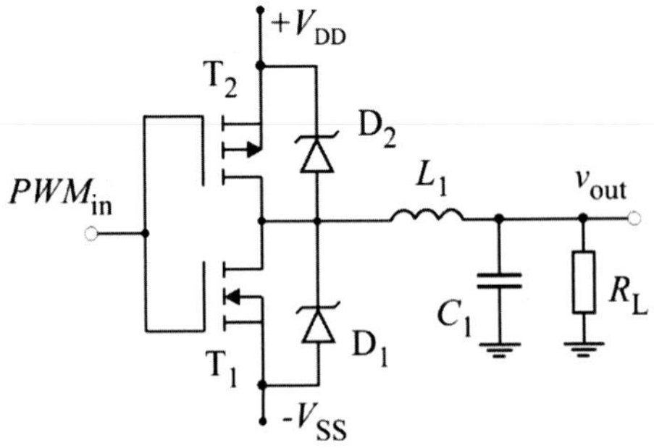

Fig. 4.3.39 Half-bridge switching output stage of Class D power amplifier

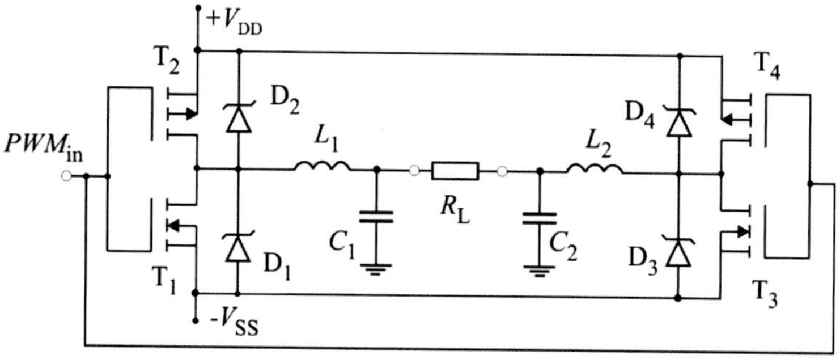

Fig. 4.3.40 Differential (H-bridge) switching output stage of Class D power amplifier

Thorough analysis of an example half-bridge amplifier is given in Appendix 4.C. The reader is advised to study the results.

Based on common experience and the Example 4.A.C.5 the following main conclusions may be drawn.

If proper filter is implemented a quality output signal may be obained. As can be seen from Example 4.A.C.5 using higher order filter will reduce the level of harmonics (in this case for about 6 dB). Here, an issue is the high-frequency electromagnetic radiation at the carrier frequency which may cause intermodulations within the system where the power amplifier is built-in.

The main advantage, i.e., work in rectangular pulse regime, brings some very important problems. Namely, during the transistion, if not protected, both transistors may be brought in conduction which creates a short circuit between the two batteries. Enormous currents can result. Consequently, current-sensing output transistor protection circuitry is needed. Alternatively, different modulation schemes are applied in place of PWM to avoid simultaneous change of the state of the transistors. Among these we may find the so-called *pulse-density modulation* (PDM) and *self-oscillating amplifiers* (which always includes a feedback loop, with properties of the loop determining the switching frequency of the modulator, instead of an externally provided clock), and *three-state modulation* schemes (implemented only to H-bridge structure). These are out of the scope of this book.

4.3.8 Class E and Class F Amplifiers

We continue now with switched mode power amplifiers but this time with the one intended to be applied in RF systems. The reader should imply that the useful signal is again impregnated into the pulse train controlling the switch with the main difference in the way how the modulation is performed (which will be not discussed here). The most promising architecture here is the Class E due to several reasons such as high efficiencies, simplicity of the load network, and satisfactory performance even with non-optimal excitation.

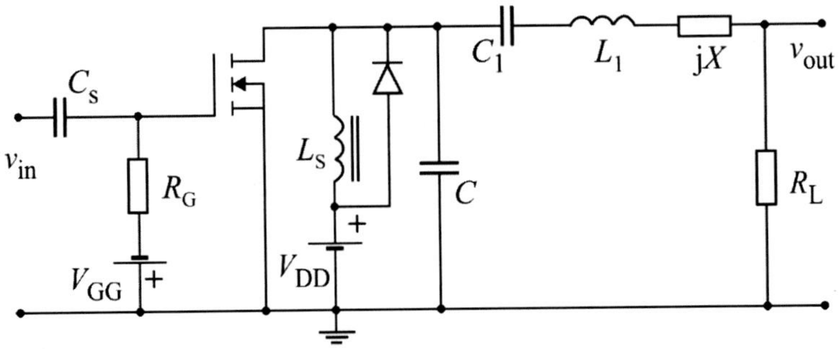

Fig. 4.3.41 Schematic of a Class E amplifier

The basic schematic of the Class E power amplifier is depicted in Fig. 4.3.41. It is built of a transistor, a shunt capacitor C, an RF choke (L_S), and a series L_1-C_1 resonant circuit. The load resistor is denoted R_L while the functionality is improved by a series-connected reactance, denoted jX, the nature and value of which will be found later. The MOSFET used can be (as is in the example given in the Appendix 4.A.C) substituted by any other type of transistor exhibiting good switching properties at RF.

For the analysis here we will use the simplified schematic (as given in Fig. 4.3.42) in which the diode is absent, and the transistor is represented by a switch. In addition, several simplifications will be implemented in order to make the closed-form analysis possible. These are.

- The transistor behaves as an ideal switch with no delay.
- 50% duty cycle of the pulse train of the carrier frequency is presumed.
- The inductance L_S is very large.
- The Q-factor of the series resonator constituted by L_1 and C_1 is high enough, so it can be considered that a purely sinusoidal current is running through the load R_L.

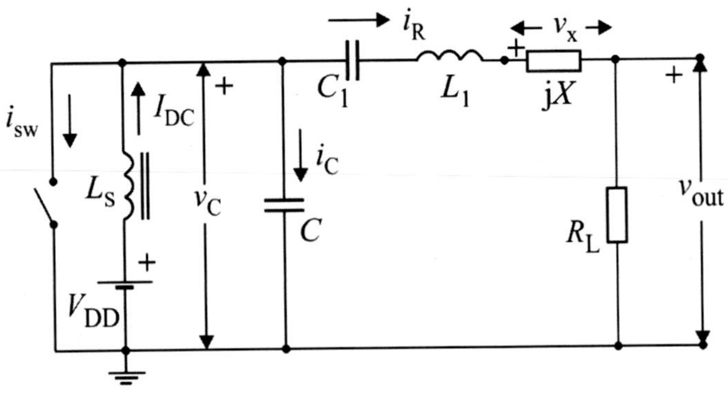

Fig. 4.3.42 Simplified schematic of the Class E amplifier

- The series resonant circuit is tuned so that $\omega_c^2 = 1/(L_1C_1)$ where ω_c is the (carrier) frequency operating the switch.

By analysis of this circuit we will create a set of expressions allowing calculation of the values of the circuit elements of the circuit.

To simplify we will use the angular time (phase) as a variable: $\theta = \omega_c t$. If so, the time period may be divided between the states as

$$0 < \theta < \pi \text{ for the Off state, and} \tag{4.3.86a}$$

$$\pi < \theta < 2\pi \text{ for the On state.} \tag{4.3.86b}$$

Additional approximation will be used based on the presumption of large L_S. Namely, we will suppose that there is no time varying component of its current. Its current will be denoted I_{DC}.

Now, since the series resonant circuit has very large Q-factor we will assume simple periodic current of fundamental frequency to flow through it. Having that in mind we can write

$$i_R(\theta) = J_m \cdot \sin(\theta + \varphi_0) \tag{4.3.87}$$

where J_m denotes the amplitude of the load current and φ_0 is the initial phase.

When the switch is open there is no current through it so that

$$i \sin(\theta) = 0 \tag{4.3.88}$$

At the same time, the current of the shunt capacitor will be

$$i_C(\theta) = I_{DC} - J_m \sin(\theta + \varphi 0) \tag{4.3.89}$$

Having in mind that in the previous half of the period the shunt capacitor was fully discharged (through the switch) for the capacitor voltage during the off-state one may write

$$v_C(\theta) = \frac{1}{\omega_c}\int_0^\theta i_c(\theta)\mathrm{d}\theta \tag{4.3.90}$$

After performing the integration for the capacitor voltage in off-state one gets

$$v_C(\theta) = \frac{1}{\omega_c C}[I_{DC} \cdot \theta + J_m \cos(\theta + \varphi_0) - J_m \cos(\varphi_0)] + V_0 \tag{4.3.91}$$

A very specific property of the Class E operation, making it different from the other switching-mode PA configurations, is the so-called "soft switching." In order to achieve soft switching, it is necessary to satisfy the Class E conditions which are

$$v_c(\theta = \pi) = 0 \tag{4.3.92}$$

$$\mathrm{d}v_c(\theta)/\mathrm{d}\theta|_{\theta=\pi} = 0 \tag{4.3.93}$$

The first equation declares that the capacitor voltage at the switching instant must be equal to zero, while the second is related to its derivative. It insists for the voltage curve to pass through an extremum which, having in mind (4.3.93), is a minimum. Since the capacitor current is related to the derivative of the capacitor voltage, the second condition may be also interpreted as a request for the switching not to produce current spikes in the switch.

From (4.3.91) and (4.3.92), we have

$$I_{DC} = \frac{2}{\pi} J_m \cos(\varphi_0) \tag{4.3.94}$$

and from (4.3.91) and (4.3.93):

$$I_{DC} = -J_m \sin(\varphi_0) \tag{4.3.95}$$

By combining the previous two equations, we find

$$\varphi_0 = \mathrm{arctg}(-2/\pi) \tag{4.3.96}$$

Noticing that I_{DC} and J_m are both positive values, the solution for the angle φ_0 must be chosen in the right quadrant, i.e., such as to satisfy (4.3.94) and (4.3.95). Therefore, we obtain the fourth quadrant solution, $\varphi_0 = -0.567$ rad. Thus,

$$\sin(\varphi_0) = -\frac{2}{\sqrt{\pi^2 + 4}} \tag{4.3.97}$$

and

$$\cos(\varphi_0) = \frac{\pi}{\sqrt{\pi^2 + 4}} \tag{4.3.98}$$

For the voltage v_x we can write

$$v_x(\theta) = J_m\sqrt{R_L^2 + X^2} \cdot \sin(\theta + \varphi_0 + \psi) \tag{4.3.99}$$

where X represents the value of the excess reactance placed between the series resonator and the load resistance, and ψ is the angle of the impedance $R_L + jX$, given by

$$\tan(\psi) = X/R_L \tag{4.3.100}$$

By expanding the sine term in (4.3.99), we can transform this equation into

$$v_x(\theta) = V_{CI}\sin(\theta + \varphi_0) + V_{CQ}\cos(\theta + \varphi_0) \tag{4.3.101}$$

where V_{CI} and V_{CQ} are the in-phase and quadrature component of the voltage $v_x(\theta)$, given by

$$V_{CI} = J_m\sqrt{R_L^2 + X^2}\cdot\cos(\psi) \tag{4.3.102a}$$

and

$$V_{CQ} = J_m\sqrt{R_L^2 + X^2}\cdot\sin(\psi) \tag{4.3.102b}$$

Therefore, we can relate the angle ψ with V_{CI} and V_{CQ} as

$$\tan(\psi) = \frac{X}{R_L} = \frac{V_{CQ}}{V_{CI}} \tag{4.3.102c}$$

We proceed with the analysis to find these two voltage components.

The series resonator L_1 and C_1 represent a short circuit for the fundamental component of the current flowing through this branch. Therefore, the two components V_{CI} and V_{CQ} can be determined by finding the Fourier components of the voltage $v_C(\theta)$ with respect to the phase of the current $i_R(\theta)$, i.e.,

$$V_{CI} = \frac{1}{\pi}\int_0^{2\pi} v_C(\theta)\cdot\sin(\theta + \varphi_0)d\theta \tag{4.3.103a}$$

and

$$V_{CQ} = \frac{1}{\pi}\int_0^{2\pi} v_C(\theta)\cdot\cos(\theta + \varphi_0)d\theta \tag{4.3.103b}$$

Since $v_C(\theta) = 0$ during the on state ($\pi < \theta < 2\pi$), the upper boundary in the integrals above can be taken to be π. By substituting (4.3.91) in (4.3.103a) and (4.3.103b), and performing the integration, we will obtain

$$V_{CI} = \frac{1}{\pi\omega C}\left\{I_{DC}[\pi\cos(\varphi_0) - 2\sin(\varphi_0)] - 2J_m\cos^2(\varphi_0)\right\} \tag{4.3.104a}$$

And

$$V_{\omega_Q} = \frac{1}{\pi\omega C}\{I_{DC}[-\pi\sin(\phi_0) - 2\cos(\phi_0)]$$

$$-2J_m\cos(\phi_0)\sin(\phi_0)-\frac{\pi}{2}J_m\bigg\} \tag{4.3.104b}$$

By substituting (4.3.94), (4.3.97), and (4.3.98) into (4.3.104a) and (4.3.104b), we obtain

$$V_{\mathrm{CI}}=\frac{J_{\mathrm{m}}}{\omega C}\cdot\frac{8}{\pi\left(\pi^2+4\right)} \tag{4.3.105a}$$

and

$$V_{\mathrm{CQ}}=\frac{J_{\mathrm{m}}}{\omega C}\cdot\frac{\pi\left(\pi^2-4\right)}{2\left(\pi^2+4\right)} \tag{4.3.105b}$$

From (4.3.102c), (4.3.105a), and (4.3.105b), it is now easy to find

$$\tan(\psi)=\frac{\pi\left(\pi^2-4\right)}{16}\approx 1.152 \tag{4.3.106}$$

Therefore, the value of the excessive reactance equals

$$X=1.152\cdot RL \tag{4.3.107}$$

Furthermore, we can derive the relationship between R_{L} and C by combining (4.3.100), (4.3.102a) and (4.3.105a), in order to obtain:

$$R_{\mathrm{L}}=\frac{1}{\omega C}\cdot\frac{8}{\pi\left(\pi^2+4\right)} \tag{4.3.108a}$$

Therefore, the nominal value of the shunt capacitance equals

$$C=\frac{1}{\omega R_{\mathrm{L}}}\cdot\frac{8}{\pi\left(\pi^2+4\right)}\approx\frac{0.1836}{\omega R_L} \tag{4.3.108b}$$

The only derivation left is to relate these parameters to the supply voltage V_{DC}, for the desired level of output power. To do so, we will evaluate the average voltage of $v_{\mathrm{C}}(\theta)$, noticing that it must be equal to the supply voltage due to the fact that the RF choke represents a short circuit for DC signal. Therefore, we can write

$$V_{\mathrm{DC}}=\frac{1}{2\pi}\int_0^{2\pi}v_{\mathrm{c}}(\theta)\mathrm{d}\theta \tag{4.3.109}$$

Again, the integration is performed only in the Off state, since in the On state $v_{\mathrm{C}}(\theta)=0$.

By combining (4.3.91), (4.3.95), (4.3.97) and evaluating the integral in (4.3.109), we obtain

$$V_{\mathrm{DC}} = R_{\mathrm{L}} I_{\mathrm{DC}} \frac{\pi^2 + 4}{8} \tag{4.3.110}$$

Thus, the DC resistance seen by the supply source is

$$R_{\mathrm{DC}} = \frac{V_{\mathrm{DC}}}{I_{\mathrm{DC}}} = R_{\mathrm{L}} \frac{\pi^2 + 4}{8} \approx 1.734 \cdot R_{\mathrm{L}} \tag{4.3.111}$$

The power delivered to the circuit by the DC power supply is then

$$P_{\mathrm{DC}} = V_{\mathrm{DC}} \cdot I_{\mathrm{DC}} = \frac{V_{\mathrm{DC}}^2}{R_{\mathrm{L}}} \cdot \frac{8}{\pi^2 + 4} \approx 0.5768 \cdot \frac{V_{\mathrm{DC}}^2}{R_{\mathrm{L}}} \tag{4.3.112}$$

and the output RF power delivered to the load R_{L} can be found as

$$P_{\mathrm{out}} = \frac{1}{2} R_{\mathrm{L}} J_m^2 = \frac{V_{\mathrm{DC}}^2}{R_{\mathrm{L}}} \cdot \frac{8}{\pi^2 + 4} \tag{4.3.113}$$

which is essentially an algebraic confirmation that the DC-to-RF power conversion efficiency ($\eta = P_{\mathrm{out}}/P_{\mathrm{DC}}$) is ideally equal to 100%. From Eq. (4.3.113), the required load resistance for the given P_{out} and V_{DC} can be calculated.

Now, we will very shortly address the Class F amplifier as depicted in Fig. 4.3.43. Three parallel resonant circuits are introduced. Their resonant frequencies are f_1, $3f_1$, and $5f_1$ which correspond to the notation of the circuit elements.

If the transistor is switched by a rectangular pulse train, as in the case of the Class E amplifier, the current waveform (excluding temporarily the delay) may be expressed in a form of a Fourier series as

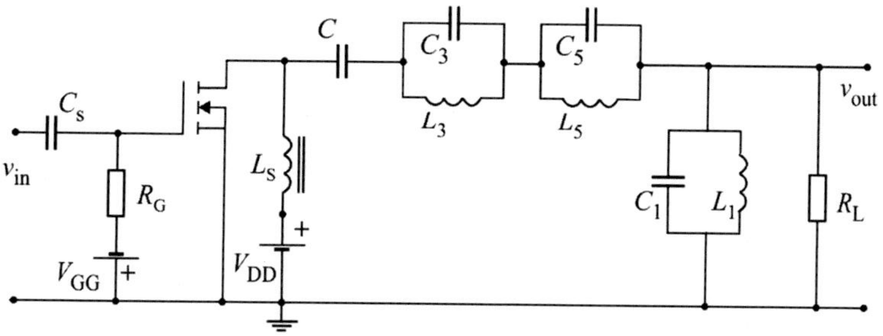

Fig. 4.3.43 Class F amplifier

$$i(t) = \frac{1}{2} + \sum_{k=1}^{\infty} \frac{\left|\sin\frac{k\pi}{2}\right|}{\frac{k\pi}{2}} \times \sin(k \cdot \omega_c t) \tag{4.3.114}$$

This means that, a part of the DC value and the main harmonic, it contains odd harmonics only. In addition, one may observe that in the Fourier series of the ideally rectangular pulse train, which has an infinite number of components, the dominant ones are the third and the fifth. Here from comes the idea implemented in the Class F amplifier. By adding two parallel resonant circuits whose resonant frequencies are $3f_1$ and $5f_1$, in series with the load one may expect better filtering of the current pulse than in Class C amplification. Of course, the parallel resonator adjusted at the carrier frequency (f_1) is retained.

The effect achieved may be observed in a different way. Namely, as can be seen from Fig. 4.3.35 the current (brown) and voltage (magenta) waveforms of a Class C amplifier have a specific property to have their maximums where the other one has its minimum. That is the key for high efficiency. The problem there was the need for the current pulse to become as short as possible with as large an amplitude as possible. That was limitting the practically achievable efficiency. Now, it may be proven that by introducing the two new parallel resonant circuit in the Class F amplifier this effect is facilitated and higher practical efficiency may be obtained.

4.3.9 Class G, Class H, and Class S Amplifiers

When listening the Overture to the Rossini's "The Barber of Seville" one would probably come spontaneously to the idea of the Class G amplification. Namely, there long periods of piano and pianissimo are suddenly interrupted by bursts of forte and even fortissimo. From the power amplifier point of view that means that, when designing, one should follow the most intensive part of the music and use as high a power supply voltage as possible. That would mean very low efficiency in most of the reproduction time.

The idea is to switch the power supply voltage to low or high value depending on the amplitude of the output signal. So, in Class G amplification two supply voltages are available and used at will. The schematic symbol of such an amplifier is the one depicted in Fig. 4.3.44a (the negative supply not shown), while Fig. 4.3.44b may be used as explanation of the functionality. The red line is playing the role of the amplitude of the output signal while the blue one the instantaneous value of the power supply voltage.

Thresholds are established (mostly using voltage reference diodes) which allow the comparator part of the circuit to switch from one to the other power supply terminal.

A simplified structure of a Class G amplifier is depicted in Fig. 4.3.45.

In this circuit DTp and DTn denote a PNP and NPN Darlington pairs, respectively. Diodes D_S (s comes from Schottkey) is a commutating diode. It supplies the rail

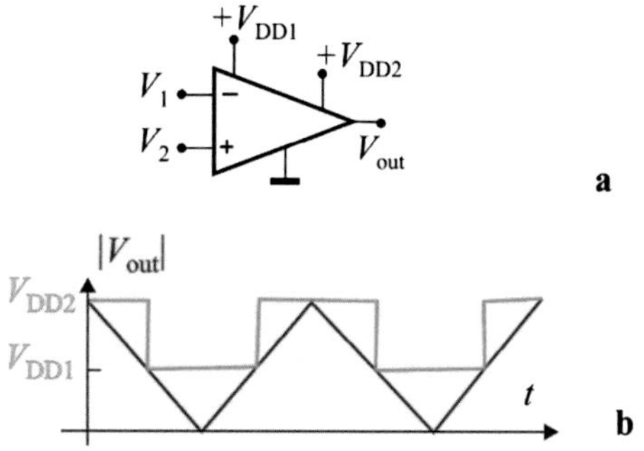

Fig. 4.3.44 The schematic symbol and the functionality of Class G amplifiers

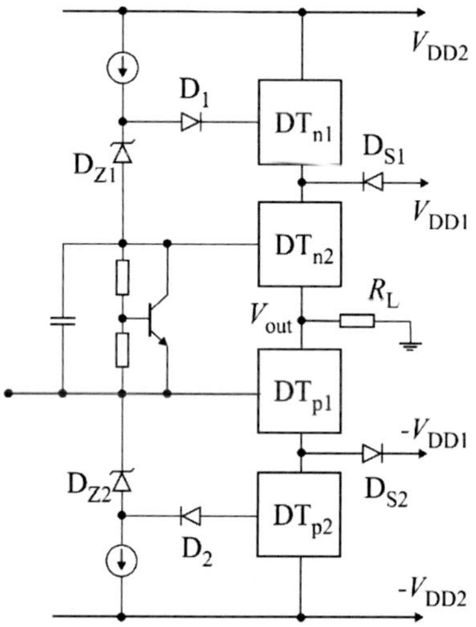

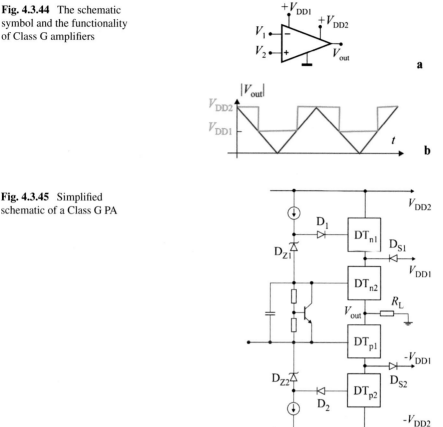

Fig. 4.3.45 Simplified schematic of a Class G PA

current from the low-voltage power supply when the signal swing is small which is most of the time. D_Z stands for voltage reference diode (z stand for Zener).

For low instantaneous values of the output signal DT_{n1} and DT_{p2} are cut-off while V_{DD1} and $-V_{DD1}$ are supplying the necessary power. When a positive-going instantaneous output signal exceeds V_{DD1}, D_1 starts conducting, DT_{n1} turns on, and D_{s1} turns off, so the entire output current is now taken from V_{DD2}. In that way, the voltage drop and hence power dissipation become shared between DT_{n1} and DT_{n2}.

We see now that, in fact, DT_{n1} and DT_{p2} are working in Class C (since they are activated during a conducting angle less than π) while DT_{n2} and DT_{p1} in Class B. Accordingly, the efficiency figures are inherited from the basic Class B amplifier.

Class H is again basically Class B, but with a method of dynamically boosting the single supply rail (as opposed to switching to another one) in order to increase efficiency. The usual mechanism is a form of bootstrapping. They are sometimes referred to as rail trackers. This is done by modulating the supply rails so that the rails are only a few volts larger than the output signal, so "tracking" it at any given time.

Class S uses a Class A stage with very limited current capability, backed up by a Class B stage connected so as to make the load appear as a higher resistance that is within the first amplifier's capability.

4.4 Discrete and Integrated Power Amplifier Circuits

Abstract This short chapter is intended to connect the low-amplitude voltage amplifiers, described in LNAE_Book 2, and the large signal amplifiers delivering large power to the load as described in Chap. 4.3.

4.4.1 Introduction

In this chapter, three complete power amplifier schematics will be presented. Two circuits will represent discrete realizations (one in MOS, and the other in bipolar technology), and the third will be integrated. Our goal here is not to design these amplifiers, but to recognize the role of individual elements and sub-assemblies and identify the design process that was described earlier. The circuits represented by no means are final designs and are used for demonstration only.

4.4.2 NMOS Power Amplifier

A power amplifier with an output stage built of NMOS power transistors will be considered first. The scheme of this amplifier is shown in Fig. 4.4.1.

When deciding on the choice of MOS technology for the design of power amplifiers, the following advantages should be kept in mind. First of all, VDMOS transistors do not exhibit a secondary breakdown, so there is no need for complex protections of exceeding the maximum power that are necessary for power amplifiers with BJT. Then, the temperature coefficient of the drain current and the transconductance of the transistor are negative, which allows parallel connection of several components, i.e., significantly facilitates the design of a temperature-stable amplifier. Of particular importance is the fact that the transfer characteristic of a MOS transistors is linear, which allows development of amplifiers with extremely small distortions. Even in its curved part, the characteristics of the MOST are favorable. Namely, the parabolic form of the transfer characteristic is generating only a second harmonic of the output current, which by its nature is easily removed in a push–pull coupling. Finally, the

V. B. Litovski, *Lecture Notes in Analog Electronics*, Lecture Notes in Electrical Engineering 958, https://doi.org/10.1007/978-981-19-6528-9_4

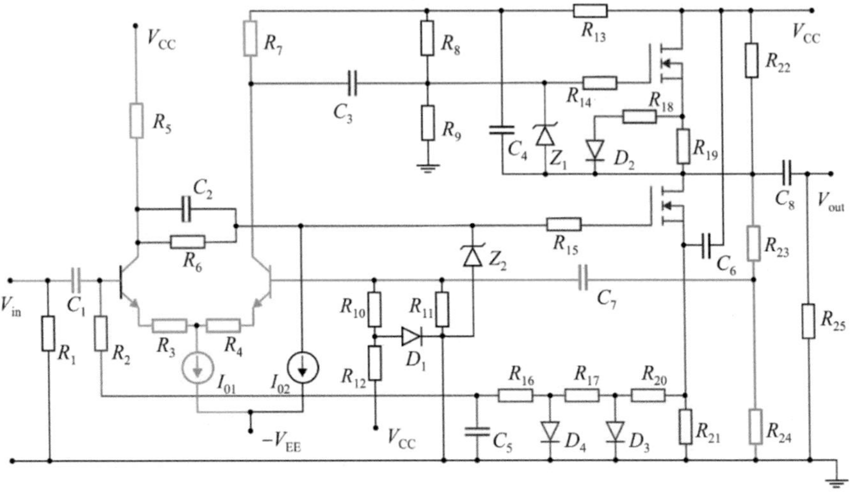

Fig. 4.4.1 Schematic of a discrete Class AB power amplifiers with NMOS power transistors in the output stage

capacitances of the MOS transistors are relatively small so that an amplifier with a high cut-off frequency of the open-loop gain can be developed. In addition, MOS transistors do not exhibit diffusion capacitance, which reduces the risk of distortion in transient mode.

The most important problem in the development of power amplifiers with MOS transistors is the provision of appropriate biasing. The DC voltage component of the transistor's gate, which is of the order of a few volts and depends on the wanted position of the quiescent operating point, should be kept constant. Unfortunately, the threshold voltage of the transistor has a temperature coefficient of about 6 mV/K, and in addition, its value varies from sample to sample. Therefore, negative feedback is needed to maintain the position of the quiescent operating point of the output transistors (and the value of the idle current) constant. In the circuit of Fig. 4.4.1, this is achieved by a circuit that generates a DC voltage proportional to the amplitude of the current pulse of the lower MOS transistor. This part of the circuit is colored in magenta. The circuit also consists of ladder built of R_{21}, R_{20}, D_3, R_{17}, D_4, $R_{16,}$ and C_5. The obtained DC voltage is fed to the amplifier input via the resistor R_2. Thus, the principle of automatic gain control (AGC) was realized. If, due to the instability of the circuit elements, the amplitude of the signal at the output increases, the gain decreases, so the amplitude returns and vice versa.

As for the amplifier itself, a differential amplifier with bipolar transistors was used in the input circuit in order to obtain as much a gain as possible (it is colored in blue). This differential amplifier is basically used as a phase splitter. Namely, the left transistor in the differential amplifier excites the lower output MOS transistor, and the right transistor in the differential amplifier excites the upper MOST. The excitation paths between the differential amplifier and the output stage are colored in brown. The output stage is colored in red.

The biasing of the input terminals of the differential amplifier is achieved by comparing the voltage drops across the resistor R_{21} and the diode D_1. In order to achieve accuracy in voltage values, the resistances in the bases of the transistors should be equal (i.e., R_{10} is compared with $R_{11} = R_2 + R_{16} + R_{17}$). Before proceeding with the analysis, it should be noted that the diode D_2 together with R_{18} and R_{19} helps the symmetry of the upper and lower halves of the output circuit.

The DC output voltage or the voltage on the lower MOST drain should be equal to half the supply voltage, and this is controlled by the distributor R_{22}-R_{23}-R_{24}. At the same time, the divider consisting of R_{13}, R_8, and R_9 determines the voltage at the gate of the upper output transistor so that its V_{GS} value is precisely determined.

From the point of view of the AC mode of operation, first of all, of interest is R_6, which with C_2 and the input capacitance of the MOST to which it is connected compensates for the decrease in gain at high frequencies. C_3 performs DC isolation only from the output of the phase splitter to the input to the power amplifier. C_4 eliminates the negative feedback that would occur through R_{13}, and C_7 forms the negative feedback for the AC signal that determines the frequency properties, non-linear distortions, and the sensitivity of this amplifier. This branch is colored in green. The capacitor C_6 together with R_{14} and R_{15} reduces the gain at very high frequencies and thus protects against self-oscillation. Finally, C_8 is necessary to DC isolate the load, and R_{25} provides a way to charge and discharge this capacitor.

Short-circuit protection of the output circuit is achieved by means of Zener diodes Z_1 and Z_2. They do not allow the voltage V_{GS} of the NMOS transistors to rise above the breakdown voltage of the diode and thus limit the value of the current of the output transistor.

4.4.3 Bipolar Power Amplifiers

A discrete amplifier with bipolar power transistors is shown in Fig. 4.4.2. The parts of the circuit are mostly easily recognizable, so we will pay relatively less attention to this circuit. Symmetrical power supply is used, which improves the response at low frequencies (there is no need for an isolating capacitor between the amplifier and the load), and at the same time shortens the time of establishing DC operating conditions in the circuit i.e., the set-up time (due to no need to charge the same capacitor).

Since a complementary pair is used at the output, the input stage does not have to be a phase splitter, so it is a differential amplifier (transistors T_1 and T_2) with an asymmetric output. For the input stage, it is extremely important to achieve a small offset so as not to form a direct current of the speaker connected to the output. Transistor T_3 is a constant current source.

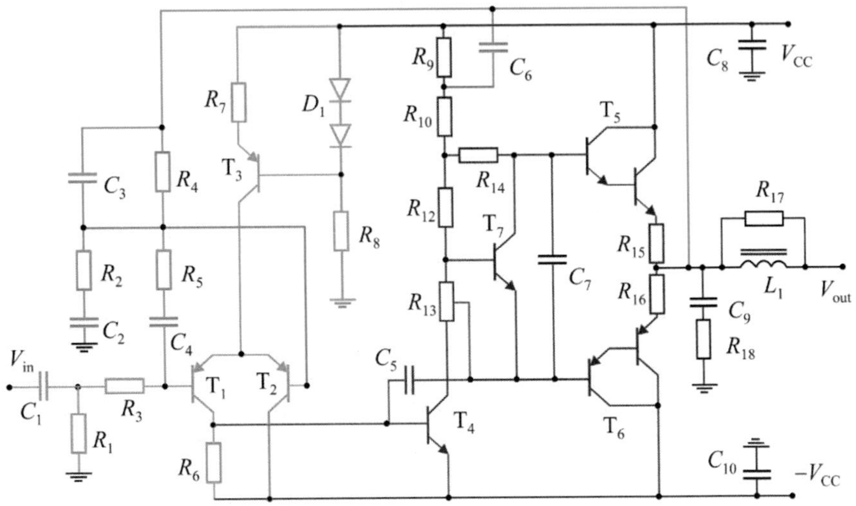

Fig. 4.4.2 Schematic of a discrete Class AB power amplifiers with BJT power Darlington pairs in the output stage

The second stage of amplification is realized by an CE amplifier (transistor T_4), and the resistor R_{13} controls the DC biasing of the output Darlington pairs of transistors. Instead of a pair of diodes that produce only $2V_{BE}$, a transistor that is so biased to provide $4V_{BE}$ is used to bias the push–pull stage with the Darlington pairs.

The characteristics of the Darlington pair have already been discussed, and here we will remind us that in the case of CC amplifiers, the current gain is practically equal to the power gain, and the Darlington pair has a very large current gain. This means that in this way an amplifier is realized whose previous stages may have a small power gain or small dissipation. R_{15} and R_{16} here serve for temperature stabilization.

The whole series of branches in the circuit is intended for stabilization of the amplifier and shaping of frequency characteristics, which we previously called compensation. Capacitors C_5 and C_7, then branch R_{18} and C_9, and finally frequency-dependent feedback circuit consisting of branch C_3-R_4 and the branch C_4-R_5 serve this purpose. The value of the capacitance C_2 is so large that it can be considered that for the AC component of the signal R_2 is connected between the base of T_2 and the ground, which is symmetric to the series connection R_3-R_1.

The capacitance of capacitor C_6 is also very high, so that negative feedback is achieved via R_{10}, which includes only the output stage and thus reduces non-linear distortions.

Finally, we mention that the branch L_1-R_{17} prevents self-oscillation due to coupling with a capacitive load such as an electrostatic speaker.

In Fig. 4.4.2 for simplicity, the circuit for protection against short circuit of the output is not shown.

A simplified diagram of an integrated power amplifier is shown in Fig. 4.4.3 This amplifier behaves as if it has a gain of 50 times or 34 dB, respectively, and its

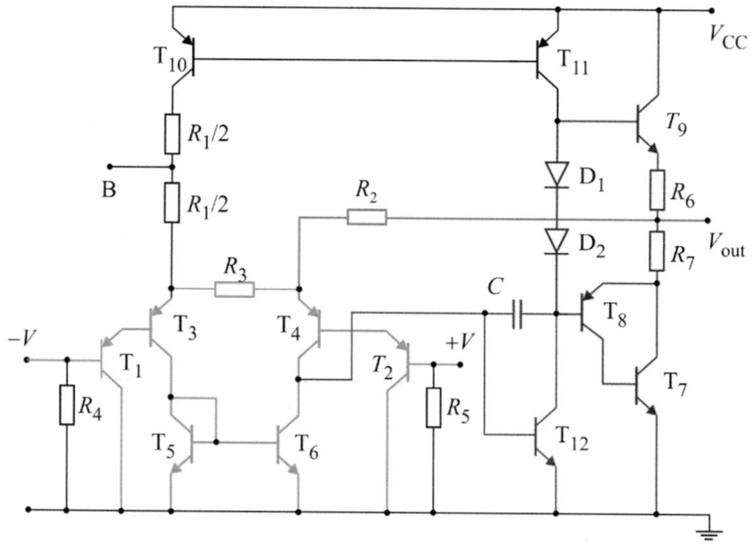

Fig. 4.4.3 Integrated power amplifier

DC output voltage is automatically adjusted to half the supply voltage. The specific construction of the input circuit allows each input connection to be directly coupled to the previous stage, DC insulated, or earthed. The output circuit is protected by both temperature and short circuit.

The input stage is composed of PNP transistors (T_1 and T_2) acting as CC amplifiers, which allows a large input impedance of the amplifier and at the same time creates the possibility of realizing a direct coupling amplifier. This is followed by a differential pair (T_3 and T_4) which is loaded by a current mirror (T_5 and T_6). The output of this stage is taken asymmetrically from the collector of T_4. Since high capacitances cannot be integrated, a terminal marked B is left, to which a high-capacitance capacitor should be connected to ground outside of the integrated circuit. Therefore, for the AC component of the signal, point B should be considered grounded.

The second stage of amplification is realized as a CE amplifier (T_{12}) which is loaded by a current source (T_{11}). Capacitor C also serves as compensation here. The output stage is a quasi-complementary pair consisting of transistors T_7, T_8, and T_9.

The numerical values of the resistances are chosen so that $R_1 = 2\ R_2$ and that $R_2 >> R_3$. Based on this, it can be shown that the output potential is equal to half the battery voltage. On the other hand, by choosing $R_2 = 25 \cdot R_3$, a fixed gain value of 50 is achieved.

The resistors R_6 and R_7 here degenerate the emitters and allow for better temperature stabilization. In addition, these two resistors may be used as sensor of rush current arising when the input becomes short-circuited to ground. The protective part of the circuit as given in Fig. 4.3.28 is not shown here.

4.5 Operational and Transconductance Amplifiers

Abstract In this chapter the structure of discrete and integrated operational (OA) and operational transconductance (OTA) amplifiers will be discussed. Attention will be paid to the stages that are making the OAs and OTAs. Then definition of a wide set of parameters will be introduced. Different technologies will be visited such as bipolar, JFET, CMOS, mixed JFET and bipolar, BiCMOS, and specialized low noise technologies in CMOS. Here, for the first time in LNAE the power supply rejection ratio calculation for a CMOS OTA will be demonstrated. As a peculiarity, a transimpedance amplifier will be shortly exemplified.

4.5.1 Introduction

The availability of an amplifier circuit in the form of a compact component of the size of the rest of passive circuit elements mounted on the printed circuit board brought a revolution in analog electronic implementation, especially in audio and measurement systems. Its properties are mostly intended to mimic an ideal voltage amplifier which means extremely large voltage gain, extremely large input impedance, and very small output impedance. Additionally, one expects for this circuit to handle large voltage amplitudes at the output which is the reason why these circuits are considered in this book. Being a real circuit, however, it suffers from imperfections such as low cut-off frequency, limited output voltage and current dynamics, existence of nonlinearities and noise, limited CMRR and some specifics such as finite Power Supply Rejection Ratio (PSRR), slew rate, and current and voltage offset which will be defined and elaborated in this chapter.

Due to its implementation in then important analog computers and generally in analog mathematic computations it was named operational amplifier (OA).

Various technologies are bringing specific advantages to OA. The bipolar circuits are known to have the largest gain while the all MOS (and CMOS) circuits are feasible for integration within modern mixed-mode chips where some analog functions (such as base-band filtering) are performed prior to the massive digital processing. Hybrid (mixed technologies) circuits were produced to meet additional requirements related to noise, input impedance, frequency band, and similar.

V. B. Litovski, *Lecture Notes in Analog Electronics*, Lecture Notes in Electrical Engineering 958, https://doi.org/10.1007/978-981-19-6528-9_5

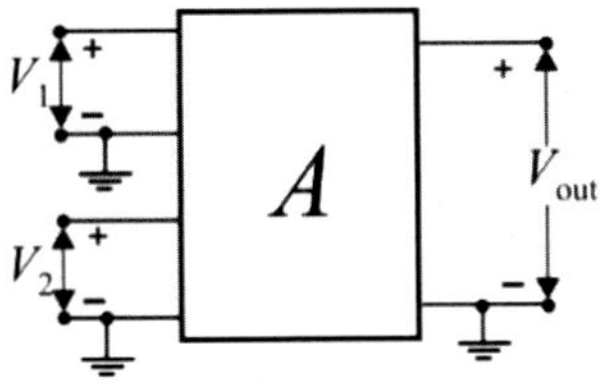

Fig. 4.5.1 Single-ended OA represented as a block

As opposed to the OA, which is a voltage amplifier, later on, the operational transconductance amplifier (OTA) was integrated with the aim to become a voltage-to-current amplifier. It is expected to have all the properties of the OA, but instead of low, a very high output resistance. Its implementation mostly goes in analog filtering where it may be used as a substitution to a resistor as well as an amplifier.

We will start here with the basic modeling and the symbols for the OA and OTA.

Both OA and OTA have differential inputs while the output may be differential or single ended. Figure 4.5.1 depicts a single-ended OA as a block. Here two input voltages are exciting the circuit while the output has one terminal grounded. It is usual for the OA to be denoted by the capital letter A. In some situations, this block is used within complex schematics.

Figure 4.5.2 depicts the simplest model of the single-ended OA. In this circuit only the "hot" input terminals are shown the ones connected to ground are considered implied. The model expresses the main function of the OA: It amplifies the difference of the input voltages with the gain being denoted as A. We will refer to this structure as an ideal OA with finite gain. The term ideal comes from the fact that this circuit has infinite input and zero-valued output resistance which means it behaves as an ideal voltage amplifier.

A more realistic model is depicted in Fig. 4.5.3 where the input (R_{in}) and output (R_{out}) resistances are included. Note the input resistance is measured between the hot input terminals so that it is frequently referred to as the differential input resistance. One can measure an input resistance from each hot terminal to ground, too.

Further improvement, or better to say, further approaching to realistic modeling includes frequency dependence of the three parameters characterizing the model. If time domain modeling is to be performed, delay created by the internal circuit of the OA should be included.

All that is in use for the case of absence of common mode gain of the circuit. If the influence of this property is to be modeled one may use the model of Fig. 4.5.4a. Here A_{d} denotes differential gain (the gain of the difference of the input signals) and

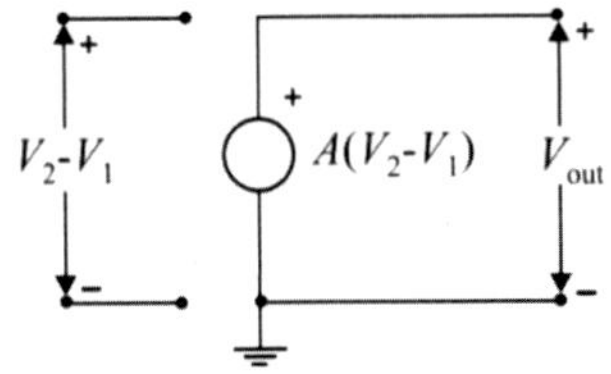

Fig. 4.5.2 Simplest model of a single-ended OA

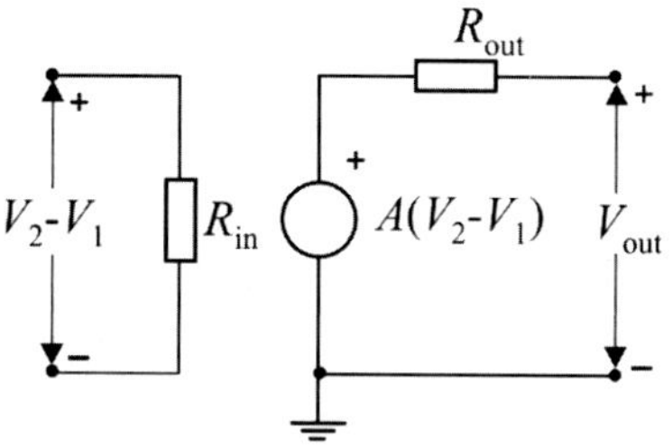

Fig. 4.5.3 More realistic model of a single-ended OA

A_{ds} the gain of the common mode input signal (arithmetical average of the input signals).

When included in complex schematic one uses simplified symbols of the OA. Some of them are depicted in Fig. 4.5.5. These are self-explanatory.

We use the opportunity here to transform the model of Fig. 4.5.4a into a circuit having a current source at the output as depicted in Fig. 4.5.4b. It is obtained by transforming the Thevenin source at the output into a Norton one. This model, however, is rarely used to model OAs. Instead, it is convenient to model OTAs. Namely the symbols denoted by "Y" in fact represent transconductances. Y_d relates the voltage difference at the input with the output current J_{out} while Y_{ds} relates the common mode voltage of the input with the output current.

Now, going backwards, on Fig. 4.5.6 the simplest model of the OTA is depicted with the common mode signal excluded. Note, Y_d is now replaced by what is normally used: g_m.

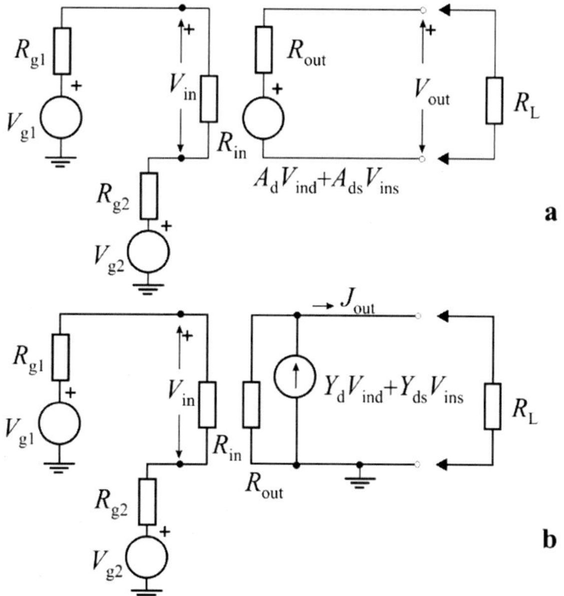

Fig. 4.5.4 Models taking account of the common mode gain. **a** Voltage output and **b** current output

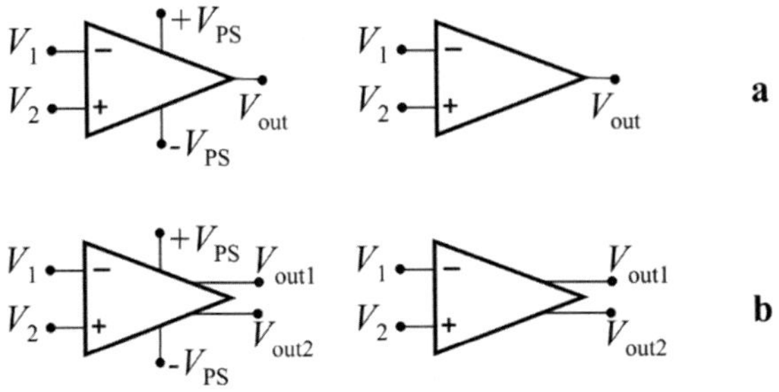

Fig. 4.5.5 Schematic symbols for OA. **a** single-ended and **b** differential output. On the left-hand side, the power supply (PS) terminals are shown

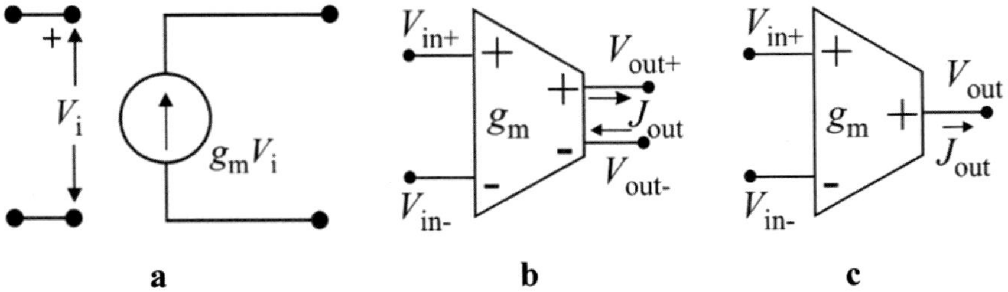

Fig. 4.5.6 **a** Simplest mode of OTA, **b** the symbol of differential output OTA, and **c** the symbol of a single ended output OTA

The simplest model of an OTA is depicted in Fig. 4.5.6a. Here an ideal voltage-to-current amplifier is considered meaning that it has infinite both the input and the output resistances. The symbols of differential output and single-ended OTAs are given in Fig. 4.5.6b and c, respectively.

In this chapter we will try to give a notion on the structure and the main characteristics of OAs and OTAs. The circuits discussed will be demonstrational only and (unless specially stated) by no means the schematics and the numerical values presented will reflect any existing circuit available on the market.

4.5.2 Bipolar Operational Amplifiers

In the practical realization of the OA, four separate blocks can most often be distinguished by their function in the integrated circuit, as shown in Fig. 4.5.7.

The input stage is a differential amplifier. It is realized as described in LNAE_Book 2 including the use of a constant current sources. Many variants of the differential amplifier are in use, starting with the basic circuit, via the differential amplifier with cascode stages, to the one with the Darlington pairs at the input.

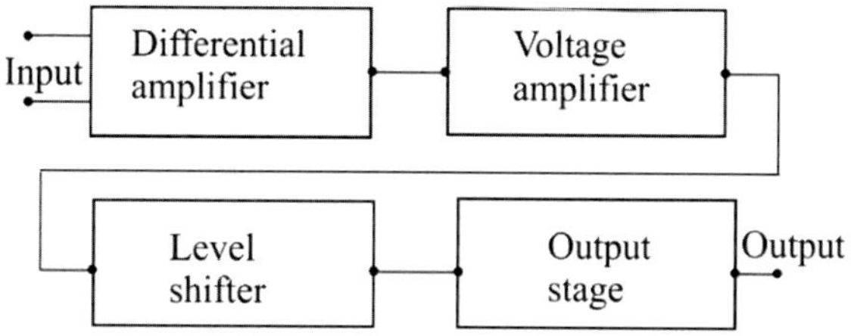

Fig. 4.5.7 Structure of an OA

The basic requirements for the differential amplifier located at the input are to provide as much gain as possible, as much input resistance as possible, as much CMRR as possible, as little noise as possible, and as little voltage and current offset as possible.

The voltage amplifier (second stage) provides for even larger gain of the operational amplifier. Usually, it is a CE amplifier with a current source as the active load as described in LNAE Book 2.

The level shifter (third stage) serves to ensure that the output voltage in the absence of excitation (the DC output voltage) is equal to zero. The DC voltage at the output of the voltage amplifier is reduced by V_{BE} by using a cascade connection with a CC amplifier.

Finally, the output amplifier is a complementary large signal amplifier described earlier in LNAE_Book 3. In addition to amplifying current (power), its main role is to provide the lowest possible output resistance.

The structure of the operational amplifier described in this way is called two-stage because it has two stages for voltage amplification: differential amplifier and voltage amplifier. There are also single-stage operational amplifiers that do not contain the voltage amplifier.

In some cases, three-stage amplification is undertaken so that the first two stages form a cascade of differential amplifiers.

In Fig. 4.5.8, the structure of a simple operational amplifier is shown as an example. Transistors are marked with T, and transistors connected as diodes are marked TD.

The input differential amplifier consists of transistors T_1 and T_2. Transistors TD_3 and T_4 form a current source as an active load to the differential amplifier. The Widlar's current source in the differential amplifier's emitter consists of transistors T_5 and TD_6.

The voltage amplifier is a CE amplifier (transistor T_8) with a current source as a collector load (transistors T_9 and TD_{10}). Since the output impedance of the differential amplifier is large (the differential amplifier acts as a transconductance stage), and the input impedance of the CE amplifier is relatively small, in order not to reduce the gain of the differential amplifier, these two stages are separated by a CC amplifier realized by the transistor T_7. Thus, the current output signal of the differential amplifier is converted into a voltage output of the CC amplifier.

This isolating amplifier in the emitter, in series with the resistor R_3, also has a constant current source (transistor T_{19}). R_3 and T_{19} act also as a voltage divider to

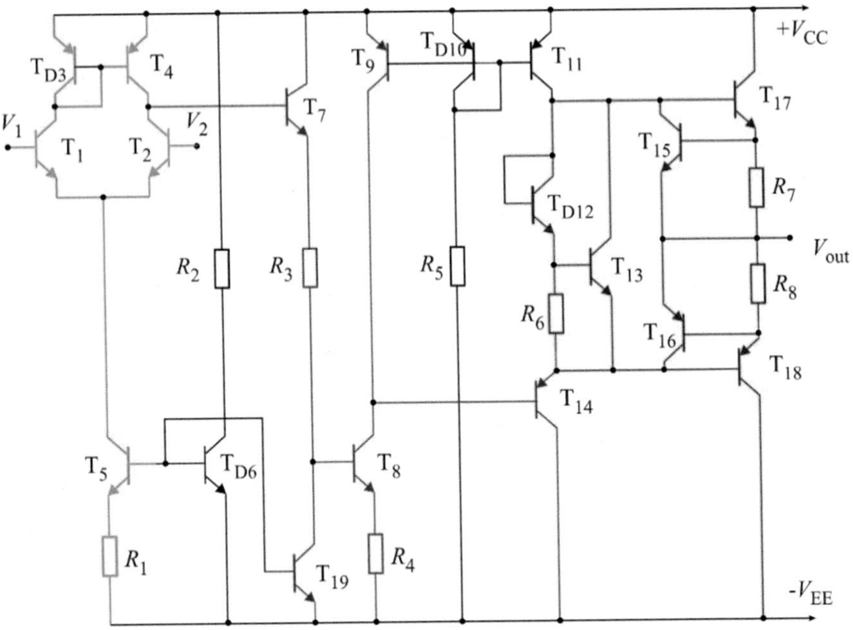

Fig. 4.5.8 Schematic of an OA

accommodate the DC voltage levels. Namely, by placing the resistor R_3, the DC voltage at the base of T_8 is reduced enough so that, in the end, in the absence of excitation, the output voltage V_{out} is equal to zero. In this way, T_7 works as a DC level shifter, too. The configuration of the voltage amplifier T_8 itself allows the DC voltage at the base of T_{14} to be low enough.

The output stage consists of transistors T_{17}, T_{18}, T_{15} and T_{16}, which is the basic circuit of a push–pull power amplifier using a complementary pair and short-circuit protection. The output complementary pair is excited by a CC amplifier (transistor T_{14}) with transistor T_{11} in the emitter acting as a current source. Transistors TD_{12} and T_{13} connected as a Darlington pair replace diodes, as previously explained in Chap. 4.3, to avoid discontinuities in the transfer characteristic. The voltage between the bases of transistors T_{17} and T_{18} is equal to the voltage drop between the bases and emitters of transistors TD_{12} and T_{13}.

Now, to exemplify, will perform a simplified circuit analysis of the circuit of Fig. 4.5.8.

4.5.2.1 DC Regimes in the Simple OA

In this section we will try to come to information about the DC values of the voltages and current of the above OA. In that we will make many presumptions since neither we have exact data for analysis nor exact analysis is possible without a computer.

First, we will use a simplified expression for the collector current as

$$I_C = \beta \cdot I_B \tag{4.5.1}$$

where the inverse saturation current of all transistors is neglected. We will also consider that the collector current does not depend on the collector voltage, i.e., that it is

$$I_C = I_s e^{V_{BE}/V_T}. \tag{4.5.2}$$

In the calculation it will be considered that when the input voltage of the OA is equal to zero, its output voltage is equal to zero, too. Of course, this cannot be true unless all voltages are stabilized by an external negative feedback circuit. Without feedback, even the slightest asymmetry at the input, for example the offset of the differential amplifier, would bring the output into saturation. This is also the reason the calculation will be done backwards from the output to the input. In the opposite case, if the calculation went in a logical order from the input to the output, even the smallest errors in the calculation of the first stages would inevitably lead to the output stage being saturated. The method of analysis will be presented for the operational amplifier from Fig. 4.5.8. Understandably, it is also applicable to other types of operational amplifiers.

At the beginning, all constant current sources should be replaced with ideal current sources. This is done in Fig. 4.5.9.

Some values must be assumed in the calculation in advance, such as the supply voltage, the value of the resistors, and the values of the current sources. The adopted values will be derived from the nature of the functionality of individual parts of the circuit. The operation of individual parts of the operational amplifier was discussed in detail in LNAE_Book 2 and in Chap. 4.3 of this book.

The following values are assumed: $V_{CC} = 15$ V, $V_{EE} = -15$ V, $R_1 = 6$ kΩ, $R_2 = 37$ kΩ, $R_3 = 40$ kΩ, $R_4 = 100$ Ω, $R_5 = 37$ kΩ, and $R_6 = 45$ kΩ. In Fig. 4.5.9 voltage and resistance values are indicated. The values of individual constant current sources are also assumed and indicated. The task of the calculation is to determine the currents and voltages of the transistor in order to determine the values of the rest of resistances. We will first determine the currents of the rest current sources.

The Widlar's current source I_5 consists of transistors T_5 and TD_6 with resistors R_1 and R_2. Based on LNAE Book 2, if the transistors are identical, we get:

$$R_E = \frac{V_T}{I_{C2}} \cdot \ln\left\{\frac{I_{C1}\left(1 + V_{CE2}/V_A\right)}{I_{C2}\left(1 + V_{CE1}/V_A\right)}\right\} \tag{4.5.3a}$$

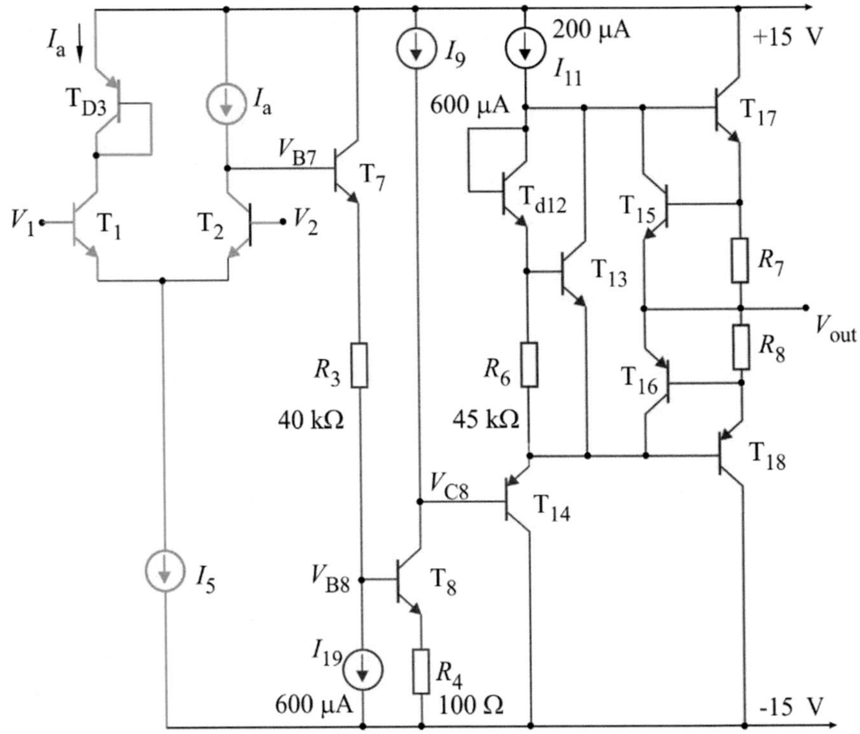

Fig. 4.5.9 Simplified schematic of the circuit of Fig. 4.5.8

where V_A stands for the Early voltage. By analogy, if the value of V_A is large enough, and the transistors are identical, we get:

$$R_E = \frac{V_T}{I_5} \cdot \ln(I_{CD6}/I_5) = R_1. \tag{4.5.3b}$$

The value of the current I_{CD6} will be determined from the relation

$$I_{CD6} = (V_{CC} - V_{EE} - V_{CED6}) \big/ R_2 = 0.789 \text{ mA} . \tag{4.5.4}$$

In the above relation, it is assumed that the voltage drop across the T_{D6} diode is equal to 0.8 V. Substituting the value of I_{CD6} in (4.5.3) gives for the room temperature ($T = 300$ K and $V_T = 26$ mV):

$$I_5 = 4.32 \times 10^{-6} \cdot \ln\left(0.789 \times 10^{-3}/I_5\right) \tag{4.5.5}$$

This transcendent equation is solved only numerically. The solution is

$$I_5 = 16.7 \text{ μA} . \tag{4.5.6}$$

This current in the differential amplifier is divided into two equal parts so that it is

$$I_a = 8.35\ \mu\text{A}. \tag{4.5.7}$$

Since resistors R_2 and R_5 are equal to each other, the currents of transistors T_{D6} and T_{D10} (Fig. 4.5.14) are equal, i.e.,

$$I_{CD10} = 0.789\ \text{mA}. \tag{4.5.8}$$

If the transistors T_9, TD_{10}, and T_{11} in the current source were equal to each other, then their collector currents would be identical. However, this current is too high. To reduce it, the transistors are not identical. Suppose that they are made so that $I_9 =$ 600 μA and $I_{11} = 200$ μA. We also assume that $I_{19} = 600$ μA. These values are entered in the schematic of Fig. 4.5.9.

The analysis of the output stage takes place in the following way. The resistances R_7 and R_8 are small, and their role is to bring the transistors T_{15} and T_{16} into the conduction if the currents through transistors T_{17} and T_{18} increase excessively. Let the conduction threshold of these transistors be $V_\gamma = 0.6$ V. If a requirement is set that the current through the output transistors T_{17} and T_{18} must not exceed 20 mA then

$$R_7 = \frac{V_\gamma}{I_{max}} = \frac{0.6}{20 \times 10^{-3}} = 30\ \Omega. \tag{4.5.9}$$

The resistor R_8 would have the same calculated value. However, a slightly lower value is taken for this resistance. This is to compensate for asymmetries in the complete output circuit. It is common in practice for the resistance R_8 to be 20% lower than the resistance R_7, so that $R_8 = 24\ \Omega$ is chosen.

In normal operation, when there is no overload, transistors T_{15} and T_{16} do not conduct and do not affect the operation of the circuit. We will also consider that the operation is not affected by the resistors R_7 and R_8 because in normal operation the voltage drops across them is negligible. If so the following will be valid

$$V_{BE12} + V_{BE13} = V_{BE17} + V_{BE18}. \tag{4.5.10}$$

Using (4.5.2) this expression reduces to

$$V_T \cdot \ln\frac{I_{C12}}{I_{s12}} + V_T \cdot \ln\frac{I_{C13}}{I_{s13}} = V_T \cdot \ln\frac{I_{C17}}{I_{s17}} + V_T \cdot \ln\frac{I_{C18}}{I_{s18}} \tag{4.5.11}$$

It should be noted that the currents I_{C18} and I_{s18} are negative because T_{18} is of the PNP type. Therefore, we will continue to take their absolute values.

Since transistors T_{17} and T_{18} are identical, it will be valid

$$I_{C17} = I_{C18}. \tag{4.5.12}$$

Substituting this value into (4.5.11) we get

$$I_{C17} = |I_{C18}| = \sqrt{I_{C12} I_{C13}} \sqrt{\frac{I_{s17} \cdot |I_{s18}|}{I_{s12} I_{s13}}}. \tag{4.5.13}$$

Transistors T_{17} and T_{18}, since they are power transistors, have higher saturation currents than transistors T_{12} and T_{13}. How big they are, depends on the transistors used. As an average value one can take three times:

$$I_{s17} = |I_{s18}| = 3I_{s12} = 3I_{s13}. \tag{4.5.14}$$

A typical voltage drop on a conducting transistor is $V_{BE12} = V_{BE13} = 0.65$ V so that the current through resistor R_6 is equal to 14.4 μA. If the transistor T_{13} has a large current gain, its base current is small and then the $I_{CD12} = 14.4$ μA. In the same way, neglecting the base currents of transistors T_{17} and T_{18} one gets

$$I_{C13} = I_{11} - I_{CD12} = 185.6\ \mu\text{A}\,. \tag{4.5.15}$$

Substituting these two currents in (4.5.13) and using (4.5.14), the following was calculated:

$$I_{C17} = |I_{C18}| = 155\ \mu\text{A}\,. \tag{4.5.16}$$

The obtained currents show that the assumption of ignoring the voltage drop across resistors R_7 and R_8 is justified. For example, the voltage drop across resistor R_7 is only 4.65 mV.

Let us now turn to the calculation of the operating conditions of the T_{14} isolating amplifier. The current through this transistor, if the base current of transistor T_{18} is neglected, is $I_{C14} = 200$ μA. With this value of current, and assuming that for the used transistors (except the output ones) it is typically $I_s = 0.3 \times 10^{-14}$ A, from Eq. (4.5.2) $V_{BE14} = 0.645$ V is calculated.

The voltage V_{BE18} between the base and the emitter of the transistor T_{18} through which a current of 155 μA flows is calculated in the same way. Considering that it is a power transistor, its inverse saturation current is three times higher and amounts to 0.9×10^{-14} A. Thus, $V_{BE18} = 0.61$ V is obtained.

These calculations were performed to determine the voltage on the collector of the voltage amplifier T_8. Namely, the value of this voltage determines the value of the DC output voltage. Since we assumed that $V_{out} = 0$ and neglecting the voltage drop across resistor R_8 we have

$$V_{C8} = -(V_{BE14} + V_{BE18}) = -1.255\ \text{V}. \tag{4.5.17}$$

The collector current of transistor T_8 is 600 μA. Ignoring the base current of transistor T_{14} and assuming that the typical value of the current gain coefficient of transistor T_8, $\beta_8 = 250$ from (4.5.1), we obtain that $I_{B8} = 2.4$ μA. Using (4.5.2), $V_{BE8} = 0.672$ V was obtained.

The voltage drop across resistor R_4 is

$$V_{R4} = I_9 R_4 = 0.06 \text{ V}. \tag{4.5.18}$$

The base voltage of T_8 is

$$V_{B8} = V_{R4} + V_{BE8} - V_{EE} = -14.27 \text{ V}. \tag{4.5.19}$$

Let us now turn to the calculation of the operating conditions of the level shifter realized by T_7. Its collector current is 600 μA and using (4.5.2) we get $V_{BE7} = 0.674$ V. The voltage at its base is

$$V_{B7} = V_{BE7} + I_{19} R_3 + V_{B8} = 10.404 \text{ V}. \tag{4.5.20}$$

This voltage is at the same time the voltage at the collector of transistor T_2. The currents through transistors T_1 and T_2 have already been calculated as $I_{C1} = I_{C2} = I_a = 8.35$ μA. For the high current gain of these transistors, their base currents are very small.

In this way, the DC operating conditions of the sub-stages built into the operational amplifier are completely calculated. It is understood that the results obtained are approximate and their purpose is not to design the OA but to make some feeling about the current and voltage levels within the circuit while having no exact layout data.

4.5.2.2 Frequency Response of the OA

We will determine the gain of the operational amplifier for low frequencies and low signals first. For that the influence of all capacitances can be neglected and the amplifier can be considered to be linear. Accurate analysis would require replacing each transistor with its own small signal model. Solving the equivalent circuit thus obtained would be a very complicated task. The obtained result even when using a computer would not be completely accurate because the exact values of the transistor parameters are not known. Therefore, we will give here an approximate analysis that serves to assess the possible gain, input, and output resistance of the OA.

The operational amplifier of Fig. 4.5.9 we are considering is two-stage because there are only two stages of voltage gain. The input differential amplifier and the voltage amplifier contribute to the voltage gain of the operational amplifier. The output stage has a voltage gain approximately equal to unity and will affect the operation of the rest of the circuit only through its input impedance. The same applies to the CC isolating amplifier.

The differential amplifier at the input is connected to an amplifier with a common collector (transistor T_7). Its input resistance is very high, usually greater than 1 MΩ, so the influence of this impedance can be neglected. Therefore, the expression

$$A'_{\mathrm{d}} = \frac{h_{\mathrm{fE}}}{2h_{\mathrm{iE}}} \cdot \frac{R_0}{1 + R_0 \cdot h_{\mathrm{oE}}/2}, \tag{4.5.21a}$$

can be used for the gain of the differential gain of differential amplifier. The collector resistance R_0 is equal to the output resistance of the current source or the transistor T_4.

$$R_{\mathrm{C2}} = R_0 = r_{\mathrm{C4}}. \tag{4.5.21b}$$

Considering that the resistance of the excitation source is small and $R_{\mathrm{g}} \ll h_{\mathrm{iE}}$ we get

$$A_1 = \frac{h_{\mathrm{fE}} r_{\mathrm{C4}}}{2h_{\mathrm{iE}}} = \frac{g_{\mathrm{m2}} r_{\mathrm{C4}}}{2}, \tag{4.5.22}$$

where the relation connecting the hybrid and the hybrid π model was used, i.e., $h_{\mathrm{iE}} = r_\pi + r_{\mathrm{B}} \approx r_\pi$ and $h_{\mathrm{fE}} = g_{\mathrm{m}} r_\pi = \beta$. The resistance of the base's body resistance r_{b} was neglected and g_{m2} is denoting the transconductance of T_2.

The resistance R_3 together with the input resistance of T_8 acts as a voltage divider and introduces attenuation into the amplified signal. This attenuation is given by the relationship

$$A_2 = R_{\mathrm{in8}} \big/ (R_3 + R_{\mathrm{in8}}) \tag{4.5.23}$$

were

$$R_{\mathrm{in8}} \approx \beta_8 R_4 + r_{\pi 8}. \tag{4.5.24}$$

The voltage amplifier gives

$$A_3 \approx \frac{-h_{\mathrm{fE}} R_{\mathrm{C8}}}{h_{\mathrm{iE}} + (1 + h_{\mathrm{fE}}) R_4} \approx -\frac{g_{\mathrm{m8}} R_{\mathrm{C8}}}{1 + g_{\mathrm{m8}} R_4}. \tag{4.5.25}$$

The resistance R_{C8} in the collector of T_8 is made of a parallel connection of three resistances: the output resistance of T_8, the input resistance of T_{14}, and the output resistance of the current source R_0:

$$\frac{1}{R_{\mathrm{C8}}} = \frac{1}{R_{\mathrm{out8}}} + \frac{1}{R_{\mathrm{in14}}} + \frac{1}{R_0}. \tag{4.5.26}$$

The input resistance of R_{in14} of the CC stage is very large, much higher than the other two resistances, so its influence can be neglected. The output resistance of T_8 is also very high due to the presence of resistor R_4 in the emitter. Therefore, in practice, the collector resistance of the transistor T_8 is reduced to the output resistance of the current source, i.e., $R_{C8} = R_0 = r_{C9}$. So, the gain of the voltage amplifier is

$$A_3 = -\frac{g_{m8} r_{C9}}{1 + g_{m8} R_4}. \tag{4.5.27}$$

The gain of the entire OA is

$$A = A_1 A_2 A_3 = \left(\frac{g_{m2} r_{C4}}{2}\right) \cdot \left(\frac{\beta_8 R_4 + r_{\pi 8}}{R_3 + \beta_8 R_4 + r_{\pi 8}}\right) \cdot \left(-\frac{g_{m8} r_{C9}}{1 + g_{m8} R_4}\right). \tag{4.5.28}$$

To illustrate the above, let us assume that all NPN transistors have $\beta = 250$ and PNP transistors $\beta = 50$. Other parameters of the transistors will differ from each other because the currents through the transistors differ. That is:

$$g_m \approx I_C / V_T = I_C / \left(26 \times 10^{-3}\right), \tag{4.5.29}$$

$$r_\pi = \beta / g_m \tag{4.5.30}$$

and

$$r_C = 1/(\eta \cdot g_m) \tag{4.5.31}$$

where $\eta = V_T/V_A$, V_A is the Early voltage. In average, for room temperatures ($T =$ 300 K), for NPN transistors it is $\eta = 2 \times 10^{-4}$, and for PNP transistors, $\eta = 5 \times 10^{-4}$.

For the adopted values of β for T_2 whose collector current is $I_{C2} = I_a = 8.35\ \mu A$ we get $g_{m2} = 0.32$ mA/V. At the same time, it is $r_{\pi 1} = r_{\pi 2} = 781\ k\Omega$. For T_4 whose current is $I_{C4} = I_a = 8.35\ \mu A$ we get $r_{C4} = 6.2\ M\Omega$. The current of T_8 is $I_{C8} = I_9 =$ 600 μA so that we have $g_{m8} = 23.1$ mA/V and $r_{\pi 8} = 10.8\ k\Omega$ and, finally, T_9 having current $I_{C9} = I_9 = 600\ \mu A$ will have $r_{C9} = 89.7\ k\Omega$. With these values, when they are introduced in (4.5.28) for the total gain one gets $A = (992)\ (0.47)\ (-626) = -291{,}866 \approx -3 \times 10^5$.

It should be constantly borne in mind that in this analysis the hybrid π model of the transistor was used where it was assumed that $r_b = 0$ and $r_\mu \to \infty$, while the influence of the capacitances was ignored.

The input resistance of the OA is determined by the input resistance of the differential amplifier. It was previously stated in LNAE_Book 2 to be

$$R_{in} = 2h_{iE} = r_{\pi 1} + r_{\pi 2} = 1.562\,M\Omega. \tag{4.5.32}$$

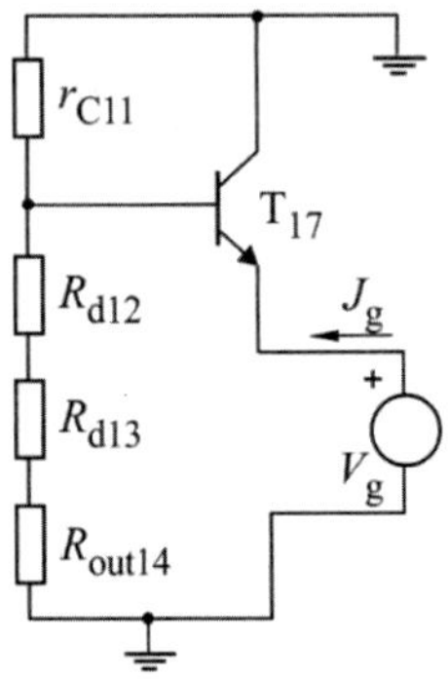

Fig. 4.5.10 Equivalent circuit used for calculation of the output resistance

To determine the output resistance, we will assume that T_{15} and T_{16} are cut-off. Let the input excitation be such that the transistor T_{17} conducts and T_{18} is cut-off. The output stage should be replaced by the equivalent circuit shown in Fig. 4.5.10. The influence of transistors TD_{12} and TD_{13} is replaced by the resistances of forward biased emitter junctions R_{d12} and R_{d13} were

$$R_d = V_T/I_C. \tag{4.5.33}$$

T_{11} and T_{14} are replaced with r_{C11} and R_{out14}, respectively.

The output resistance at the emitter will be

$$R'_{out} = V_g/J_g. \tag{4.5.34}$$

The total output resistance will be obtained by summing this resistance and the R_7.

For the currents of T_{D12} and T_{13} which are $I_{CD12} = 14.4\,\mu A$ and $I_{C13} = 185.6\,\mu A$ at room temperature, by application of (4.5.33) we get $R_{d12} = 1.8\,k\Omega$ and $R_{d13} = 140\,\Omega$. For the current $I_{11} = 200\,\mu A$ of T_{11} using (4.5.29) and (4.5.31) and considering that it is a PNP transistor, we get $r_{C11} = 260\,k\Omega$.

The output resistance of T_{14} acting as a CC amplifier is

$$R_{out14} = (r_{\pi 14} R_{B14})\big/(1+\beta), \tag{4.5.35a}$$

which comes from

$$R_{out} = \frac{r_\pi + \frac{(r_b+R_B)r_\mu}{r_b+R_B+r_\mu}}{1+g_m r_\pi} \approx \frac{r_\pi + r_b + R_B}{1+g_m r_\pi}. \tag{4.5.35b}$$

where $r_b = 0$ was used. The resistance R_{B14} in the base of T_{14} simultaneously represents the collector resistance of T_8 for which we already calculated $R_{B14} = R_8 = r_{C9} = 89.7\,k\Omega$. Also, for the calculated current of $I_{C14} = 200\,\mu A$ we get $r_{\pi 14} = 32\,k\Omega$. Putting together we get $R_{out14} = 0.49\,k\Omega$.

The resistance R_{B17} in the base of T_{17} represents a parallel connection of r_{C11} with the series connection $R_{d12} + R_{d13} + R_{iz14}$ and amounts $R_{B17} = 2.37\ k\Omega$.

The output resistance of T_{17} is

$$R'_{out} = (r_{\pi 17} + R_{B17}) / (1 + \beta). \tag{4.5.36}$$

The current of the output transistor T_{17} was calculated to be $I_{C17} = 155\ \mu A$. There from we have $r_{\pi 17} = 41.66\ k\Omega$, so that $R'_{out} = 176\ \Omega$. The overall output resistance of the OA is

$$R_{out} = R'_{out} + R_7 = 206\ \Omega. \tag{4.5.37}$$

It should be noted that if lower output impedance is desired, it would be necessary for the direct current through the output stage to be higher.

At higher frequencies, the influence of the transistor's capacitance on the functionality of the operational amplifier can no longer be neglected. Then, in the analysis of the amplifier, the capacitance C_π and C_μ should be included in the model. As a result of the analysis, the transfer function of the circuit would be obtained, where the gain, in the general case, would be given as

$$T(s) = A \frac{\prod_{i=1}^{n} (s/z_i + 1)}{\prod_{i=1}^{m} (s/p_i + 1)}, \tag{4.5.38a}$$

where z_i and p_i are the zeros and the poles of the transfer function, respectively. It is understandable that such an analysis is very difficult and almost impossible to perform in closed form. Therefore, we will give here only the approximate frequency characteristic of the operational amplifier. To this end, we will recall the structure of the operational amplifier. It consists of a cascade of a differential amplifier, a CC stage, a CE stage, and an output CC stage. As mentioned earlier, only the first and third stages act as voltage amplifiers.

The differential amplifier, according to the analysis in LNAE_Book 2, acts as a CE with a halved gain and therefore it can be considered that the analysis from LNAE_Book 2 can be applied to its frequency response at high frequencies. It can generally be assumed that the frequency response is determined by the gain as given by

$$A_v = A \frac{1 - s\tau_\mu}{1 + s\tau_1 + s^2\tau_2^2} \tag{4.5.38b}$$

where $s = j\omega$ and

$$\tau_\mu = C_\mu r_\mu / (g_m r_\mu - 1) \approx C_\mu / g_m. \tag{4.5.38c}$$

We will not repeat here the complete data for the time constant for sake of brevity.

The input impedance, as seen from (4.5.32), in this case, can be considered very large so that its frequency dependence, in the first approximation, can be neglected. Thus, the influence of the first stage can be calculated as a function with single pole, i.e., as a function that introduces an asymptotic slope of −6 dB/oct.

The second stage, as stated, also serves as a separation stage in the sense of the discussion in LNAE_Book 2. Due to the very high input impedance, due to the negligible frequency dependence of the gain and due to the very low (inductive) output impedance, the influence of the common collector amplifier that makes the T_7 transistor on the frequency response can be neglected. This conclusion is also correct because due to the presence of R_4, the input resistance of the third stage is very high.

The third stage is a CE amplifier with a gain of

$$A_{\mathrm{v}}^{'} = A_{\mathrm{v0}}^{'} \cdot \frac{1}{1 + j\omega/\omega_{\mathrm{v}}} \cdot \frac{1}{1 + \frac{j\omega\tau_3 R_{\mathrm{g}}}{R_{\mathrm{g}} + R_{\mathrm{in}}}} \tag{4.5.39a}$$

were

$$A_{\mathrm{v0}}' = A \cdot \frac{R_{\mathrm{in}}}{R_{\mathrm{in}} + R_{\mathrm{g}}}. \tag{4.5.39b}$$

The frequency characteristic of the CE amplifier that makes up the T_8 transistor is similar to the frequency characteristic of a conventional CE amplifier, except that due to the negative feedback (due to the presence of R_4), the upper cut-off frequency is increased. Therefore, this stage will also contribute for one pole (−6 dB/oct) but at a slightly higher frequency than the first amplifying stage.

Finally, if the output stage, which is in fact a CC amplifier is assumed to be loaded with a large capacitive load (which is a possible unfavorable case), its upper cut-off frequency can be expected to approach the value of the bandwidth of the OA.

Based on all this, we conclude that the frequency characteristic of the operational amplifier as depicted in Fig. 4.5.11 can be approximated by the following expression

$$A(j\omega) = \frac{A_0}{(1 + jf/f_1)(1 + jf/f_2)(1 + jf/f_3)} \tag{4.5.40}$$

where A_0 is the gain at low frequencies, and f_1, f_2, and f_3 are frequencies of the poles. The corresponding amplitude characteristic is shown in Fig. 4.5.11 in asymptotic form.

The cut-off frequency at which the gain decreases for 3 dB is determined by the frequency (f_1) of the dominant pole. It is attributed to the cut-off frequency of the first stage. The frequency at which the gain becomes equal to unity is denoted by f_{T}.

When the OA is in open loop, i.e., when no negative feedback is applied, the cut-off frequency f_1 is very low. Its value differs significantly for different types of operational amplifiers and, in average, is of the order of ten kHz. The pole frequency f_2 is usually one decade higher. The frequency f_{T} is of the order of MHz, and this

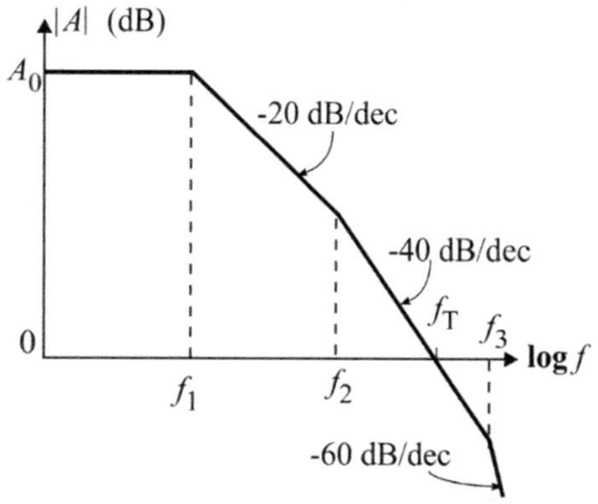

Fig. 4.5.11 Approximative amplitude characteristic of an OA

information is most often given by manufacturers and referred to as gain-bandwidth product.

In OA applications when negative feedback circuit is added, the cut-off frequency f_1 is shifted to higher frequencies by an amount equal to the amount of gain reduction.

4.5.2.3 OA with Differential Output

OAs are also realized with a symmetrical output; one example is depicted in Fig. 4.5.12. In such a case, the model of the OA would correspond to the one shown in Fig. 4.5.4a.

It is a three-stage amplifier in which the first two stages are differential amplifiers, followed by CC output stages. It is clear that the output signals V_{out1} and V_{out2} are equal in amplitude and opposite in phase.

In this circuit, different values of voltage gain can be achieved by tying the A and A', the B and B' terminals, or by leaving these terminals open. Figure 4.5.13 shows the amplitude and phase characteristics of the voltage gain for all three possibilities of connecting of terminals A, A′, B, and B′. Note the ultra large value of the open loop gain shown by a full line. It is above 200 dB which is larger than 10^6. In that the cut-off frequency is of the order of several dozens of MHz.

Fig. 4.5.14 shows the amplitude and phase characteristics for three different values of the compensating capacitance C_{k}. For $C_{\text{k}} = 1$ pF, the amplifier is still unstable. The limiting case of stability (when the phase margin: $M_{\phi} = 0$. Note, the phase margin was defined in LNAE_Book 2) is reached for the capacitance $C_{\text{k}} = 4$ pF. For $C_{\text{k}} = 7$ pF the simulation results indicate good stability of the amplifier.

The introduction of the compensating capacitor C_{k} fulfills the basic goal which is the stability of the amplifier even at maximum feedback. The price paid is reflected in two aspects. First of all, the value of the capacitance of the C_{k} is so large that this capacitor occupies disproportionately largest area on the silicon chip as compared with any of the rest of devices. In some cases, it occupies up to a third of the whole.

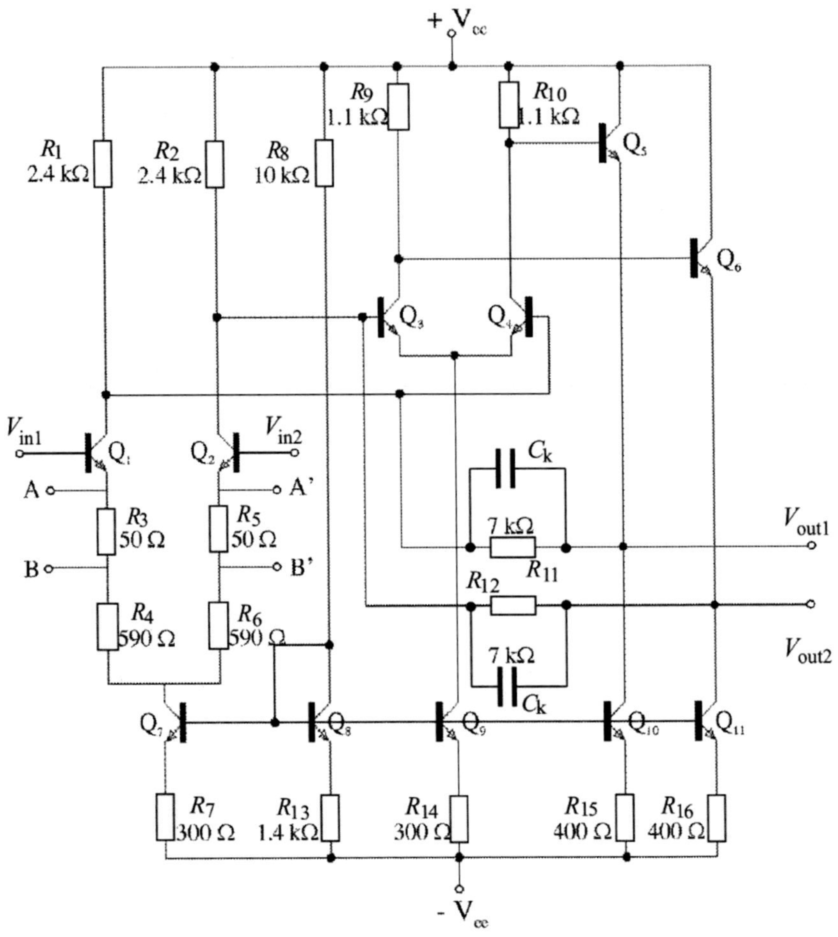

Fig. 4.5.12 OA with a symmetric (balanced) output (The schematic and simulation results shown in this, and the following two figures were produced at the Faculty of Electronic Engineering, University of Niš, Serbia, by Prof. Nebojša Janković and associates. Here we use the original drawing of the schematic diagram so that the transistors are denoted Q instead of T)

A more significant problem than the increase in chip area due to the introduction of C_k capacitor, however, is the emergence of a new limitation to the capabilities of the operational amplifier known as slew rate. The concept of the slew rate will be very shortly presented here on the example of the OA of Fig. 4.5.12.

It is related to the response of the output voltage of the OA to sudden and large change of the input signal. For such signals some of the transistors become (current) saturated while other become cut-off. In any case the capacitor C_k gets charged or discharged. Due to the large time constant of this capacitor, the output voltage has difficulties to follow fast changes of large amplitude of the input signal. Its slope (in

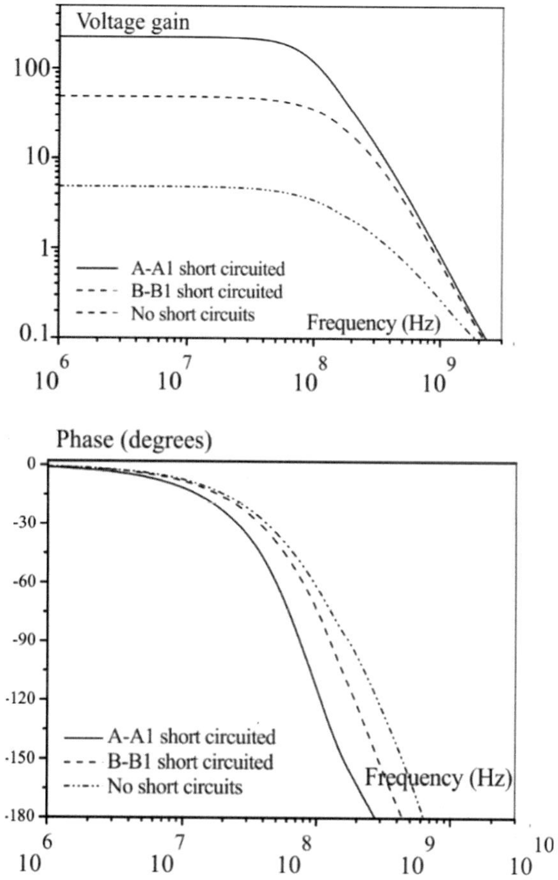

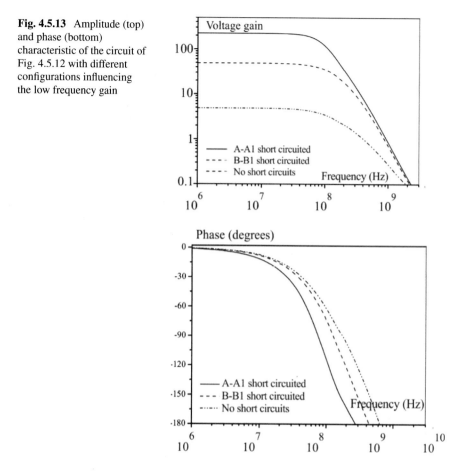

Fig. 4.5.13 Amplitude (top) and phase (bottom) characteristic of the circuit of Fig. 4.5.12 with different configurations influencing the low frequency gain

time) is much smaller than the slope of the input signal. The maximum slope of the output signal defined for a given value of C_k is named slew rate.

For the circuit of Fig. 4.5.12 imagine that due to a sudden change of the input signals (both V_{in1} and V_{in2}) the output transistors Q_5 and Q_6 go cut-off. In such a case the complete currents of the current sources Q_{10} and Q_{11} run through their corresponding capacitor C_k, which for sudden changes, behaves as a short-circuit as compared with R_{11} or R_{12}.

Since the current of the capacitor is $i_C = C \cdot (dv_C/dt)$, for this case we will have

$$I_0 \approx C_k(dv_{out}/dt) \tag{4.5.41a}$$

where I_0 is the current of the current source (Q_{10} or Q_{11}). Since C_k and I_0 are constants, they determine the maximum value of the slope (derivative) of the output voltage. In other words

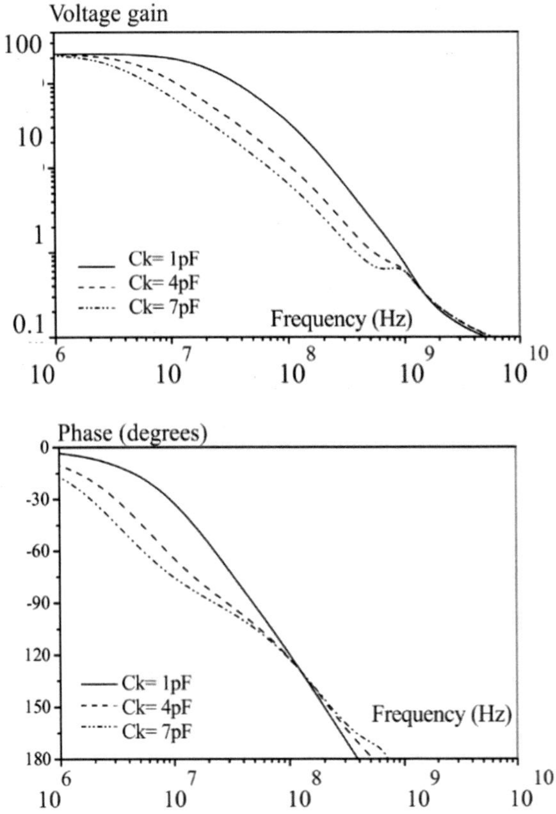

Fig. 4.5.14 Amplitude (top) and phase (bottom) characteristic of the circuit of Fig. 4.5.12 for different values of the compensating capacitance

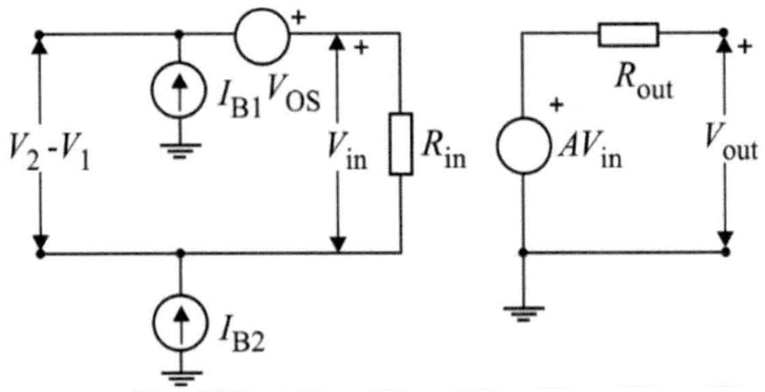

Fig. 4.5.15 Model of the OA with input currents and input voltage offset sources shown

$$\frac{dv_{\text{out}}}{dt}_{|\max} = \frac{I_0}{C_k} \tag{4.5.41b}$$

This maximum value is named slew rate. For the example circuit under consideration, it amounts 30 μA/30 pF = 1 V/μs. The slew rate is usually expressed in volts per microseconds as was done here.

It is now important to keep in mind that the slew rate is a substantially different limiting property from the upper cut-off frequency of the amplifier. Namely, a signal of relatively low frequency, if its amplitude is large, shows a high rate of change and can exceed the slew rate, while belonging to the bandwidth of the amplifier. On the other hand, a low-amplitude high-frequency signal can have a small slope, and thus not exceed the slew rate, but exceed the upper cut-off frequency.

Bearing in mind, however, that the value of C_k is determined from the stability requirements, i.e., from the upper cut-off frequency, the cut-off frequency and the slew rate are interdependent. To determine this dependence, we will assume that the cut-off frequency of the amplifier is f_0, and that the nominal gain is A_0. The gain is now

$$A = A_0 / (1 + s/\omega_0). \tag{4.5.42}$$

If the circuit is excited by a step function whose Laplace transform is V_m/s, for the output voltage we have

$$\frac{v_{out}(s)}{V_m} = \frac{1}{s}\frac{\omega_0 A_0}{s+\omega_0} = \frac{A_0}{s} - \frac{A_0}{s+\omega_0}. \tag{4.5.43}$$

The output voltage in the time domain is now obtained by the inverse Laplace transform as

$$v_{out}(t) / V_m = A_0\left(1 - e^{-\omega_0 t}\right), \tag{4.5.44a}$$

and its derivative is

$$\frac{dv_{out}}{dt} = \left(A_0\omega_0 e^{-\omega_0 t}\right) \cdot V_m, \tag{4.5.44b}$$

which at $t = 0$, when the derivative is the largest, becomes $A_0\omega_0 V_m$. This result suggests that the value of the slew rate can be expressed over the cut-off frequency and vice versa. In this case, for a given cut-off frequency, we can identify the maximum amplitude of the step function that can be fed to the input, while maintaining the linearity of the amplifier.

4.5.2.4 Voltage and Current Offset

In this section some attention will be paid to the offset of the operational amplifier, which is basically due to the offset of the input differential amplifier.

If for equal excitation voltages at the input terminals the output offset voltage of the OA is ΔV_{out}, then the voltage offset at the input is

$$V_{os} = \Delta V_{out}/A, \tag{4.5.45}$$

where A is the open loop gain of the OA. For example, if $A = 10^5$, and if at the output we measure $\Delta V_{out} = 5$ V, the offset voltage is $V_{os} = 5 \times 10^{-5}$ V. In this way, the offset of the OA is represented by the voltage offset at the input. This representation is more convenient because the exact value of the gain is not known in advance.

If the input voltages (of a BJT-based differential amplifier) are equal to zero and if identical resistors are connected from the input to ground, due to the unavoidable asymmetries there will be $I_{B1} \neq I_{B2}$. The difference in the input currents:

$$I_{os} = I_{B1} - I_{B2} \tag{4.5.46}$$

is named a current offset of the OA.

As current offset always exists, when it comes to the biasing current of input inputs, it is defined as

$$I_B = (I_{B1} + I_{B2})/2. \tag{4.5.47}$$

To include the offset into the model of the OA one needs to modify the one given in Fig. 4.5.3. Figure 4.5.15 represents the modification. I_{B1} and I_{B2} denote the biasing currents. The offset voltage V_{os} is connected in series with one of the inputs and represents the potential difference of the input terminals when $V_1 = V_2$. In that the current flowing through R_{in} is considered negligible. The difference in the currents I_{B1} and I_{B2} models the current offset.

The voltage and current offsets also depend on temperature as transistor parameters depend on temperature. Typical values of voltage offset in the OA are several mV, and its temperature dependence is several μV/K. The current offset is about 10% of the mean value of the biasing current of the input base terminals. The dependence of the current offset on temperature is determined from the dependence of the input currents of the bases on the temperature. In more complete datasheets of OA manufacturers, this dependence is shown graphically.

The influence of offset on the operation of the operational amplifier will be considered on the example of the non-inverting amplifier from Fig. 4.5.16a. Applying the model from Fig. 4.5.15, in order to calculate the influence of the offset, the input source V_g should be short-circuited. However, its resistance R_g must be considered. In this way, the equivalent circuit shown in Fig. 4.5.16b is obtained. We will consider that the gain of the operational amplifier is large enough that the inverting and non-inverting input terminals are at the same potential.

Using the superposition method, we will calculate the output offset voltage.

If $I_{B1} = 0$ and $I_{B2} = 0$ the output voltage is

$$V'_{outos} = V_{os}(1 + R_2/R_1). \tag{4.5.48}$$

For $I_{B1} = 0$ and $V_{os} = 0$ the output voltage is

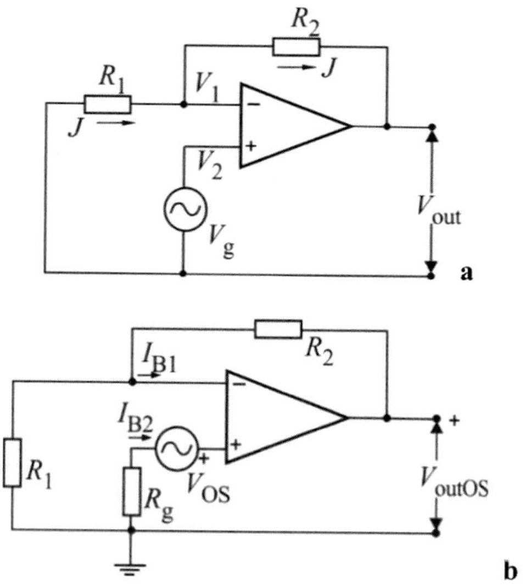

Fig. 4.5.16 Offset analysis of the non-inverting amplifier. **a** The amplifier and **b** offset voltage source inserted at the input

$$V''_{\text{outos}} = -R_g I_{B2}(1 + R_2/R_1). \tag{4.5.49}$$

Finally, if $I_{B2} = 0$ and $V_{os} = 0$ the output voltage is

$$V'''_{\text{ouos}} = R_2 I_{B1}. \tag{4.5.50}$$

Accordingly, by superposition, the output offset voltage will be

$$V_{\text{outos}} = V'_{\text{outos}} + V''_{\text{outos}} + V'''_{\text{outos}}, \tag{4.5.51}$$

or

$$V_{\text{outos}} = V_{os}(1 + R_2/R_1) + R_2 I_{B1} - R_g(1 + R_2/R_1)I_{B2}. \tag{4.5.52}$$

In order to eliminate the influence of the voltage offset, i.e., to compensate for V_{os}, we connect a voltage battery V_k in series with R_1. This circuit is shown in Fig. 4.5.17, with the current offset neglected. The analysis of this circuit gives

$$V'_{\text{out}} = -R_2 V_K/R_1. \tag{4.5.53}$$

If this output voltage, caused by the used battery, is by modulo equal to the output voltage caused by the offset voltage as

$$\left|V'_{\text{out}}\right| = V''_{\text{outos}}, \tag{4.5.54}$$

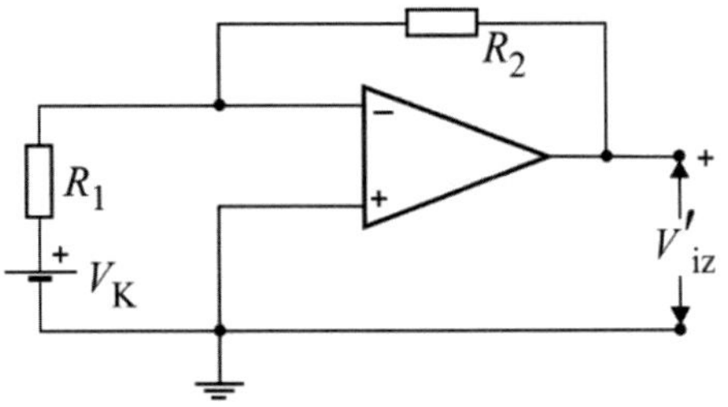

Fig. 4.5.17 Circuit, in principle, for compensation of the offset voltage

the output offset voltage will be zero. So, the condition for compensation of the offset voltage is

$$R_2 V_K / R_1 = V_{os}(1 + R_2/R_1) \tag{4.5.55}$$

or

$$V_K = V_{os}(1 + R_1/R_2) \tag{4.5.56}$$

The compensating voltage V_K is very small of the order of millivolts. To obtain it, a potentiometer connected to the circuit of the operational amplifier is used as shown in Fig. 4.5.18.

It should be noted that in the circuit of Fig. 4.5.18 the compensation is not easy. Due to the large gain of the operational amplifier, it is difficult to adjust the required value of the voltage V_K, i.e., the position of the potentiometer P. Therefore, operational amplifiers are produced where the compensation of the voltage offset is realized in the input differential amplifier in the manner shown earlier in LNAE_Book 2. There, the compensation is facilitated by smaller value of the gain of the differential amplifier. Such an operational amplifier has two additional external terminals used to connect the potentiometer P.

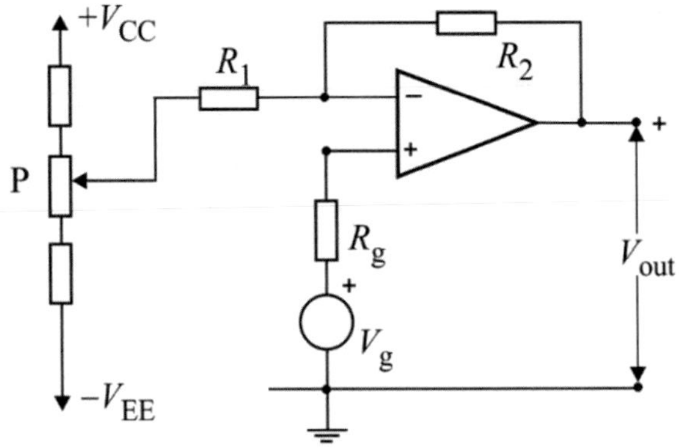

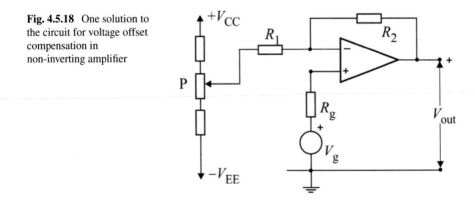

Fig. 4.5.18 One solution to the circuit for voltage offset compensation in non-inverting amplifier

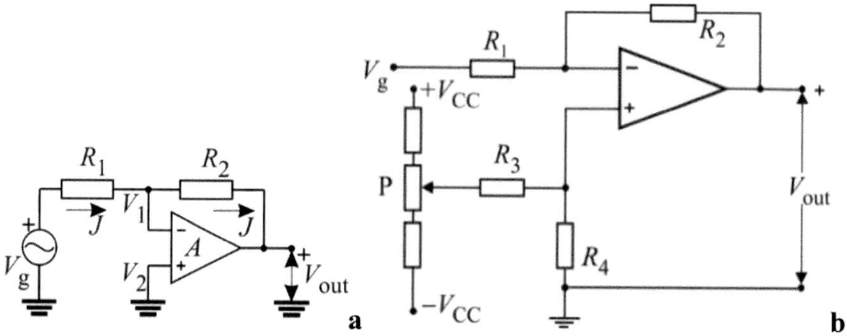

Fig. 4.5.19 **a** Inverting amplifier and **b** compensation of the voltage offset

Equation (4.5.52) indicates how the current offset should be compensated. It can be reduced by appropriate choice of the resistance of the source R_g. Namely, if it is

$$R_2 = R_g(1 + R_2/R_1), \tag{4.5.57}$$

or

$$R_g = R_1 R_2 \big/ (R_1 + R_2), \tag{4.5.58}$$

the output offset voltage caused by the current offset at the input is

$$V_{outos} = V'_{outos} + V''_{outos} = R_2(I_{B1} - I_{B1}) = R_2 I_{os}. \tag{4.5.59}$$

This current offset compensation satisfies practical applications, especially since the currents I_{B1} and I_{B2} are extremely small. The procedure for reducing these currents is explained in LNAE_Book 2.

The same conclusions would be reached in the case of an inverting amplifier depicted in Fig. 4.5.19a. The voltage offset compensation circuit is shown in Fig. 4.5.19b.

The explained offset compensation is static. If for $V_1 = V_2$ the total output offset in the compensated circuit is equal to zero, it will not be when $V_1 \neq V_2$ because the operating points of the transistor have changed. The same will apply to the effect of temperature changes. However, static offset compensation significantly reduces offset even under dynamic operating conditions.

4.5.2.5 Definitions of the OA's Parameters

Here we will give a short overview of the characteristic parameters of the OAs which allow for making choices when application comes in fore. The diagrams displayed are demonstrational only and, unless explicitly stated, are by no means related to any specific OA available on the market. That stands for the numerical values for the parameters given below.

4.5.2.5.1 Ratings

Maximum allowable values (or ratings) are those parameters whose value must not be exceeded. Otherwise, the operational amplifier would be damaged.

Supply voltage. Most operational amplifiers use two supply sources, one positive voltage, the other negative, relative to ground. The maximum allowable values of variations of these voltages are usually specified. Typical values are 5 V, but there are others such as 22 V, 15 V, 9 V, 3 V, etc.

Power supply current is a quantity that indicates the consumption of the OA in static mode, i.e., in the absence of excitation and without a load. For illustration, the supply current as a function of temperature for two values of the supply voltage of one OA is shown in Fig. 4.5.20.

Short-circuit duration. This is the maximum time for which the output can be short-circuited to ground and the maximum time for which the output can be short-circuited to one of the power sources. When proper current protection is implemented in the output circuit of the OAs, these times are unlimited.

Differential input voltage is the maximum difference of input voltages V_1-V_2 that can be amplified. It is given for pre-assumed values of supply voltage.

Absolute input voltage is the maximum allowable values of voltage V_1 and V_2 individually, at the assumed value of the voltage of the power supply.

The maximum output voltage is the allowable change in the output voltage up to which the operational amplifier operates linearly and does not enter saturation (e.g., ±15 V for a supply voltage of ±20 V).

Internal dissipation is dissipation in a chip in the absence of a signal and it depends on the supply voltage. For the OA known as μA 741E at supply voltages of ±20 V,

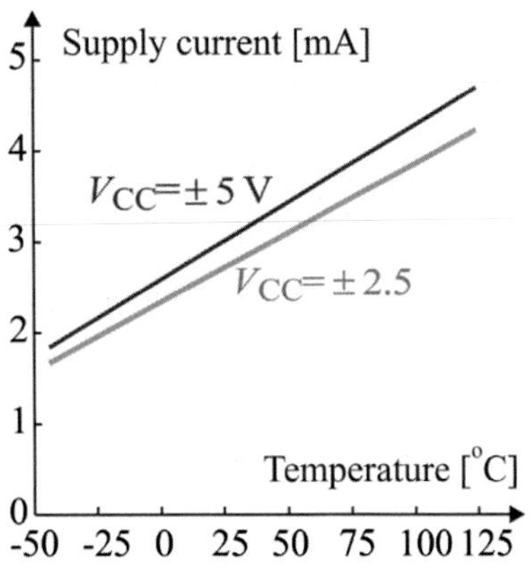

Fig. 4.5.20 Supply current as function of temperature for two values of the supply voltage

the dissipation is typically 80 mW. When the excitation is introduced, dissipation increases mainly due to dissipation at the output transistors. The maximum dissipation on the chip depends on the way the chip is made, but also on the way it is cooled, i.e., on the way the casing in which the chip is located is made.

The operating temperature range is the maximum permissible ambient temperatures in which the operational amplifier operates. The maximum temperature of the soldering agent for external leads and the duration of soldering for that temperature are also given. Examples of storage temperatures are (−60, +150) °C, (−55, + 125) °C, (−40, +125) °C or (0, +70) °C.

4.5.2.5.2 Electrical Parameters

Electrical parameters do not refer to maximum values, but to normal operating conditions. Therefore, they are given for pre-specified values of supply voltage and for a given value of the resistance of the load connected to the OA. This value is usually 2 kΩ. It is also considered that no external feedback is connected to the OA, that is, that it works in an open loop.

Under the above conditions, *the voltage gain* is defined for small input signals and low frequencies, below the frequency of the dominant pole. Due to the large variation of the gain from sample to sample of the same type of OA, the minimum, mean, and maximum values of the gain are given in datasheets. In Fig. 4.5.21 the voltage gain is given for a wide range of load resistances at two values of the supply voltages. This amplifier has relatively small gain.

Cut-off frequency f_T is the one at which the gain decreases to unity (0 dB). This value should be distinguished from the cut-off frequency f_{3dB}, which is defined in the classical way. Typical values are $f_T = 2$ MHz and $f_{3dB} = 8$ Hz (for the μA741 OA). However, there are also circuits that are manufactured in more advanced technologies, so that amplifiers with a much higher value of f_T can also be found (such as the one depicted in Fig. 4.5.12).

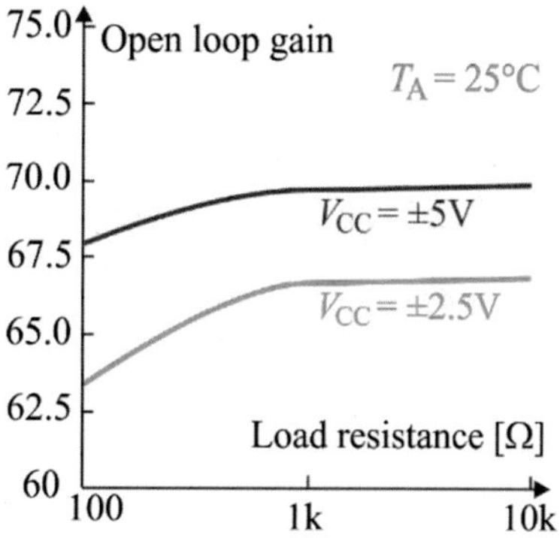

Fig. 4.5.21 Open loop voltage gain of an OA as a function of the load resistance for two values of the supply voltage

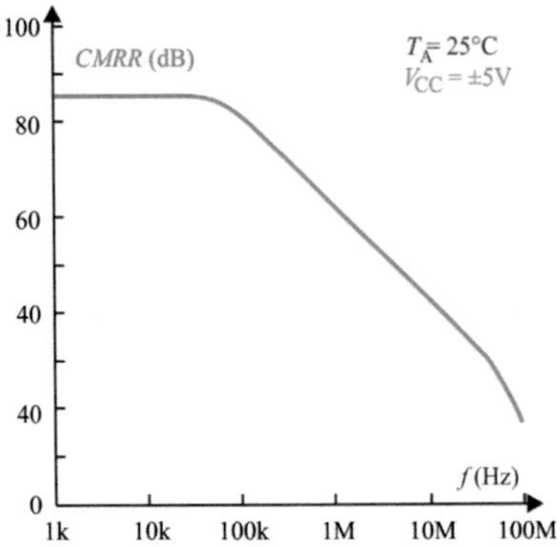

Fig. 4.5.22 *CMRR* as a function of frequency

The amplitude margin and the phase margin are quantities that indicate the stability of the amplifier when the feedback is applied and are defined in LNAE_Book 2.

The input resistance is the resistance seen between the input terminals for alternating excitations (The value is about 6 MΩ for μA 741). This quantity is also recognized as symmetric input resistance. Of course, the resistance can be measured from each input terminal to ground (asymmetrically).

Input capacitance is the symmetrically measured capacitance between the input terminals.

Output resistance is the resistance between the output and ground for alternating signals (approx. 75Ω for μA741).

The common mode rejection ratio (CMRR) was explained in detail in LNAE_Book 2 and is given as the ratio of the differential gain and the common mode gain (about 95 dB for μA741). Figure 4.5.22 shows a dependence of a CMRR on the signal frequency.

The maximum response rate for large signals (*Slew rate*) is the maximum rate of change of output voltage when the OA is excited by a large input signal in the form of a step function and is given in V/μs (about 0.7 V/μs, for OA μA 741 which is connected so that its gain is equal to unity and the input signal is ≈10 V).

Settling time is the time interval required for decaying the oscillations in response to a low-amplitude step function.

Noise voltage and current are defined at the input of the OA as equivalent noise voltage and equivalent noise current. Manufacturers often give these values in a form of diagrams expressing the frequency dependence of the equivalent input noise. These quantities will be discussed in particular in LNAE_Book 5.

The total harmonic distortion (*THD*) speaks on the distortions that are generated inside the operational amplifier and is usually given together with the noise data. A typical example would be $THD = 0.08\%$.

The input polarization current is the mean value of the base input currents $I_B = (I_{B1} + I_{B2})/2$ when the output voltage is equal to zero (about 80 nA for μA 741).

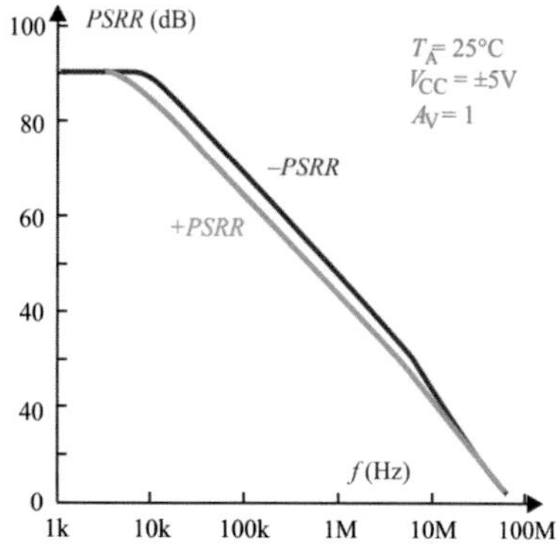

Fig. 4.5.23 Frequency dependence of typical *PSRR*

The current offset represents the difference of the input excitation currents for the output voltage equal to zero (about 30 nA for μA 741).

The temperature coefficient of the current offset (or drift of the input offset current) is the dependence of the offset current on the temperature, $\alpha_{\text{Ios}} = \text{d}I_{\text{os}}/\text{d}T$ (about 0.5 nA/K for μA 741).

Voltage offset is the value of the voltage that needs to be connected between the input terminals so that the output voltage is equal to zero (about 0.8 mV for μA 741).

The temperature coefficient of the voltage offset is the dependence of the offset voltage on the temperature, $\alpha_{\text{Vos}} = \text{d}V_{\text{os}}/\text{d}T$ (about 15 μV/K for μA 741). For this parameter, as above, the term temperature drift of the offset voltage is used.

The Power Supply Rejection Ratio (*PSSR*) is a measure of the effect of the supply voltage variations on changes in output voltage. A frequency dependence of typical *PSRR* is depicted in Fig. 4.5.23. Typical values of these factors are 80 dB. The *PSRR* determines the required DC voltage stability. He also talks about the danger of inter-coupling of the circuit elements (in this case OAs) through the internal resistance of the supply voltage. The process of calculating *PSRR* when the computer is not in use is too complex for bipolar amplifiers, so it will be illustrated later on the example of MOS OA.

4.5.3 Realization of MOS OA and OTA

Whenever comparison of electronic circuits with BJT and with MOS transistors is mentioned, a much higher gain can always be pointed out, which can be achieved by using BJT. Therefore, the operational amplifiers that are implemented with BJTs, which were discussed in the previous section, have no equal in terms of the size of the gain and they have the most massive use when the OA is used as a component to build a system on a PCB.

However, with the increase of applications of microelectronics, there is an increasing need for embedded integrated MOS OAs. This need is especially noticeable in circuits that receive measured (analog) signals from various sensors including antennas, and such signals are delivered to microcomputer systems for further processing. In the stages of analog signal processing that can be characterized as receiving part of the system, amplification, filtering, and conversion of the analog to digital signal is performed wherefrom the need to use OAs. An earlier concept of designing such circuits was to separate the signal processing system into analog and digital parts, which allowed each of them to be produced in a different technology. Thus, the analog part was mainly realized by bipolar subassemblies, and the digital part, in order to save low consumption or maintain a high scale of integration, in MOS (or CMOS) technology. The printed circuit board that performs the entire function was serving as a technological intermediary between bipolar and MOS technology. Recently, with the advent of nano-electronics, it has become possible to integrate the analog part on the same chip with the digital. In this way, the overall scale of integration of the system was dramatically increased as well as its reliability, having in mind the reduction of the number of soldering points and testing capabilities. One is not to forget the dramatic fall in the price of this kind of mixed signal (analog and digital) circuits.

The application of MOS OAs within the integrated circuit has enabled the development of two classes of MOS operational amplifiers. Namely, in a large number of applications, the operational amplifier is loaded with a very small load (a small load, even here, is the one through which a small current flow, i.e., whose resistance is high). This means that it is not necessary to install an output stage that performs only current amplification.

If the output stage of the OA is eliminated, however, the remaining circuit has a high output resistance and, at the output, can be considered as a current source. Therefore, it is said that such an operational amplifier converts voltage into current and is called a transconductance amplifier or OTA. In addition to OTAs, MOS operational amplifiers with a complete output stage are also being developed, which, of course, still contains an OTA as a basic building block. We will call such amplifiers complete.

OTAs can be realized as single-stage and as two-stage. The single-stage would contain only a differential amplifier (preferably cascode), while the two-stage would contain an additional stage with a CS, similar to the CS amplifier we had in bipolar OAs. The second stage in a two-stage amplifier can also be realized as a cascode. Finally, in CMOS, the second stage may be the well-known CMOS inverter working as a linear amplifier.

In the following text, several circuits of OAs with MOS transistors will be presented. In order not to go into a very detailed analysis, only CMOS amplifiers will be considered. Two versions of OTAs will be shown first, followed by a complete one.

A two-stage OTA is shown in Fig. 4.5.24. The first stage consists of a differential pair of transistors T_1 and T_2, which has a current mirror T_3 and T_4 as a load. The constant current source of the differential amplifier is the transistor T_5 which forms a

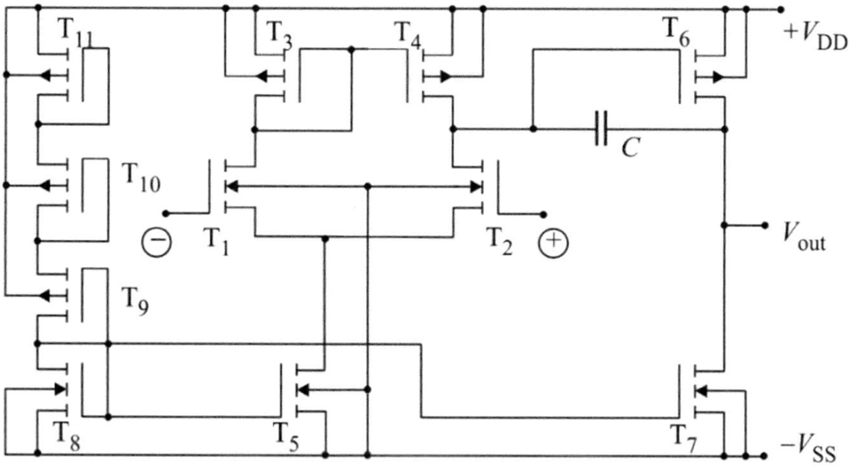

Fig. 4.5.24 Transconductance amplifier using Miller's compensation

current mirror with T_8, while the series connection T_9, T_{10}, and T_{11} forms a resistor R which determines the current of the constant current source. The common source output amplifier is the P-channel transistor T_6, and T_7 is its dynamic load. The amplifier is frequently compensated by a Miller's capacitor C.

This amplifier, realized (in 0.5 μm technology) as a discrete circuit for supply voltages of ±5 V, exhibits the following features:

- Dynamics of the common mode signal at the input is ±3 V,
- Dynamics of the output voltage is ±4 V,

Gain (open loop) $A = 15000, f_T = 1$ MHz, $P_{dis} = 1.6$ mW.

For a load capacitance of $C_L = 20$ pF, and with a Miller capacitance $C = 4.4$ pF, its slew rate is 2 V/μ.

This circuit, being the simplest two-stage amplifier, will be used to describe the procedure for calculating the *PSRR*. To this end, Fig. 4.5.26 shows a simplified equivalent circuit of the OTA depicted in Fig. 4.5.24 provided that it is connected in a circuit as in Fig. 4.5.25a.

The *PSRR* is calculated as the product of the differential gain and the quotient of the supply voltage and output voltage increments, provided that there is no excitation. If with V_{dd} we denote the amplitude of the alternating component of the supply voltage, and with V_{out} the corresponding value of the output voltage which is its consequence, for the gain from the power supply to the output we have

$$A_{dd} = \frac{V_{out}}{V_{dd}}_{|V_{in}=0}. \tag{4.5.60a}$$

Now, for the power supply rejection ratio we have

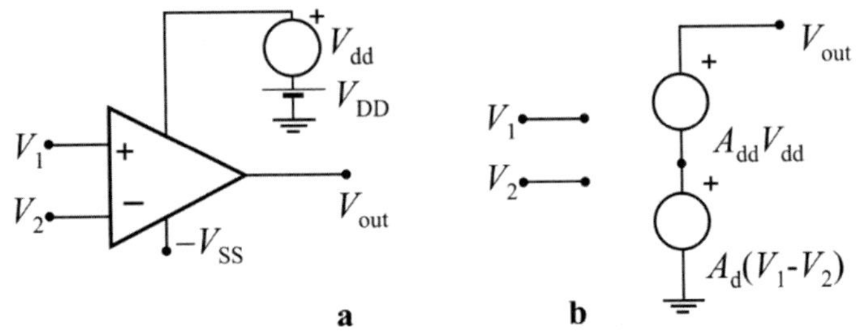

Fig. 4.5.25 On the definition of the *PSRR*

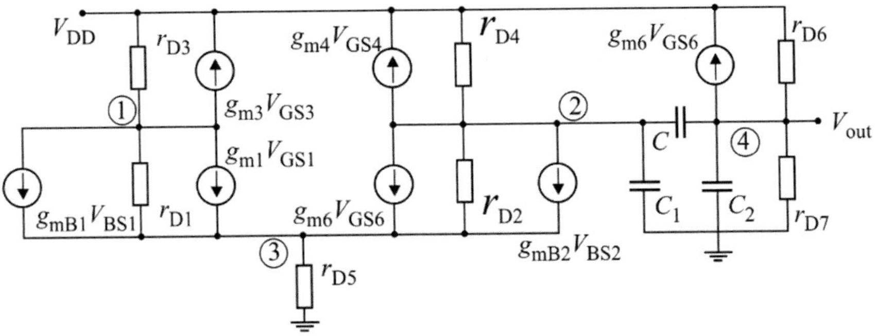

Fig. 4.5.26 Simplified circuit for evaluation of the *PSRR*

$$\mathrm{PSRR} = A_d/A_{dd}, \tag{4.5.60b}$$

where A_d is the differential gain.

For circuits using two power supply sources we will need two numbers for *PSRR*. So, we would have $PSRR_{dd}$ and $PSRR_{ss}$ or $PSRR_{cc}$ and $PSRR_{ee}$, for MOS and BJT amplifiers, respectively.

The circuit of Fig. 4.5.25a, in which an ideal voltage amplifier is used, illustrates the case where only the V_{DD} is subject to change. The output voltage of the amplifier is, of course, a function of the potential of the input terminals and the voltage V_{dd}, which represents the variations of the supply voltage. With this in mind, the amplifier model has been extended with the source $A_{dd}V_{dd}$. By substituting this model into the circuit of Fig. 4.5.25a the circuit 4.5.25b is obtained. The analysis of this circuit yields the following significant results:

$$V_{out} = A_d(V_1 - V_2) + A_{dd}V_{dd} = A_d\left[(V_1 - V_2) + \frac{1}{\mathrm{PSRR}_{dd}}V_{dd}\right]. \tag{4.5.61}$$

This result makes it possible to map the variation of the supply voltage onto the input of the amplifier and compare it with the input signal. Obviously, if the *PSRR* is higher, the influence of the supply voltage is smaller.

To calculate the *PSRR* for the circuit of Fig. 4.5.24 we will suppose the following: $r_{D1} = r_{D2} = 1/G_1$, $r_{D3} = r_{D4} = 1/G_3$, $g_{m1} = g_{m2}$, $g_{m3} = g_{m4}$ and $g_{mB1} = g_{mB2} = g_{mB}$. In addition, based on the circuit structure we have

$$\begin{aligned}
&V_{GS1} = V_4 - V_3, \quad V_{GS2} = -V_3, \\
&V_{GS3} = V_1 - V_{dd}, \quad V_{GS4} = V_1 - V_{dd}, \\
&V_{GS6} = V_2 - V_{dd}, \quad V_{BS1} = -V_3 \\
&\text{and } V_{BS2} = -V_3,
\end{aligned} \tag{4.5.62}$$

and, to simplify the analysis, we suppose $V_{GS5} = V_{GS7} = 0$. In that, the transfer of the signal from the supply terminal via the connection T_{11}–T_{10}–T_9–T_8–T_5 (T_7) and beyond, we will consider negligible in comparison to the transfer via the rest of the paths.

For the circuit of Fig. 4.5.26 the following system is valid

$$\begin{aligned}
&(G_1 + G_3 + S_3)V_1 + 0 \cdot V_2 \\
&\quad - (G_1 + g_{m1} + g_{mB})V_3 + g_{m1}V_{out} = (G_1 + g_{m3})V_{dd} \\
&g_{m3}V_1 + [(G_3 + G_1 + s(C_1 + C)]V_2 \\
&\quad - (G_1 + g_{m1} + g_{mB})V_3 - sCV_{out} = (G_3 + g_{m3})V_{dd}
\end{aligned} \tag{4.5.63}$$

$$\begin{aligned}
&- G_1V_1 - G_1V_2 + (G_1 + g_{m1} + g_{mB} + G_1 + g_{m1} + g_{mB} \\
&\quad + G_5)V_3 - g_{m1}V_{out} = 0 \cdot V_1 + (g_{m6} - sC) + 0 \cdot V_3 \\
&\quad + [G_6 + G_7 + s(C + C_2)]V_{out} = (G_6 + g_{m6})V_{dd}
\end{aligned}$$

This system shows that the *PSRR* is a very complex function of the value of the circuit elements and the frequency.

Since our goal is not to deal here with details about the *PSRR* of this circuit, but to point out the method of its calculation, we will introduce some additional simplifications. First, the frequency dependence will be omitted. So, we put $s = 0$. This simplification is not justified from the point of view of the actual frequency spectrum of the signal generated by the battery and is introduced only to shorten the proceedings. Then, the transconductance of any transistor will be considered to be significantly higher than the reciprocal of the internal resistance of any transistor. Finally, $g'_{m1} = g_{m1} + g_{mB}$ will be introduced so that the following system is created

$$\begin{bmatrix} g_{m3} & 0 & -g'_{m1} & g_{m1} \\ g_{m3} & G_1 + G_3 & -g'_{m1} & 0 \\ -G_1 & -G_1 & 2g'_{m1} & g_{m1} \\ 0 & g_{m6} & 0 & G_6 + G_7 \end{bmatrix} \cdot \begin{bmatrix} V_1 \\ V_2 \\ V_3 \\ V_{out} \end{bmatrix} = \begin{bmatrix} g_{m3}V_{dd} \\ g_{m3}V_{dd} \\ 0 \\ g_{m6}V_{dd} \end{bmatrix}. \tag{4.5.64}$$

The solution of this system, i.e., the *PSRR* is

$$\frac{V_{\mathrm{dd}}}{V_{\mathrm{out}}} = PSRR_{\mathrm{dd}} \approx \frac{g_{\mathrm{m1}}}{G_1} = \mu_1. \tag{4.5.65}$$

The obtained value approximately corresponds to the voltage gain of a single-stage amplifier. This means that the signal coming to the output of the differential amplifier (to the drain of transistor T_2) from the power supply (via transistor T_4) is further amplified equally as the useful signal. Therefore, the useful signal is amplified only A_{d} (the gain of the differential amplifier) times more than the unwanted signal.

The amplifier we have described is practically the most commonly used CMOS OTA circuit. However, it is also characterized by some shortcomings, among which we should mention a small open loop gain at zero frequency, the inability to precisely control the position of high-frequency poles and a relatively small value of the *PSRR*. The use of cascoded stages offers a possibility to eliminate these shortcomings. Since our goal is not to go into details about the design of CMOS OAs, only one of the many possible cascode structures will be discussed here. It will enable a significant increase in circuit gain.

The increase in gain that is achieved by the use of the cascode differential amplifier built into the OA depicted in Fig. 4.5.27, is based on increase of the output resistance of the differential amplifier.

Namely, by the analysis of the differential amplifier built in the OTA depicted in Fig. 4.5.27 it can be shown that the output resistance of the cascode amplifier can be expressed as

$$R_{\mathrm{out}} \approx (\mu_{14} r_{\mathrm{D4}}) || (\mu_{15} r_{\mathrm{D5}}). \tag{4.5.66}$$

This can be a very high output resistance, which makes the differential amplifier much better current adapted to the next stage and thus increases the overall transconductance.

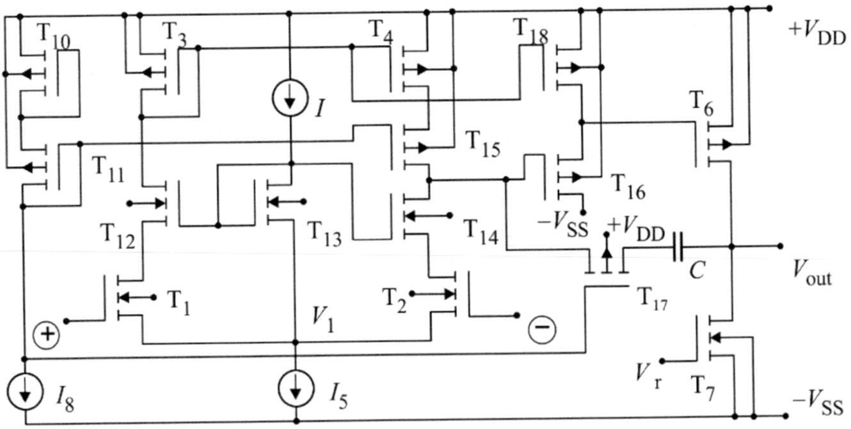

Fig. 4.5.27 Transconductance amplifier using a cascode differential amplifier as a first stage. The substrates of T_1, T_2, T_{12}, T_{13} and T_{14} are connected to $-V_{SS}$

The next stage (transistor T_{16}) allows the adjustment of the DC levels, and the second stage of amplification is the transistor T_6. It should be noted that the compensation is achieved by a series connection of a capacitor and a MOS transistor that serves as a dynamic resistor (T_{17}). This compensation scheme corresponds to the one shown LNAE_Book 2. If the second stage of amplification were performed as a cascode amplifier, an increase in the *PSRR* would be achieved.

The schematic of a complete OA is shown in Fig. 4.5.28. For this circuit we say that in addition to the two-stage OA, it also has an output stage which is realized in the form of a push–pull stage. The basic differential pair consists of transistors T_1 and T_2. The second stage of amplification is the transistor T_7, which is loaded with a large impedance T_{12} and thus produces large DC gain. The capacitance C is used for compensation. The output stage is specific in that T_{18} and T_{19} are used to adjust the DC levels and operate as CD amplifiers, and to avoid phase inversion, the output complementary transistors have swapped places. Thus, T_{22} is a P-channel and T_{21} is an N-channel transistor.

The following values were measured for this amplifier: $A\,(\mathrm{dB}) = 65, f_\mathrm{T} = 60\,\mathrm{MHz}$ at a capacitive load of $C_\mathrm{L} = 1$ pF, maximum amplitude of alternating (dynamics) output voltage 0.65 V, and dynamics of the common mode input voltage 3 V. Since the output transistors can provide relatively high current, it can be expected that this circuit has a relatively large slew rate. Of course, since the output transistors in the push–pull stage work as CD amplifiers the output resistance of the OA is dramatically reduced as compared with any OTA.

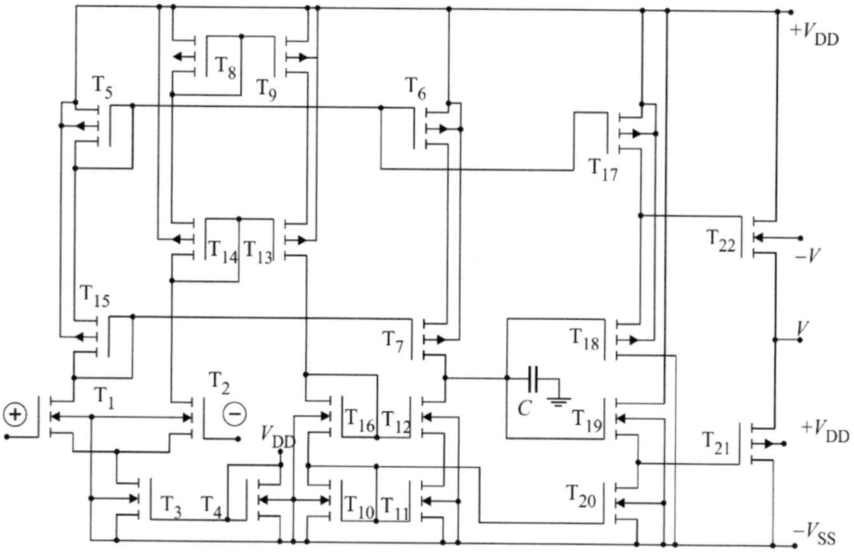

Fig. 4.5.28 Schematic of a CMOS OA exhibiting small output resistance

4.5.4 Realization of the OA Using JFETs

As already mentioned, the differential amplifier with JFETs, and thus the operational amplifier based on it, has significantly smaller voltage gain than the corresponding bipolar amplifier circuit. However, there are features of the JFET that make it suitable for the construction of operational amplifiers with specific properties. Among such features, we would mention the high input resistance, low noise, and the square transfer characteristic of the transistor. Namely, in the push–pull stage in Class A, where the even harmonic is easily neutralized, if the component has a parabolic transfer characteristic, no higher harmonics remain.

These properties of the JFET were used in a discrete operational amplifier depicted in Fig. 4.5.29. The circuit has three gain stages.

First, transistors T_1 and T_2 form a differential pair. T_1, T_2, T_3, T_4, and T_5 are N-channel, while T_6 and T_7 are P-channel JFETs. It should be borne in mind that discrete components with the same designations, as a rule, show a certain difference in the values of the parameters, so when forming a differential pair, it is always necessary to take a larger number of transistors and (by measurements) find two that have close characteristics. T_3 (as well as T_4) is connected as a constant current diode, which is a replacement for a constant current source using a current mirror. Such a solution is not possible when BJT or MOSFET technology is solely implemented. As with

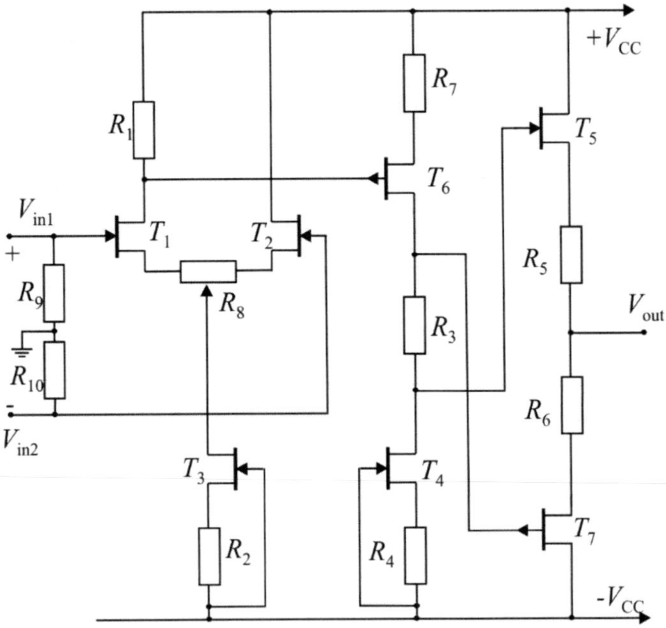

Fig. 4.5.29 Discrete OA using JFETs

the CS amplifier, the gates are held at zero DC potential by means of high-resistance resistors (R_9 and R_{10}).

The output of the differential amplifier is asymmetric. The second amplifier stage is a CS amplifier (T_6) which has a resistor for temperature stabilization R_7 in the source. This resistor also enables the formation of a larger voltage drop across R_1, and thus a larger gain of the differential amplifier.

Since the voltage drop across the R_3 is very small (in principle R_3 can be replaced by a short-circuit) the push–pull stage consisting of T_5 and T_7 is working in Class A, and its voltage gain is less than unity. R_3 reduces the DC component of the current at the quiescent operating point of T_5 and T_7 and thus improves the efficiency by leading the operating point toward Class AB. Here, too, the symmetry of the output stage must be ensured by careful selection of the pair of complementary transistors.

The resistors R_5 and R_6 as well as the potentiometer R_8 degenerate the sources where they are connected and increase the temperature stability of the circuit. R_8, of course, also serves for offset compensation.

4.5.5 Realization of OA Using Mixed Technologies

There is often a need to combine the advantages of individual technologies in order to obtain the most successful solution. The bipolar transistor is characterized by high transconductance, and the JFET by high input resistance and low noise. The MOS transistor has a high noise at low frequencies, and the BJT an input resistance, which in some cases can be unacceptably low. The combination would take advantage of individual technologies. In LNAE_Book 2 it is shown, however, that this is not an easy task, especially given the completely different isolation procedures of the active components on the chip. Therefore, that is done by developing a circuit that is mostly realized in single technology, and a smaller part of it is created by using some specific structures that can be implemented in the same technology, but with some additional interventions perform as if they are in some other technology. Here we will consider three circuits that are specific to use two types of transistors.

Figure 4.5.30 shows the input stage of an integrated OA that is manufactured in CMOS technology, but the input circuit is realized as a differential amplifier with Darlington pairs of BJTs. The PNP BJTs are realized as lateral in the N-well which is connected to the lowest potential in the circuit (V_{SS}). In addition, a conductive (polysilicon) electrode is placed over the collector junction of the BJT. It is physically realized like any gate of a MOS transistor. Its role is to deplete the surface at the junction, to reduce the surface current, and thus to reduce the $1/f$ noise of the PNP transistor.

At a nominal gain of 56 dB, this OA exhibits a cut-off frequency of 4.5 MHz, an input current of 15 μA, a supply current of 2.1 mA, and the amplitude of the noise voltage spectrum at medium frequencies of 7.3 nV/(Hz)$^{1/2}$ with f_L (as defined in LNAE_Book 5) for the noise voltage of only 19.3 Hz. Typically, the amplitude of the noise current spectrum at room temperature is 203 fA/(Hz)$^{1/2}$, with $f_L = 22.3$ Hz.

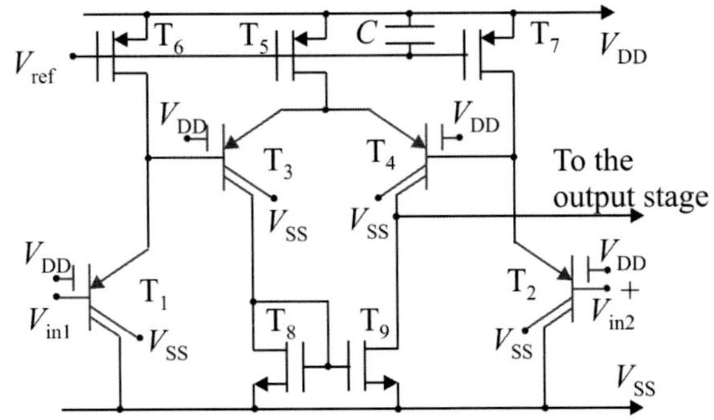

Fig. 4.5.30 Input stage of a CMOS OA using lateral PNP BJTs

The *CMRR* of this OA is about 100 dB, and the *PSRR* of the positive supply voltage is about 70 dB. Finally, the circuit exhibits a slew rate of about 40 V/μs.

In the previous application, structures that are present in the given technology and which, sometimes, are even considered parasitic, were used. The circuit of Fig. 4.5.31 is similar. An OA is realized, which uses bipolar technology, but improved so that the input circuit is now realized with the help of JFETs. In this way, the input resistance of the circuit was dramatically increased to 10^{12} Ω. The input polarization current is only 1 nA. By other measures undertaken within the bipolar part, a large linearity of the circuit was achieved, so that the *THD* is 0.01% only.

Another JFET was used in this circuit, which is connected as a constant current diode and is located in the reference voltage source in the part of the circuit which is not shown.

The key effect achieved, however, is to reduce the noise at low frequencies. As can be seen from Fig. 4.5.32 the equivalent noise voltage is reduced to about 23 nV/(Hz)$^{1/2}$ with f_L below, the almost unbelievable, 0.1 Hz.

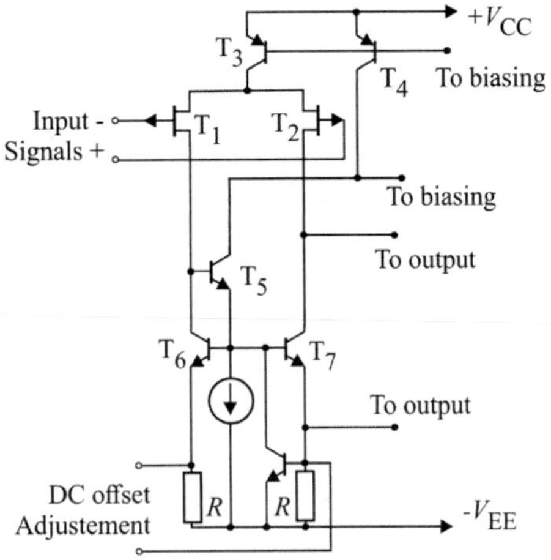

Fig. 4.5.31 OA with a differential pair at the input implementing JFETs

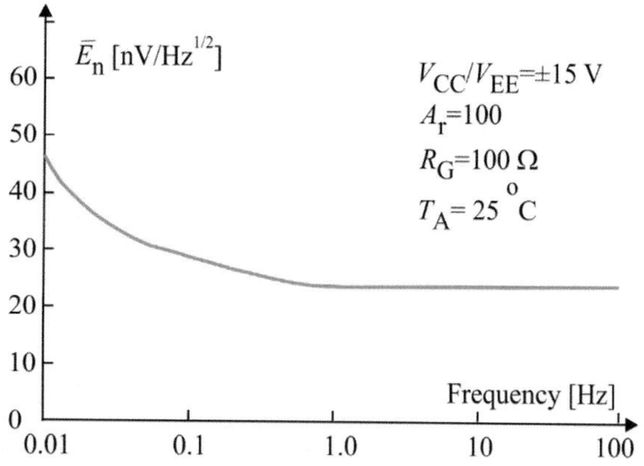

Fig. 4.5.32 Equivalent noise voltage of the circuit of Fig. 4.5.31

The following example will be used to show that sometimes "new" technology is literally generated in order to incorporate the specific properties of individual components. Namely, in recent times, starting from the ideas created by the application of parasitic structures in CMOS technology, a new technology has emerged that allows simultaneous use of NPN BJT and CMOS transistors.

This eliminates the difficulties associated with the use of PNP BJT. This is how BiCMOS technology was created. The example of an operational amplifier in BiCMOS technology will be used to describe a specific circuit that is characterized by low input and low output resistance. Such circuits are called transimpedance amplifiers. Of course, the input resistance is reduced by introducing parallel feedback. The amplifier circuit is shown in Fig. 4.5.33.

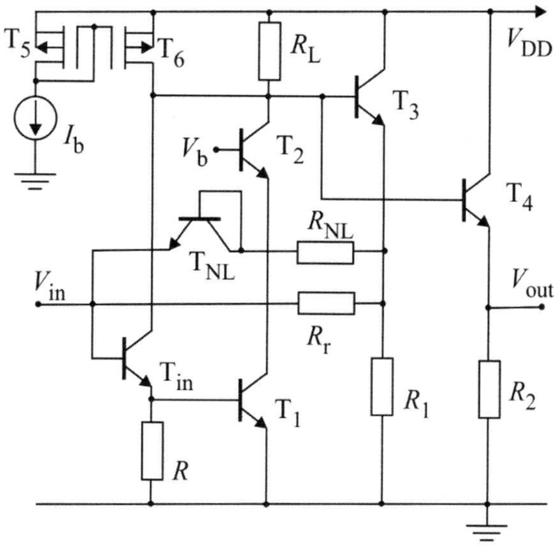

Fig. 4.5.33 Transimpedance amplifier in BiCMOS technology

The basic amplifier is single-stage and consists of a Darlington pair (T_u and T_1) which is implemented as CE amplifier. This reduces the base current and thus the noise. The load of this pair contains the cascode transistor T_2 and the load resistance R_L in parallel with the output resistance of the MOS current mirror (transistors T_5 and T_6). The additional direct current obtained from this current mirror increases the transconductance of the Darlington pair and thus increases the loop gain of the circuit.

The output CC stage (T_4—R_2) is outside the feedback loop and provides the loading capacity of the amplifier, i.e., a small output resistance. The feedback circuit consists of three parts. First, we have an isolation amplifier made of the transistor T_3 and the resistor R_1. Then it is R_r that transfers the signal in the inverse direction and, finally, it is the nonlinear branch R_{NL}–T_{NL} which is activated at large signals and increases the transfer coefficient of the feedback circuit so improving the linearity of the circuit.

4.6 Analog Computations

Abstract This chapter is devoted to the implementation of the OA mostly in linear and partly in nonlinear analog computation. We will start with simple inverter and end with electronic filters going via logarithmic and other nonlinear applications.

4.6.1 Introduction

The OA is a basic integrated circuit in linear analog electronics. It is characterized by very high gain, high input, and low output impedance. It is usually realized as an integrated circuit on a silicon wafer—a chip. One of the applications of the OA is realization of electronic circuits that perform mathematical operations on electrical signals. This is how the name OA came about. It is understandable that there are several different types of OAs intended for different applications.

In this chapter, the ideal and realistic OA will be defined first, and their applications will be considered. These applications will apply to linear analog circuits. It should be mentioned that the OA has significant applications in both pulse and digital electronics, which will not be considered here.

4.6.2 The Ideal OA

Figure 4.6.1a is a block diagram of an OA. Here, his idealized version will be considered first, for which the above block diagram, being the most general, also applies. The model of an ideal OA is shown in Fig. 4.6.1b. It amplifies the difference between the input signals; i.e., the output voltage is:

$$V_{\text{out}} = A \cdot (V_2 - V_1), \tag{4.6.1}$$

where it is assumed that A is a positive number.

Unless otherwise indicated, the ideal OA has infinite gain, i.e., $A \to \infty$. Infinite gain does not mean infinite output signal. It should be understood that when V_{out}

V. B. Litovski, *Lecture Notes in Analog Electronics*, Lecture Notes in Electrical Engineering 958, https://doi.org/10.1007/978-981-19-6528-9_6

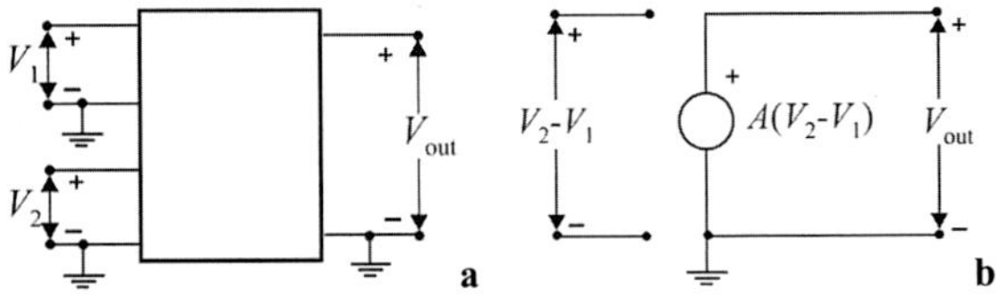

Fig. 4.6.1 **a** Block diagram of the OA and **b** model of the ideal OA

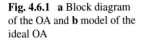

is infinite, the difference between the input signals becomes infinitively small or $(V_2 - V_1) \rightarrow 0$. The input resistance is infinite, $R_{in} \rightarrow \infty$, i.e., no current flows in the input terminals. The output resistance is equal to zero, $R_{out} \rightarrow 0$. The relation (4.6.1) is valid regardless of the size of the difference $(V_2 - V_1)$; i.e., the range of linearity of the transfer characteristic of an ideal OA is not limited. And, finally, the bandwidth of the ideal OA is infinite; i.e., the amplitude and phase characteristics do not depend on the frequency. Such an ideal OA can be represented by the model as in Fig. 4.6.1b. Each of the input terminals can be grounded independently.

More will be said later about the features of real OAs. Here we will only indicate some of the most important features of the same in order to better understand the idealizations that were made when defining the ideal OA. The real OA has finite input and output impedances and a finite gain. This is a consequence of its practical realization. The input impedance will, truth be told, be very large, depending on the type of differential amplifier at the input of the OA, so that in most applications it can be considered infinite. The output impedance is of the order of several hundred ohms. If negative feedback is applied to the OA, which is always the case, it can be extremely reduced, depending on the size of the feedback.

The gain of the OA varies from type to type, but the average values (for bipolar circuits) could be considered to be of the order of 10^5 or 100 dB, respectively. In Fig. 4.6.2, a simplified model of a real OA is presented.

The real OA, like any amplifier, is not linear; i.e., the output voltage is a nonlinear function of the input. Namely, at large input signals, the transfer characteristic of the amplifier is curved (exhibits saturation) so that the value of the gain decreases. However, there is always one interval of the input voltage value when the characteristic can be considered linear. The transfer characteristic of the OA can be roughly represented as in Fig. 4.6.3. When considering this characteristic, special attention should be paid to the difference in the scale used to represent the input and output voltages.

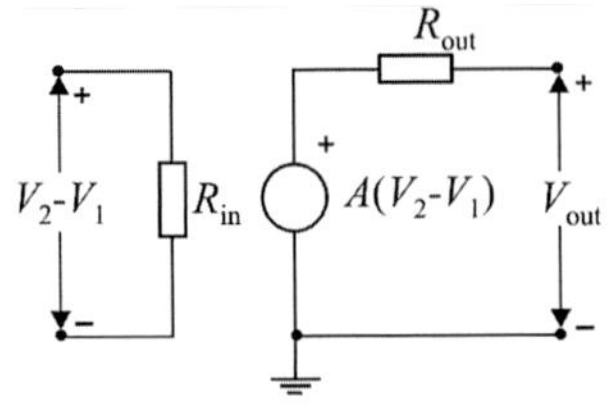

Fig. 4.6.2 Simplified model of the real OA

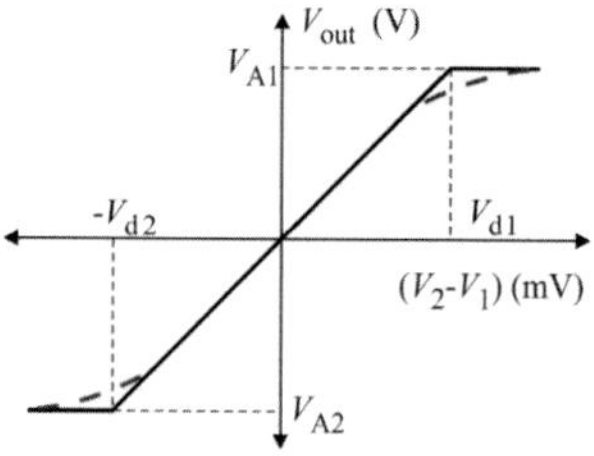

Fig. 4.6.3 Transfer characteristic of an OA

The voltages V_{A1} and V_{A2} are those values for which linearity is lost and when the output voltage is in saturation they determine the dynamics of the output voltage. Their values depend on the components used and the value of the voltage of the power supply. Typical values of these voltages are from 5 to 15 V. Most often, the OA is symmetrical, so $V_{A1} = V_{A2}$.

The range of values of input voltages at which the linearity is maintained is very small, which is understandable given the high gain. With a symmetric OA $V_{d1} = V_{d2} = V_d$, this value is always less than a few millivolts.

The simplified model of the real OA of Fig. 4.6.2 is used to obtain basic indicators in individual applications. Even the input resistance is often considered to be infinite. However, the practical performance of the OA leads to the occurrence of other effects such as the finite *CMRR* so that the output voltage also depends on the average of the input voltages; presence of voltage and current offset, very small bandwidth, influence of the variations of the power supply, etc. We will talk about these characteristics and their influence later.

The OA block of Fig. 4.6.1a is most often schematically represented as in Fig. 4.6.4. The difference in representation in the two-variant depicted here is in the absence of ground connections, which are frequently omitted for simplicity but are implied.

The input terminal marked with "−" is called inverting input, because when $V_2 = 0$ (tied to zero)

$$V_{out} = -A \cdot V_1, \tag{4.6.2}$$

so the input and output voltages are in the counter phase.

Similarly, when $V_1 = 0$

$$V_{out} = A \cdot V_2 \tag{4.6.3}$$

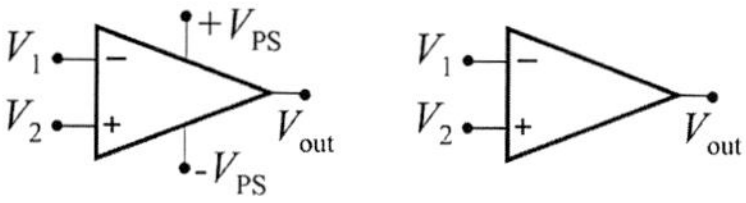

Fig. 4.6.4 Schematic symbols for OA. V_1 and V_2 are excitation signals, V_{PS} is the power supply voltage while V_{out} is the output voltage

so the input and output voltages are in phase. The input terminal marked with a "+" is called a non-inverting input.

4.6.3 Basic Applications of the OAs

As already pointed out, the applications of OAs in linear analog circuits will be described here. Given the large number of applications, only the most basic ones will be considered. Some others will be discussed later, for example with oscillators and AC-to-DC converters.

The OA is always used with external additional components, most often passive. Of these additional external circuits, the one that produces negative feedback from the output of the OA to its input is mandatory. Therefore, the number A is most often called open loop gain. In the case of passive feedback circuits, the inverting input is used to connect the feedback signal. The main reason for the introduction of a negative feedback is the large tolerances that occur in the production of OAs. For example, if a type of OA has a nominal gain of 10,000, during production, under the same conditions, amplifiers whose gain varies from 5000 to 15,000 will be obtained. The reason for this is the great complexity of the OA circuit and the impossibility to keep all technological parameters strictly constant during production. Subsequent selection would take a long time and raise the price of the component. LNAE_Book 2 shows that these problems can be successfully solved by applying negative feedback. The same applies to the reduction of the sensitivity of the gain to changes in the voltage of the power supply, changes in temperature and aging of components (instability of the gain, etc.). In addition, the negative feedback increases the bandwidth of the OA.

4.6.3.1 Inverting Amplifier

One of the most commonly used circuits with an OA called an inverting amplifier, or just an inverter, is shown in Fig. 4.6.5.

An inverting input terminal is used, so the output and input voltages will be counter-phase. Hence the name of this amplifier–inverter. The second input, the non-inverting one, is connected to ground, i.e., $V_2 = 0$. The resistors R_1 and R_2 form a negative feedback circuit. Let us first look at the gain of the inverter as

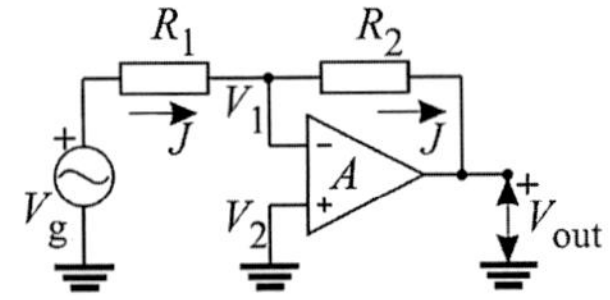

Fig. 4.6.5 Inverting amplifier

$$A_v = V_{out}/V_g, \tag{4.6.4}$$

in the case when the OA is an ideal voltage amplifier ($R_{in} = 0$ and $R_{out} = \infty$) with finite gain A.

Since $V_2 = 0$ the circuit is described by the following system

$$\frac{V_1 - V_g}{R_1} + \frac{V_1 - V_{out}}{R_2} = 0. \tag{4.6.5}$$

$$V_{out} = -A \cdot V_1 \tag{4.6.6}$$

where the model of Fig. 4.6.1b was used.

The system of Eqs. (4.6.5) and (4.6.6) is a complete set of equations for determining the node voltages in this circuit. Since the inverter is a representative of a very wide class of linear electronic circuits that contain one or more operational amplifiers, it is extremely important that the reader notices the procedure for formulating equations for this circuit. Namely, despite the fact that the circuit of Fig. 4.6.5 contains three nodes, only one equation is written according to Kirchhoff's current law and that is Eq. (4.6.5). No equation was written for the input node because the voltage at that node is known and is V_g. No equation was written for the output node for the same reason, although this is not explicitly visible. Namely, an ideal voltage source is connected to the output node, the current of which is determined by the external circuit, and the voltage value is determined by the controlling voltage V_1. If V_1 is known the output voltage is known too. Therefore, two unknowns remain that are related by Kirchhoff's relation for the input node of the operational amplifier (4.6.5) and the equation that represents the model of the operational amplifier (4.6.6). This way of equation formulation will be followed in the following text.

By eliminating the voltage V_1 from the last two equations, we obtained

$$\frac{R_2}{R_1} V_g = -V_{out} \left[1 + \frac{1}{A}(1 + R_2/R_1) \right]. \tag{4.6.7}$$

Due to the large gain of the OA, A, and considering practical values for resistances R_2 and R_1, the following will apply:

$$\frac{1}{A}(1 + R_2/R_1) \ll 1. \tag{4.6.8}$$

Using this inequality, for an inverter in which an ideal OA with infinite gain is built, one gets

$$V_{out} = -\frac{R_2}{R_1} V_g. \tag{4.6.9}$$

Fig. **4.6.6** Equivalent circuit of the inverter with $A \to \infty$

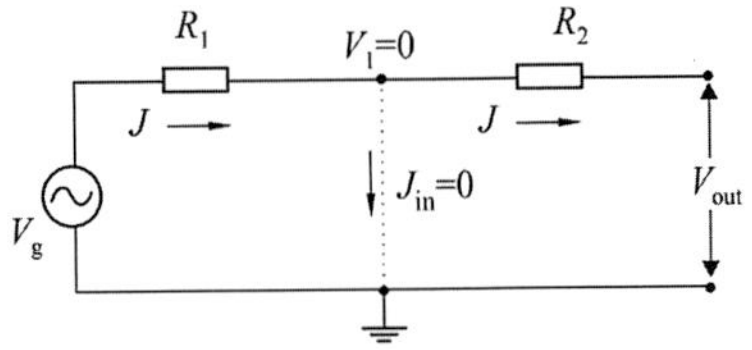

Thus, the gain of the inverter for the case when the gain of the OA is infinite becomes

$$A_v = -R_2/R_1. \tag{4.6.10a}$$

The gain depends only on the resistance ratio R_2/R_1, and not on the gain of the basic OA. This result is expected because it is an amplifier with large negative feedback. This is valid only under the condition that A is large enough, which makes the loop gain large. Since A decreases with increasing frequency, the importance of expression (4.6.10) is limited to low frequencies. Also A_v is considerably less than A. It follows from this

$$V_1 \ll V_g. \tag{4.6.11}$$

As the gain of the OA, A, increases, the voltage V_1 decreases and approaches zero. In case $V_1 = 0$, the inverting and non-inverting inputs are at the same potential. Since the non-inverting input is at the potential of the ground, it is said that the inverting input is at the "virtual ground." In this way the inverter of Fig. 4.6.5 can be represented by the equivalent circuit given in Fig. 4.6.6.

In the equivalent circuit, the input of the operational amplifier is short-circuited, only no current flows through this short-circuit.

When formulating equations for circuits in which the input voltage is on the virtual ground, one unknown variable is eliminated (in this case V_1), so it remains to write only the equation for the input node of the operational amplifier:

$$-V_g/R_1 - V_{out}/R_2 = 0, \tag{4.6.10b}$$

wherefrom the gain given by (4.6.10a) comes.

The derived expressions are approximate. It is of interest, given the large changes in the gain of the OA, to establish the impact of these changes on the changes in the gain of the inverter A_v.

From the expression given in (4.6.7) we get:

$$A_v = \frac{V_{out}}{V_g} = -\frac{R_2/R_1}{1 + (1 + R_2/R_1)/A}. \tag{4.6.12}$$

Before we look at the effect of gain variations, consider the following example.

Table 4.6.1 Gain of the inverter

A	10^5	10^3	100	10	1
A_v	9.999	9.891	9.009	4.762	0.833

Example 4.6.3.1 Calculate the gain of the circuit from Fig. 4.6.5 in which an ideal operational amplifier with infinite gain and $R_1 = 2$ and $R_2 = 20$ is used. Then determine the inverter's gain if the gain of the OA is finite and gets the values: (a) $A = 10^5$, (b) $A = 10^3$, (c) $A = 100$, (d) $A = 10$ and (e) $A = 1$.

Solution From (4.6.10), when $A \to \infty$, we get $A_V = -R_2/R_1 = -10$.

When the gain of the OA is finite, the values from Table 4.6.1 are obtained.

By analyzing the obtained results, we conclude that due to the very large loop gain of the OA in this circuit, the decrcasc in gain becomes influential only when its value decreases hundred times (40 dB).

Let us now consider a more general case. Let A change to $A + \Delta A$. Then the gain of the inverter will change and will be:

$$A_v + \Delta A_v = -\frac{R_2/R_1}{1 + \frac{1+R_2/R_1}{A+\Delta A}}. \tag{4.6.13}$$

Elimination of the gain A_v from the last two equations gives

$$\Delta A_v = \frac{-\frac{R_2}{R_1} \cdot \frac{1}{\beta} \cdot \frac{\Delta A}{A(A+\Delta A)}}{\left[1 + \frac{1}{A} \cdot \frac{1}{\beta}\right]\left[1 + \frac{1}{A+\Delta A} \cdot \frac{1}{\beta}\right]}. \tag{4.6.14}$$

where $\beta = (R_1 + R_2)/R_1$

If the gain A is large enough and ΔA not too large, the following will apply:

$$\frac{1}{A + \Delta A}(1 + R_2/R_1) \ll 1 \tag{4.6.15a}$$

and

$$\frac{1}{A}(1 + R_2/R_1) \ll 1 \tag{4.6.15b}$$

If we implement this inequalities for Eq. (4.6.14) and if we put $R_2/R_1 = -A_{vo}$ we get

$$\frac{\Delta A_v}{A_{v0}} = \frac{\Delta A}{A} \frac{1 - A_{v0}}{A(1 + \Delta A/A)}. \tag{4.6.16}$$

Based on this, a relationship can be determined as

$$\frac{A_v/A_{v0}}{\Delta A/A} = \frac{1 - A_{v0}}{A\left(1 + \frac{\Delta A}{A}\right)} = \frac{1 + R_2/R_1}{A\left(1 + \frac{\Delta A}{A}\right)} \quad (4.6.17)$$

which is called logarithmic sensitivity and shows the relationship between the relative changes of the inverter's gain and the gain of the OA.

Example 4.6.3.2 An OA is given, the gain of which varies by 50%. It is necessary to make an inverter whose nominal gain is $A_{v0} = -100$ with an acceptable variation of its gain up to 1%. Determine the minimum acceptable value of the gain of the OA.

Solution From the given values we calculate first: $\Delta A_v / A_{v0} = -0.01$ and $\Delta A/A = -0.5$. Here, too, we should keep in mind that A is a positive number and that it is the largest possible value of the gain. Therefore, ΔA must be negative. On the other hand, $A_{v0} = -R_2/R_1$ is a negative number. Decreasing the gain A results in a decrease in the absolute value of A_v, which means that ΔA_v is a positive number. Therefore, both relative increments are negative. Now from (4.6.17) we easily get $A = 3367$.

4.6.3.1.1 Input Impedance of an Inverter

The input impedance of the inverter "seen" by the source is

$$Z_{in} = V_g/J. \quad (4.6.18)$$

If we apply the equivalent circuit of Fig. 4.6.6 we get

$$Z_{in} = R_1. \quad (4.6.19)$$

This allows the resistor R_1 to be determined according to the nature of the source at a given desired inverter gain value (A_v). For example, if one wants $R_{in} = R_1 = 1$ kΩ and the gain $A_v = -100$, one gets the value $R_2 = 100$ kΩ.

The calculation of the input impedance assumed that the input of the inverter was at virtual ground, i.e., that the input impedance of the OA was infinite. We will show now that the finite input impedance of the OA will not affect the input impedance of the inverter.

Figure 4.6.7 shows the part of the inverter behind resistor R_1. The impedance Z_1 represents the input impedance of the rest of the circuit. Suppose first that $R_{in} \to \infty$. The input impedance Z_1 is:

$$Z_1 = V_1/J \quad (4.6.20a)$$

where

$$J = \left(V_g - V_1\right)/R_1 = (V_1 - V_{out})/R_2 \quad (4.6.20b)$$

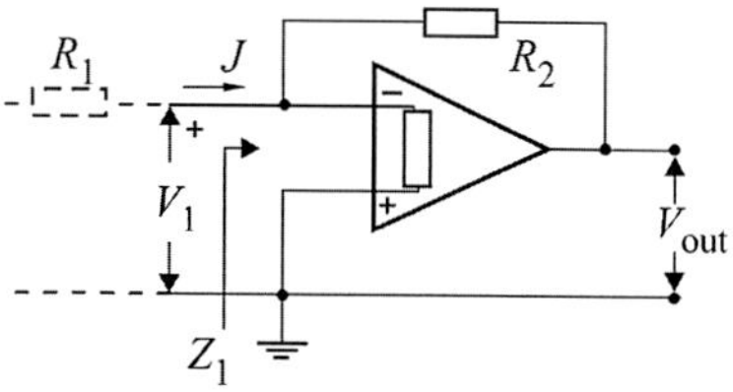

Fig. 4.6.7 Circuit for evaluation of the input impedance of a part of the inverter

After substitution of J one gets

$$Z_1 = R_2/(1 - V_{out}/V_1) = R_2/(1 - A) \tag{4.6.21}$$

This expression would otherwise be obtained by direct application of Miller's theorem.

Sine A is a large number, Z_1 is small. For example, for the amplifier with $R_2 =$ 100 kΩ and $A = 10{,}000$, one gets $Z_1 \approx 10\ \Omega$.

When $R_{in} \neq \infty$, we consider its parallel connection with Z_1. Since it is always valid that $R_{in} \gg Z_1$, it has practically no effect on the value of the input resistance behind the resistor R_1. As for the input resistance of the inverter: $Z_{in} = Z_1 + R_1$, it remains practically equal to R_1 due to the small value of Z_1.

To calculate the output impedance of the inverter, we will use the circuit of Fig. 4.6.8. In this circuit, the input source is short-circuited, and the model of the operational amplifier of Fig. 4.6.2 is applied. It is assumed that the input impedance of the OA is infinite. The source J_o is connected to the output of the amplifier, and the voltage on it defines the output resistance

$$Z_o = V_o/J_o. \tag{4.6.22}$$

From Fig. 4.6.8 we have

$$V_1/R_1 + (V_1 - V_o)/R_2 = 0 \tag{4.6.23}$$

$$(V_o - V_1)/R_2 + (V_o + A \cdot V_1)/R_{out} = J_o. \tag{4.6.24}$$

By elimination of V_1 we get

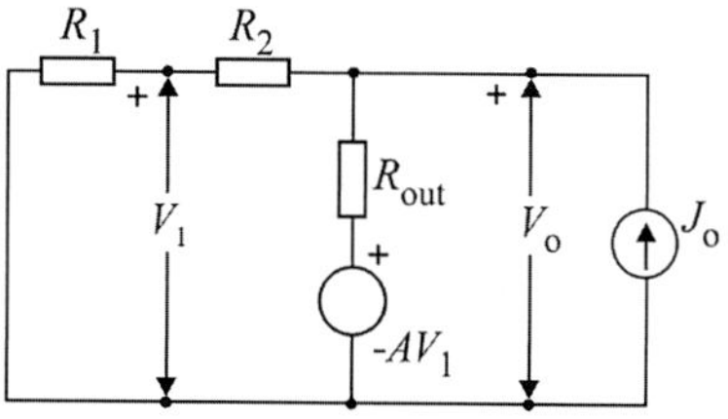

Fig. 4.6.8 Inverter's output impedance calculation circuit

$$Z_o = R_{out} \Big/ \left(1 + \frac{R_{out}}{R_1 + R_2} + \frac{A \cdot R_1}{R_1 + R_2}\right). \quad (4.6.25)$$

Since the output resistance of the OA, R_{out}, is small (of the order of several hundred ohms), the following will apply:

$$\frac{R_{out}}{R_1 + R_2} + 1 \ll \frac{A \cdot R_1}{R_1 + R_2}. \quad (4.6.26)$$

Based on this inequality (4.6.22) becomes

$$Z_o = R_{out} \frac{1 + R_2/R_1}{A}. \quad (4.6.27)$$

If $A \gg (1 + R_2/R_1)$ the output resistance of the inverter is significantly reduced compared to the output resistance of the OA. For the example above it is $A = 10{,}000$, $R_1 = 1\ \text{k}\Omega$, $R_2 = 100\ \text{k}\Omega$ and, even for $R_{out} = 500\ \Omega$, for the output impedance of the inverter one gets $Z_o = 5\ \Omega$. For $A = 10^5$, one gets $Z_o = 500\ \text{m}\Omega$.

4.6.3.2 Non-inverting Amplifier

In a non-inverter amplifier, the excitation signal is fed to the non-inverting input so that the output and input signals are in phase. At the same time, for the reasons already mentioned, the circuit of negative feedback was maintained. So, the non-inverting amplifier is shown in Fig. 4.6.9.

For the circuit of Fig. 4.6.9 the following system of equations can be written

$$V_{out} = A \cdot (V_2 - V_1) \quad (4.6.28)$$

$$V_2 = V_g \quad (4.6.29)$$

$$V_1/R_1 + (V_1 - V_{out})/R_2 = 0. \quad (4.6.30)$$

Fig. 4.6.9 Non-inverting amplifier

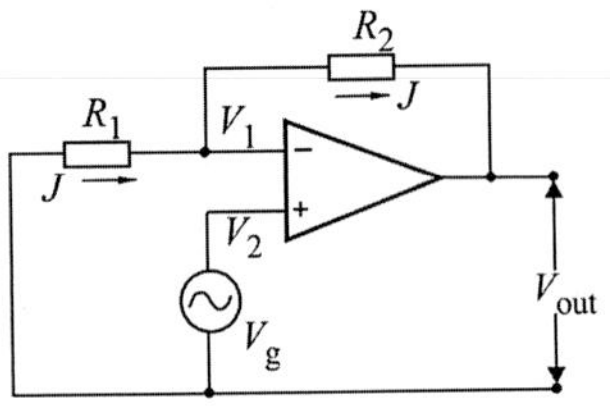

Equation (4.6.28) represents the model of the OA, (4.6.29) is the loop equation that is imposed here since ideal voltage source at the input is used, and (4.6.30) is the node equation written for the inverting input.

These equations give the gain as

$$A_v = V_{out}/V_g = A/[1 + A \cdot R_1/(R_1 + R_2)]. \tag{4.6.31}$$

If A is a large number the following is valid

$$A \cdot R_1/(R_1 + R_2) \gg 1 \tag{4.6.32}$$

so the approximate value of the gain of a non-inverter amplifier is

$$A_v = 1 + R_2/R_1. \tag{4.6.33}$$

The same result would be obtained if a large gain, A, were assumed in advance, which means $V_1 = V_2 = V_g$. The following node equation would be used

$$V_g/R_1 + (V_g - V_{out})/R_2 = 0. \tag{4.6.34}$$

In any case, for a sufficiently large gain of the OA, the gain of the non-inverting amplifier does not depend on A. It is always larger than unity.

The input impedance of a non-inverter amplifier is equal to the input impedance of the OA.

The same calculation procedure as for the inverter for the sensitivity of the non-inverting amplifier would give

$$\frac{\Delta A_v}{A_{v0}} = \frac{\Delta A}{A} \frac{1 + R_2/R_1}{A_{v0} + A(1 + \Delta A/A)}. \tag{4.6.35}$$

Having in mind that $A(1 + \Delta A/A) \gg A_{v0}$ the relative sensitivity of the gain of both amplifiers (inverting and non-inverting) are approximately equal if their gains are equal in modulus.

It is possible to realize a non-inverting amplifier by putting $R_1 = \infty$ and $R_2 = 0$. That would be a unity gain voltage amplifier or isolation amplifier ($A_V = 1$, $R_{in} = \infty\,\Omega$, and $R_o = 0\,\Omega$). It is depicted in Fig. 4.6.10.

By solving equation

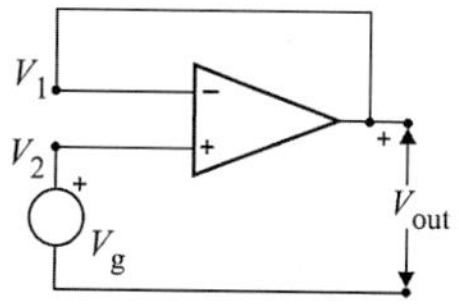

Fig. 4.6.10 Unity gain (isolation) amplifier

$$V_{out} = A \cdot (V_2 - V_1) = A \cdot \left(V_g - V_{out}\right) \tag{4.6.36}$$

we get the gain

$$A_v = \frac{A}{1 + A}. \tag{4.6.37}$$

If $A \gg 1$ the gain of the circuit of Fig. 4.6.10 becomes

$$A_v = 1 \tag{4.6.38}$$

The input impedance of this amplifier is equal to the input impedance of the OA. The output impedance for $A \to \infty$ would be zero. The practical values for the output impedance R_o are of the order of ten parts ohm.

The unity gain voltage amplifier has a very high input resistance and a very low output resistance and is used as an isolating amplifier between a high output resistance source (previous amplification stage) and a low resistance load.

4.6.3.3 Differential Balanced Amplifier

The OA has essentially the same role as the differential amplifier since its output voltage is equal to the amplified difference of the input voltages. Due to the extremely high gain, however, it cannot be used without negative feedback. At the same time, the configuration of the external circuit should continue to ensure the dependence of the output signal only on the difference of the input ones. One such circuit, called a differential balanced amplifier, is shown in Fig. 4.6.11.

For the circuit with Fig. 4.6.11 the following system of equations can be set

$$\frac{V_{g1} - V_1}{R_1} = \frac{V_1 - V_{out}}{R_2} \tag{4.6.39}$$

$$\left(V_2 - V_{g2}\right) / R_3 + V_2/R_4 = 0. \tag{4.6.40}$$

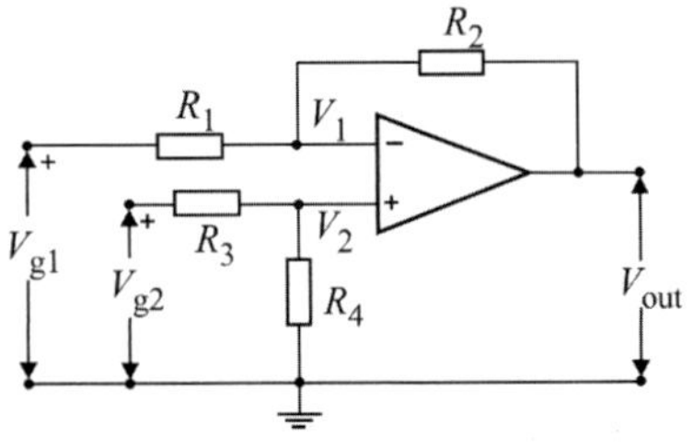

Fig. 4.6.11 Differential balanced amplifier

$$V_{\text{out}} = A \cdot (V_2 - V_1), \tag{4.6.41}$$

where it is taken that A represents only the differential gain of the OA, i.e., that the common mode gain is equal to zero. This assumption is usually justified.

From these equations we get

$$\frac{V_{g1}}{R_1} - \frac{V_{g2}}{R_1}\left(\frac{1 + R_1/R_2}{1 + R_3/R_4}\right) = -\frac{V_{\text{out}}}{R_2}\left[1 + \frac{1}{A}\left(1 + \frac{R_2}{R_1}\right)\right]. \tag{4.6.42a}$$

Having in mind that

$$\frac{1}{A}\left(1 + \frac{R_2}{R_1}\right) \ll 1, \tag{4.6.43}$$

(4.6.42a) becomes

$$\frac{V_{g1}}{R_1} - \frac{V_{g2}}{R_1}\left(\frac{1 + R_1/R_2}{1 + R_3/R_4}\right) = -\frac{V_{\text{out}}}{R_2}. \tag{4.6.42b}$$

If the resistances R_1, R_2, R_3, and R_4 are set so that the expression in parentheses of the above equation is equal to unity, or, if the condition

$$R_1 R_4 = R_2 R_3, \tag{4.6.44}$$

is met, one gets

$$V_{\text{out}} = \frac{R_2}{R_1}\left(V_{g2} - V_{g1}\right). \tag{4.6.42c}$$

This fulfills the requirement that only the difference of the input signals is amplified and that the gain can be controlled. The reader is recommended to repeat the analysis of this circuit in case the open loop gain is infinite by putting $V_1 = V_2$ instead of (4.6.41).

Let us now consider the case when a precise differential amplifier is considered and when the influence of the common mode gain of the OA cannot be ignored, i.e., when the *CMRR* of the OA is not infinite. Of interest are the values of the differential gain and the common mode gain as well as the *CMRR* of the differential balanced amplifier. To this end, we write for the output voltage

$$V_{\text{out}} = A_{\text{d}}(V_2 - V_1) + A_{\text{ds}}(V_1 + V_2)/2, \tag{4.6.45a}$$

where A_{d} and A_{ds} are the differential and the common mode gain of the OA, respectively, and it is still assumed that its output resistance is negligibly small. On the other hand for a differential balanced amplifier we can introduce

$$V_{\text{out}} = A_{\text{db}}\big(V_{\text{g2}} - V_{\text{g1}}\big) + A_{\text{sb}}\big(V_{\text{g1}} + V_{\text{g2}}\big)\big/2, \tag{4.6.45b}$$

where, now, A_{db} and A_{sb} are the differential and common mode gain of the differential balanced amplifier. Having in mind the new expressions for the output voltage, by analysis of the circuit from Fig. 4.6.11, the following expressions for the required gains are obtained

$$A_{\text{db}} = \frac{\alpha(A_{\text{d}} - A_{\text{ds}}/2) + \gamma(A_{\text{d}} + A_{\text{ds}}/2)}{2 \cdot [1 + \beta(A_{\text{d}} - A_{\text{ds}}/2)]} \tag{4.6.46a}$$

and

$$A_{\text{sb}} = \frac{\alpha(-A_{\text{d}} + A_{\text{ds}}/2) + \gamma(A_{\text{d}} + A_{\text{ds}}/2)}{1 + \beta(A_{\text{d}} - A_{\text{ds}}/2)} \tag{4.6.46b}$$

where $\alpha = R_2\big/(R_1 + R_2)$, $\beta = R_1\big/(R_1 + R_2)$ and $\gamma = R_4/(R_3 + R_4)$.

Now, for the common mode rejection ratio of the differential balanced amplifier we have

$$\rho_b = \frac{A_{\text{db}}}{A_{\text{sb}}} = \frac{1}{2}\frac{A_{\text{d}}(\gamma + \alpha) + A_{\text{ds}}(\gamma - \alpha)\big/2}{A_{\text{d}}(\gamma - \alpha) + A_{\text{ds}}(\gamma + \alpha)\big/2} \tag{4.6.47a}$$

This expresses at the same time the influence of the OA and the mismatch of the resistors on the *CMRR* of the whole circuit. If we now again assume that the *CMRR* of the OA ($\rho = A_{\text{d}}/A_{\text{ds}}$) is infinite, i.e., that $A_{\text{ds}} = 0$, we can estimate the effect of resistor's asymmetry on the *CMRR* of the differential balanced amplifier (ρ_b) as

$$\rho_r = \frac{1}{2}\frac{2R_2R_4 + R_2R_3 + R_1R_4}{R_1R_4 - R_2R_3} \tag{4.6.47b}$$

where ρ_r is the requested *CMRR* under the condition $\rho \to \infty$. Now, for ρ_b we have

$$\rho_b = \frac{\rho_r\rho + 1/4}{\rho_r + \rho} \tag{4.6.47c}$$

or, if $\rho_r \cdot \rho \gg 1/4$,

$$\frac{1}{\rho_b} \approx \frac{1}{\rho_r} + \frac{1}{\rho}. \tag{4.6.47d}$$

This result suggests that if we want to make a differential amplifier with a large *CMRR*, we should simultaneously use an amplifier that has a high *CMRR* on his own, and resistors with very low tolerances to keep ρ_r big enough. Bearing in mind that ρ is frequency dependent ρ_b will follow its frequency dependence when the *CMRR* of the OA becomes dominant in relation to ρ_r.

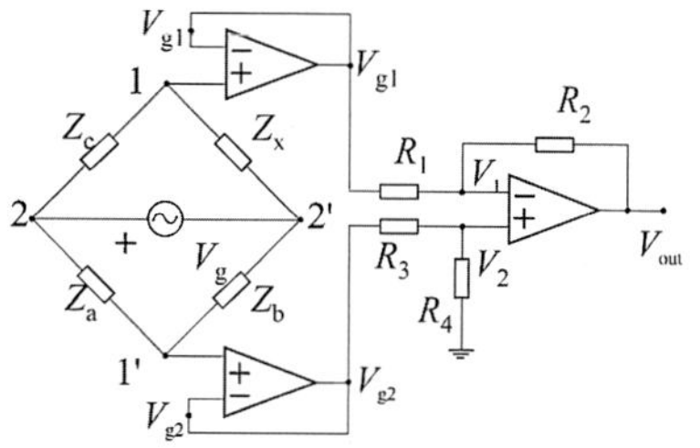

Fig. 4.6.12 One application of the differential balanced amplifier

The differential balanced amplifier is often used in electronic measurement circuits. That is why it is frequently named instrumentation amplifier. One such application used to measure unknown impedances is shown in Fig. 4.6.12.

The circuit consists of three stages. First we have a bridge which is a common circuit for measuring impedance. The problem with it is that the measuring instrument loads the circuit, and at the same time as we approach equilibrium, the value of the signal representing the mismatch of the bridge becomes smaller and more difficult to measure. By installing two isolation amplifiers, the bridge is separated from the measuring circuit, and by installing an instrumentation amplifier, the error signal is amplified and in this way the circuit becomes much more sensitive. It is also ensured that one end of the sources may be connected to ground without affecting the measurement. In the measuring bridge, the unknown impedance Z_x is determined from

$$Z_x = \frac{Z_b}{Z_a} Z_c, \tag{4.6.48}$$

if the other impedances are adjusted so that the voltage between the points $1\text{–}1'$ is equal to zero, i.e., $V_{g1} = V_{g2}$. A much more precise determination of the moment of equalization of these two voltages is provided by a differential balanced amplifier thanks to its gain.

4.6.3.4 Voltage-to-Current Converter

There are applications, when it is necessary to obtain current proportional to the voltage of the source. One such circuit is shown in Fig. 4.6.13 which is in fact a non-inverting amplifier with the impedance Z being used in place of the resistor R_2.

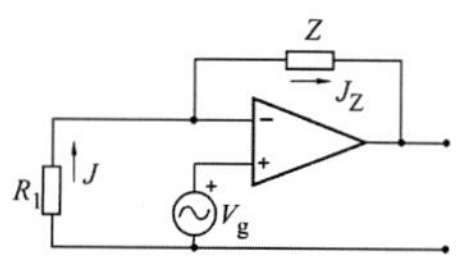

Fig. 4.6.13 Voltage-to-current converter

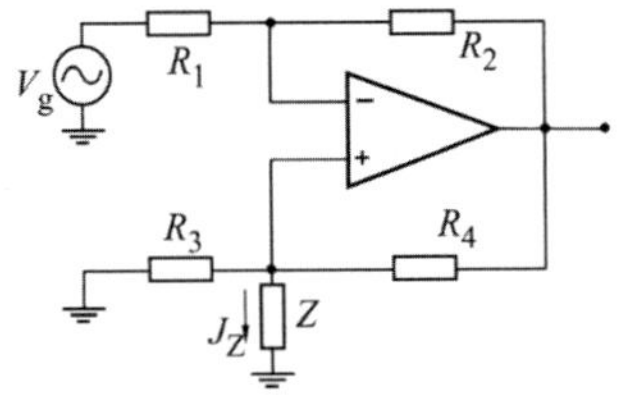

Fig. 4.6.14 Voltage-to-current converter with grounded Z

Since the inputs of the operational amplifier are short-circuited when its gain is very large the following is valid

$$J_Z = V_g/R_1. \tag{4.6.49a}$$

The current through the load impedance Z is proportional to the source voltage and will not depend on the value of the impedance Z.

It should be noted that if the voltage V_g were constant, this circuit would act as a constant current source (CCS).

In Fig. 4.6.13 neither terminal of the impedance Z is grounded. If that is necessary, one can use the circuit of Fig. 4.6.14.

For the circuit of Fig. 4.6.14, if the condition

$$R_1/R_2 = R_3/R_4 \tag{4.6.50}$$

is fulfilled, one may prove that

$$J_Z = V_g/R_3. \tag{4.6.49b}$$

4.6.3.5 Current-to-Voltage Converter

Figure 4.6.15 shows a circuit for converting current into voltage. If the input terminals of the OA are on the virtual ground, then the current through the resistor R_1 will be equal to zero, so the current through R_2 will be equal to the current of the source. That is why the output voltage will be

$$V_{out} = -R_2 J_g. \tag{4.6.51}$$

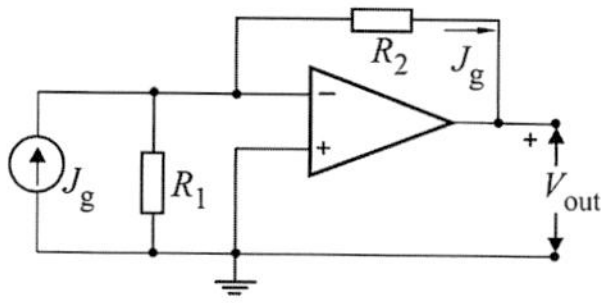

Fig. 4.6.15 Current-to-voltage converter

4.6.3.6 Analog Computational Circuits

An analog signal in electronics is considered to be any signal that is expressed as a function of time and receives continuous values. The term analog is used because other physical quantities such as speed, force, pressure, lighting, temperature, and the like can have the same waveform in another scale. The reader is probably aware that there are a virtually unlimited number of electronic components that can be considered a converter of some physical quantity into some electrical. Therefore, the processing of analog electrical signals can be identified with the processing of arbitrary analog signals. Having this in mind, any signal in nature can be considered an analog signal, and hence the importance of performing computational operations with such signals. The usual procedure for working with signals is to record, process, and communicate their values to the user. If that user is a machine, then it is possible that all three phases of processing that we mentioned are done electronically. At the same time, it is possible that operations on signals are performed without conversion into another domain (say a digital word) directly using specialized electronic circuits that act as operators. Electronic circuits that process analog signals are called analog circuits.

There are practically no restrictions regarding the type of operator that can be realized with an analog electronic circuit. Multiple signals can be added, subtracted, multiplied by a constant, or multiplied by each other. It is also possible to integrate, differentiate and logarithmize the input signals. In this section, only some principles and some basic circuits used for analog signal processing will be presented.

For greater precision in the processing of analog signals, the OAs used are of higher quality. This is especially true for the requirement for extremely large gain, which in these applications is greater than 10^6. Therefore, in the analysis that follows, unless otherwise indicated, the gain A will be considered infinite. In addition, temperature stability may be of crucial importance when evaluating the feasibility of the analog computation, as such.

4.6.3.6.1 The Adder Circuit

Figure 4.6.16a shows the circuit by which voltage values $v_1, v_2, \ldots, v_n$ can be summed while using a set of weighting coefficients $k_1, k_2, \ldots, k_n$. Here k_i, $i = 1, 2, \ldots, n$ are positive real numbers. The output voltage is thus obtained as

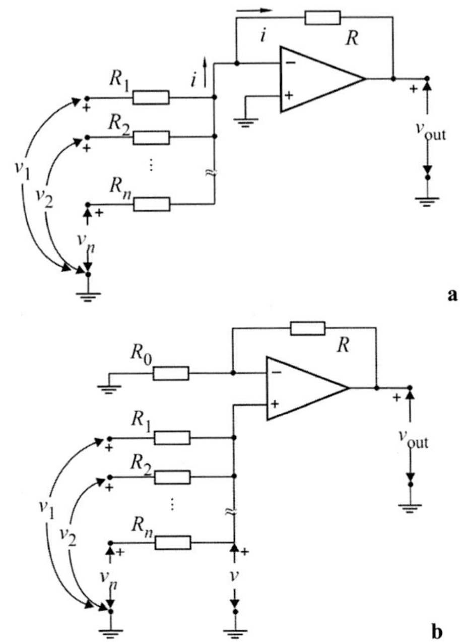

Fig. 4.6.16 Adder circuits. **a** Inverting and **b** non-inverting

$$v_{\text{out}} = \sum_{i=1}^{n} k_i v_i. \tag{4.6.52}$$

In the special case $k_1 = k_2 = \cdots = k_n = k$ one gets

$$v_{\text{out}} = k \sum_{i=1}^{n} v_i, \tag{4.6.53}$$

where k may be equal to unity.

For the circuit of Fig. 4.6.16a we can write

$$i = \frac{v_1}{R_1} + \frac{v_2}{R_2} + \cdots + \frac{v_n}{R_n} = -\frac{v_{\text{out}}}{R}, \tag{4.6.54}$$

wherefrom

$$v_{\text{out}} = -R\left(\frac{v_1}{R_1} + \frac{v_2}{R_2} + \cdots + \frac{v_n}{R_n}\right). \tag{4.6.55}$$

For $R_1 = R_2 = \cdots = R_n = R$ one gets

$$v_{\text{out}} = -\frac{R}{R_1}(v_1 + v_2 + \cdots + v_n). \tag{4.6.56}$$

When designing this popular circuit, the dynamics of the signal at the output of the amplifier should be considered seriously. Namely, since the output voltage is equal to the weighted sum of the input ones, care should be taken that the amplitudes and phases of the input signals are such that the output signal does not exceeds the maximum output voltage (rating) of the operational amplifier shown in Fig. 4.6.3. The same is true for current restrictions. The current i given by (4.6.54) must not be higher than the maximum output current of the operational amplifier or the maximum emitter current of the output transistor. From these considerations, the value of the resistance R is determined.

For example, consider the case when it is necessary to add five signals whose maximum amplitude does not exceed one volt. An OA is available with a maximum output voltage of 14 V and a maximum current of 2.5 mA. The values of R and R_1 of the formula (4.6.56) are determined from the following considerations.

When the voltage at the output is at maximum, the output current is closed through R, with the left end of this resistor on the virtual ground. The minimum value of the resistance is $R = 14/(2.5 \times 10^{-3}) = 5.6\ \text{k}\Omega$. In the worst case, when all the input signals are in phase and have a maximum value, the output current is divided into five equal parts that flow through each excitation source. Since the voltage drop across R_1 is equal to the input signal (or 1 V), and the current is $2.5 \times 10^{-3}/5 = 0.5$ mA, for R_1 we get $1/(0.5 \times 10^{-3}) = 2\ \text{k}\Omega$. Therefore, the gain of this circuit for each excitation signal is individually $A_v = -5.6/2 = -2.8$.

The negative sign in (4.6.56) indicates that the output voltage is opposite in phase with the linear combination of the input voltages. If we want to have a positive sign, we can do it in two ways. The simplest way is to cascade one inverter, i.e., to ensure that the next stage inverts the phase once more. This, however, can be an expensive solution. Another possibility is to use the circuit of Fig. 4.6.16b. The following relation applies to this circuit

$$v_{\text{out}} = (1 + R/R_0) \cdot v \tag{4.6.57}$$

and

$$v = \frac{R_e}{R_1} v_1 + \frac{R_e}{R_2} v_2 + \cdots + \frac{R_e}{R_n} v_n, \tag{4.6.58}$$

where

$$1/R_e = 1/R_1 + 1/R_2 + \cdots + 1/R_n. \tag{4.6.59}$$

For $R_1 = R_2 = \cdots = R_n = R$ one gets

$$v_{\text{out}} = (1 + R/R_0) \cdot \frac{1}{n} (v_1 + v_2 + \cdots + v_n). \tag{4.6.60}$$

4.6.3.6.2 The Integrator Amplifier

If in the circuit of Fig. 4.6.5 instead of resistor R_2 one puts a capacitor, the circuit shown in Fig. 4.6.17a will arise. It is often called an integrator amplifier or simply integrator. The circuit from Fig. 4.6.17b will be used for the analysis of the integrator. We can write

$$i = v_g/R = -C\frac{dv_{out}}{dt}, \tag{4.6.61}$$

which gives

$$v_{out} = v_{out}(0) - \frac{1}{RC}\int_0^t v_g(\tau)\cdot d\tau. \tag{4.6.62}$$

In this way, the output signal is an integral of the input signal. For example, if the excitation signal is sinusoidal [$v_g = V_g \sin(\omega t)$] we get

$$v_{out} = \frac{1}{RC}\frac{V_g}{\omega}\cos(\omega t) + v_{out}(0). \tag{4.6.63}$$

In case $v_{out}(0) \neq 0$ the circuit of Fig. 4.6.17a should be modified by connecting a battery of value $E = v_{out}(0)$ in parallel to the capacitor (active before the moment $t = 0$), and the input terminal should be grounded as shown in Fig. 4.6.18. When starting ($t = 0$), the excitation is switched on and the battery is switched off. This is, in fact, about setting the boundary conditions for integration.

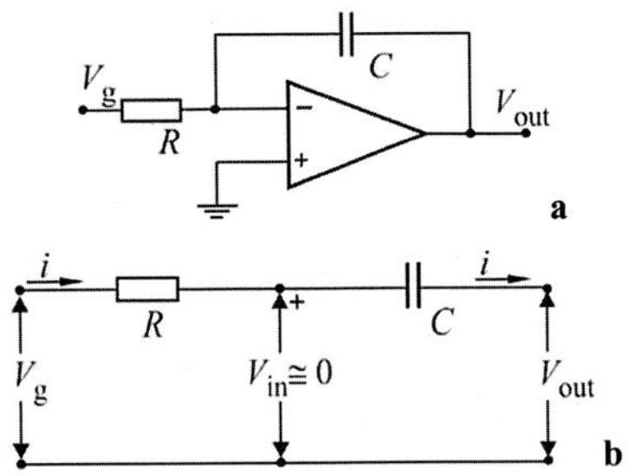

Fig. 4.6.17 Integrator amplifier **a** schematic and **b** equivalent circuit

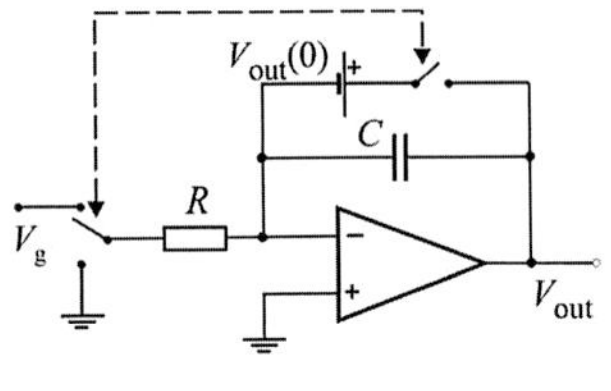

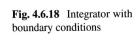

Fig. 4.6.18 Integrator with boundary conditions

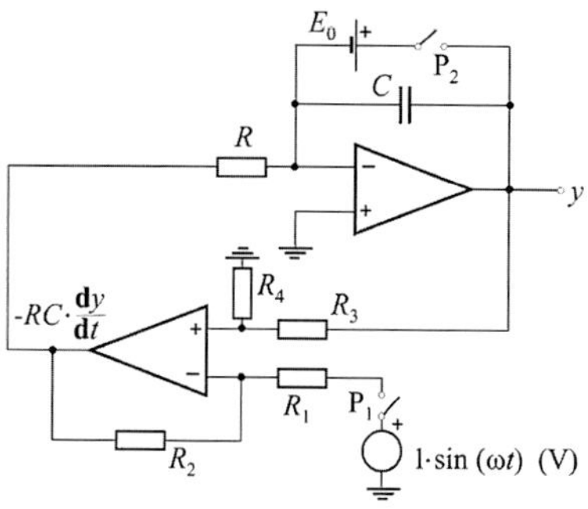

Fig. 4.6.19 Circuit for solving a differential equation. At time instant $t = 0$, P_1 is switched On and P_2 is switched Off

The integrator shown in Fig. 4.6.18 can be easily used to solve (integrate) differential equations. Consider, for example, the equation

$$\frac{dy}{dt} + ay = k \cdot \sin(\omega t) \quad (4.6.64)$$

with the condition $y(0) = 1$. The circuit that generates the solution of this differential equation is shown in Fig. 4.6.19. To make it easier to understand the operation of the circuit, the differential Eq. (4.6.64) will be rewritten as

$$\frac{dy}{dt} = k \cdot \sin(\omega t) - ay. \quad (4.6.65)$$

This indicates that the derivative dy/dt is obtained as a linear combination of the excitation and the solution. In the circuit of Fig. 4.6.19, first, we have a subtraction circuit that forms the aforementioned linear combination. The only thing missing is the coefficient for the time scale, RC, obtained by the integrator. Thus for the constants we have $k = R_2 / (R \cdot R_1 C)$ and $a = R_4(R_1 + R_2) / [R \cdot R_1 C(R_3 + R_4)]$. Here we should pay attention to the fact that the circuit is valid only for the positive values of the constants a and k. If k is negative, an adder circuit should be used instead of the subtraction circuit, and if a is negative, an additional inverter with unity gain should be installed in the loop of Fig. 4.6.19. The figure itself shows analog values of the voltages and the variables.

When the integrator is observed in the frequency domain for the circuit of Fig. 4.6.17a the following applies

$$V_g/R = -sCV_{out}, \quad (4.6.66)$$

or

$$A_{int} = -1/(sCR). \quad (4.6.67)$$

By analyzing this expression, we conclude that the gain of an integrator with an ideal operational amplifier has a pole at the origin. This means that its gain for zero frequency is infinite but, at the same time, its upper cut-off frequency is equal to zero. The passive integrator based on a simple RC circuit had similar characteristics. It was a low-pass filter whose upper cut-off frequency was lower if the capacitance of the capacitor was larger. For integrators with an ideal operational amplifier, the Miller capacitance is infinite, and the upper cut-off frequency is zero.

The reader should not be confused by the fact that the integrator exhibits infinite gain for a DC signal. Namely, if $v_g = C^{\underline{te}}$ is set, which corresponds to a signal whose frequency is equal to zero, the output voltage would actually grow linearly (integral of the constant) to infinity if it were not limited by the maximum dynamics of the output signal of the actual OA.

4.6.3.6.3 Differentiator Amplifier

If the resistor and capacitor in the integrator change places, a differentiatiator circuit is obtained, shown in Fig. 4.6.20a.

For the corresponding equivalent circuit of Fig. 4.6.20b we can write

$$i = C\frac{\mathrm{d}v_g}{\mathrm{d}t}. \tag{4.6.68}$$

and

$$v_{out} = -R \cdot i \tag{4.6.69}$$

leading to

$$v_{out} = -RC\frac{\mathrm{d}v_g}{\mathrm{d}t}. \tag{4.6.70}$$

The output signal is equal to the derivative of the input signal.

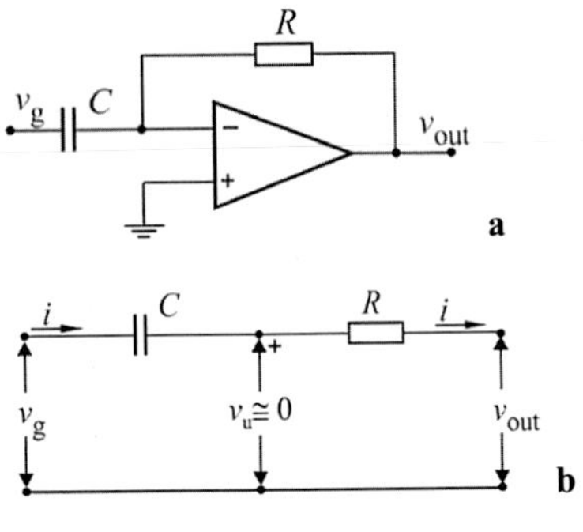

Fig. 4.6.20 **a** Differentiator amplifier and **b** corresponding equivalent circuit

When observed in the frequency domain for the circuit of Fig. 4.6.20a it is easily obtained

$$A = \frac{V_{out}}{V_g} = -sRC, \tag{4.6.71}$$

which means that the circuit function has zero at the origin. Of course, the passive RC differentiation circuit had a similar frequency response. In doing so, its function had a pole at the finite frequency, too. Due to the Miller effect, however, the pole now moves to infinity so that (4.6.71) remains valid.

At this point, the opportunity will be used to point out how the imperfection of the operational amplifier can affect the characteristics of a circuit. Namely, if it is assumed that the operational amplifier in the circuit with Fig. 4.6.20a has a finite gain and a finite upper cut-off frequency and if it is assumed that the slope of its amplitude characteristic is only 6 dB/oct, for the gain we can write the following

$$A = \frac{A_0}{1 + s/\omega_0}. \tag{4.6.72}$$

Now by substituting (4.6.72) into (4.6.12) and putting R instead of R_2, and $1/(sC)$ instead of R_1, for the gain of the differentiatiator circuit we get

$$A_v = -\frac{s\omega_0 A_0}{s^2 + \frac{1+RC\omega_0}{RC}s + \frac{(1+A_0)\omega_0}{RC}}. \tag{4.6.73a}$$

At relatively low frequencies, the third term in the denominator of the above function is dominant because A_0 is a large number, so that the circuit acts as an ideal differentiator. For high-frequency signals, the poles of the function (4.6.73a) are given by

$$s_{1,2} = -\frac{1 + RC\omega_0}{2RC}\left(1 \pm j\sqrt{1 - \frac{4RC\omega_0(1 + A_0)}{(1 + RC\omega_0)^2}}\right)$$
$$\approx -\frac{1 + RC\omega_0}{2RC} \pm j\sqrt{\frac{A_0\omega_0}{RC}}. \tag{4.6.73b}$$

From this expression we can see how high are the frequencies at which the imperfections of the operational amplifier come to the fore. One thing is obvious, if the gain A_0 is infinite, the poles will also migrate to infinity. The same is true for the influence of ω_0 to the pole.

For example, let $A_0 = 10^5$, $f_0 = 100$ Hz or $\omega_0 = 2\pi \cdot 100\ \text{s}^{-1}$ and let $R \cdot C = 10\ \mu\text{s}$. The poles of the function (4.6.73a) are $s_{1,2} = -50 \times 10^3(1 \pm j \cdot 50.4) \approx \pm j2.5 \times 10^6\ \text{s}^{-1}$. The corresponding pole frequency is $f_{1,2} = \pm 404$ kHz. Below this frequency, the circuit acts as a differentiator.

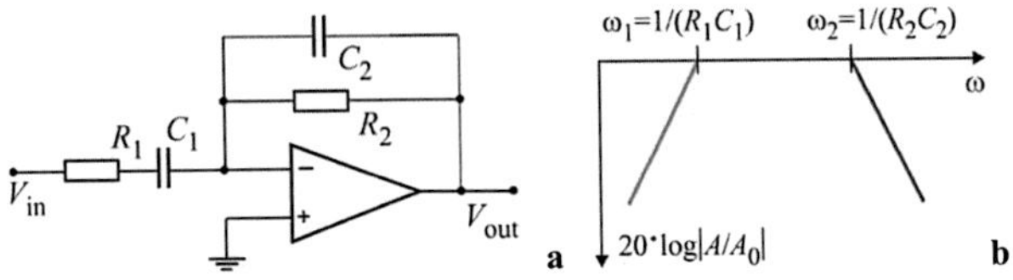

Fig. 4.6.21 **a** Practical differentiator and **b** its frequency response

As well as the passive RC differentiation circuit, based on (4.6.71) the circuit of Fig. 4.6.20a acts as a high-pass filter. In ideal conditions, for an infinite frequency, the gain is infinite. In practical circuits, as we have seen, infinite gain will not be achieved for the simple reason that the amplifier itself has a finite cut-off frequency so that (4.6.71) ceases to be valid. However, the upper cut-off frequency of the circuit with Fig. 4.6.20a, as we have seen above, can be very high which can also affect the stability of the amplifier. Bearing in mind that the signal to be differentiated usually lacks spectral components in the very high-frequency range, the property of the differentiatiator circuit to have a high upper cut-off frequency can become a significant weakness. Namely, in addition to the danger of self-oscillation, the noise power of the system is proportional to the bandwidth of the amplifier. If the bandwidth is wider than necessary, the noise will be higher than it needs to be. Therefore, instead of the original circuit, the circuit with Fig. 4.6.21a is used for differentiation. Its frequency response is shown in Figs. 4.6.21b. In that $R_2C_2 \ll R_1C_1$ is selected. In this way, the choice of the time constant R_2C_2 controls the bandwidth of the amplifier. The gain of this "inverter" is given by

$$A = -\frac{sR_2C_1}{(sR_1C_1 + 1)(sR_2C_2 + 1)}. \tag{4.6.74}$$

Up to the frequency $\omega_1 = 1/(R_1C_1)$, the circuit acts as a differentiator since at very low frequencies the denominator of the circuit function can be approximated by unity. At frequencies between ω_1 and $\omega_2 = 1/(R_2C_2)$ the circuit behaves as an ordinary amplifier, and above ω_2 the circuit behaves like an integrator. It is understandable that ω_1 is chosen so that it is above the highest frequency of the signal spectrum, and that ω_2 is chosen so as to ensure minimal noise or stability of the amplifier.

4.6.3.6.4 Logarithmic and Anti-logarithmic Amplifier

Consider the circuit in Fig. 4.6.22. For a forward biased diode we have

$$i_d = I_s\left(e^{v_d/V_T} - 1\right) \approx I_s e^{v_d/V_T}. \tag{4.6.75}$$

Having in mind that

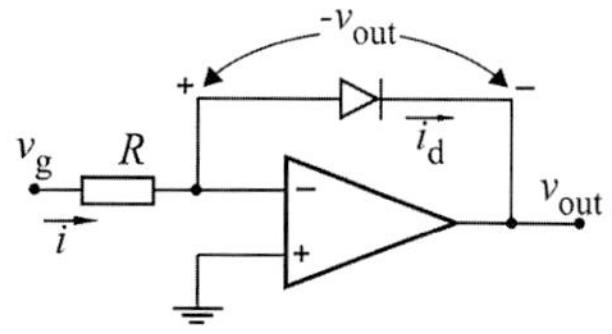

Fig. 4.6.22 Logarithmic amplifier

$$i = i_d = v_g/R \tag{4.6.76}$$

And

$$v_d = -v_{out} \tag{4.6.77}$$

we get

$$v_g/R = I_s e^{-v_{out}/V_T} \tag{4.6.78}$$

or

$$v_{out} = -V_T \cdot \ln\left[v_g/(R \cdot I_s)\right], \tag{4.6.79}$$

which means that the output voltage is proportional to the negative logarithm of the input voltage.

The logarithmic function itself is defined only for the positive values of the argument and takes positive and negative values. This means that the calculation is performed in the first and fourth quadrants. The logarithmic amplifier considered so far performs a logarithmic function also in the first and fourth quadrants, except that, due to the negative sign, the quadrants swap places.

From a practical point of view, it is important that the approximation made in (4.6.75) is valid in a relatively narrow range of changes in diode currents. This range is about six decades and is in the vicinity of a current of 1 mA. At low currents, errors occur due to the leakage current of the diode as well as due to the offset of the OA. At high currents, again, the diode becomes linear, not a logarithmic element. If we assume that the diode current is about one milliampere and that the maximum input voltage is 10 V, we get $R = 10\ \text{k}\Omega$. If a bipolar transistor is used instead of a diode in the feedback branch, due to the gain of the transistor, logarithmization in a wider current interval can be expected.

If the diode and resistor in the circuit with Fig. 4.6.22 swap places for the output voltage we get

$$v_{out} = -RI_s \cdot e^{v_g/V_T} \tag{4.6.80}$$

which means we got an anti-logarithmic or exponential amplifier.

Although at first glance (4.6.80) is valid for every v_g, we can easily conclude that this is not true. The circuit will operate as an exponential amplifier only if the

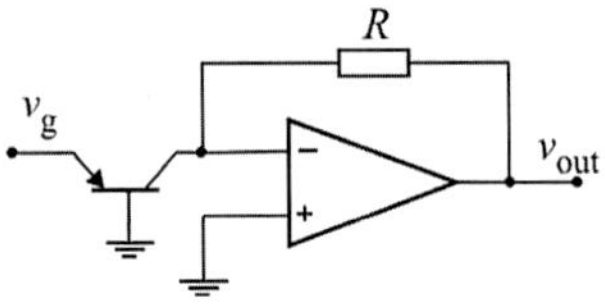

Fig. 4.6.23 Anti-logarithmic amplifier using BJT

diode is forward biased or if the output voltage is negative. Therefore, the input voltage must be positive. The circuit calculates the exponent of a positive number. Calculations are performed in the fourth quadrant. The third quadrant is crossed if the diode electrodes are swapped. The anti-logarithmic circuit performing on the entire axis of real numbers is more complex and will not be considered here.

The resistance R determines the current of the diode as in the case of the logarithmic circuits, so a similar calculation is applied to it. In order to expand the dynamic range of the circuit, instead of a diode, a BJT can be used, where the exponential dependence of the collector current on the base-emitter voltage is also used. One such anti-logarithmic amplifier is shown in Fig. 4.6.23.

Equations (4.6.79) and (4.6.80) also apply to the logarithmic or anti-logarithmic amplifiers using a BJT, with the current I_s being the collector saturation current.

The logarithmic and anti-logarithmic circuits shown are only in principles. Namely, they are very temperature unstable due to the temperature dependence of V_T and I_s. Additional temperature stabilization circuits are used for practical application.

The combination of logarithmic and anti-logarithmic amplifiers enables construction of a multiplication circuit. This circuit is shown in Fig. 4.6.24a. The principle of operation of the circuit is that the two signals are first logarithmized separately, and then their logarithms are added together. This gives the logarithm of the product. This quantity is anti-logarithmized, so the product is obtained at the output. With minor modifications from this circuit, a division circuit can be synthesized. It is shown in Fig. 4.6.24b. Now the logarithmic signals are subtracted, which makes it possible to obtain the logarithm of the quotient, which is reduced to the quotient by anti-logarithmization.

The multiplication and the division circuit are complementary in the sense that only one operation is replaced (subtraction instead of addition of logarithms). Therefore, it is reasonable that they can be generated from each other. Figure 4.6.24c shows the use of a multiplication circuit to make a divider, and Fig. 4.6.24d gives the reverse situation.

It should be borne in mind that the exponential characteristics of transistors can be used in other ways to multiply signals. In that we think of the sub-threshold characteristic of the MOSFET which is modeled as

$$I_D = I_{D0} \cdot \frac{W}{L} \cdot \exp\left(\frac{V_{GS}}{n \cdot kT/q}\right). \tag{4.6.81}$$

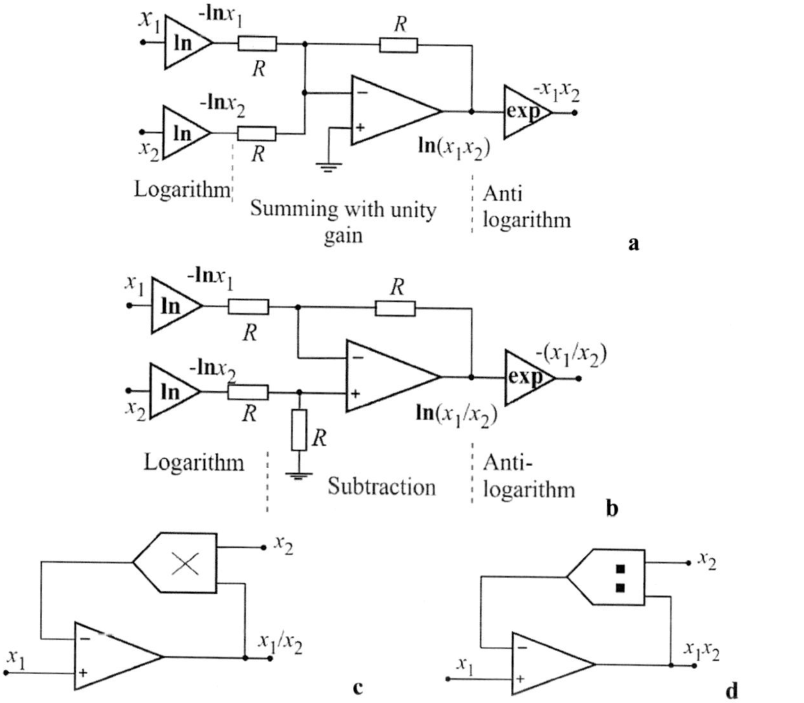

Fig. 4.6.24 **a** Multiplier in principle, **b** divider in principle, **c** divider realized using a multiplier (×), and **d** multiplier realized using a divider (:)

This means that the MOS transistor can also be used for multiplication. Thus, in nonlinear integrated circuits, transistors are used for multiplication. We will not go into the physical realization of this principle here.

4.6.3.7 Active RC Filters

In LNAE_Book 2 some passive electric circuits are shown, which have the property of behaving differently against signals with different frequencies. We have identified low-pass, high-frequency, and band-pass circuits. A common feature of these circuits is that the spectrum of the complex periodic signal that is fed to the input will be changed. Those spectral components that belong to the passband of the circuit will be transmitted to the output, and those components whose frequencies do not belong to the passband will be not. The signal will therefore be filtered from unwanted spectral components. Therefore, these circuits are called electric filters. The electric filter behaves selectively against signals of different frequencies.

The main disadvantage of passive RC circuit is the fact that they have finite and frequency-dependent output impedance. That means that, when loaded, these circuit, as any passive circuit, will change their fundamental selective properties. Special actions must be implemented in order for their frequency response to be preserved.

The properties of RC circuits can be significantly improved by using a combination of RC circuits and operational amplifiers. Such circuits are called active RC filters. In that, the property of the OA to have small output impedance, i.e., to behave as an ideal voltage source is exploited. That is further improved when negative feedback is implemented. The load will not affect the transfer function of the filter. In that way filtering circuits may be cascaded to produce extremely complex transfer functions.

Here are four of the simplest circuits that perform basic filtering functions. They have in common that they use only one operational amplifier and that the function of the circuit is of the second order (the function of the circuit contains two poles).

We will first show a low-pass filter. The structure is shown in Fig. 4.6.25a. The analysis of this circuit yielded

$$T(s) = \frac{V_2}{V_1} = \frac{1}{1 + sC_2(R_1 + R_2) + s^2 C_1 C_2 R_1 R_2}. \tag{4.6.82}$$

It is usual to set $R_1 = R_2$, so that

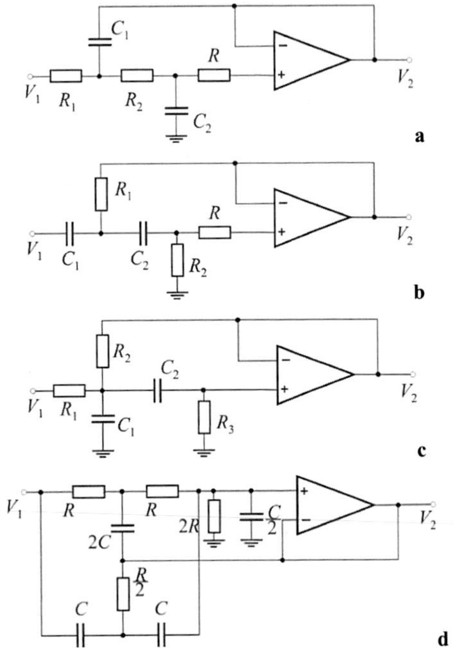

Fig. 4.6.25 Second-order active RC filters using single OA **a** low-pass, **b** high-pass, **c** band-pass, and **d** band-stop

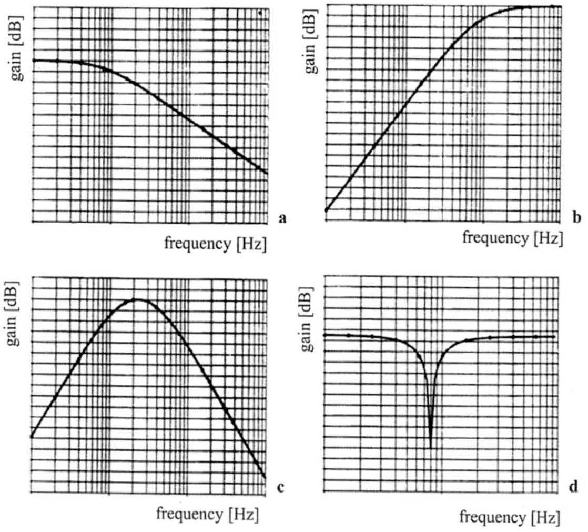

Fig. 4.6.26 Frequency characteristics of the simplest filter cells **a** low-pass, **b** high-pass, **c** band-pass, and **d** band-stop

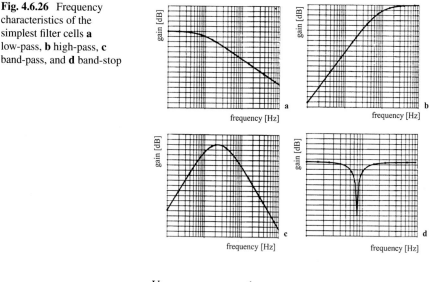

$$T(s) = \frac{V_2}{V_1} = \frac{1}{1 + 2sC_2R_1 + s^2C_1C_2R_1^2}. \tag{4.6.83}$$

By choosing different values for C_1 and C_2, it is possible for the transfer function to have a pair of different poles. If $C_1 = C_2$ is chosen, the denominator becomes a double square, so the double pole of the circuit function is $s = -1/(R_1C_1)$, and the cut-off frequency is $\omega_v = \left(\sqrt{\sqrt{2} - 1}\right) \Big/ (R_1C_1)$.

It is not necessary to prove that the maximum gain of this circuit is at zero frequency and that the gain at high frequencies has an asymptote with a slope of 12 dB/octave (Fig. 4.6.26a).

The second-order high-pass filter is shown in Fig. 4.6.25b. The transfer function of this circuit is

$$T(s) = \frac{V_2}{V_1} = \frac{s^2C_1C_2R_1R_2}{1 + sR_1(C_1 + C_2) + s^2C_1C_2R_1R_2}. \tag{4.6.84}$$

This function has a maximum value for $s \to \infty$ and a double zero at $s = 0$ which means that at low frequencies the gain decreases with a slope of 12 dB/oct. (Fig. 4.6.26b). Here again, if $R_1 = R_2$ and $C_1 = C_2$ the function of the circuit becomes a double square, and the double pole is $s = -1/(R_1C_1)$, while the lower cut-off frequency is $\omega_n = 1 \Big/ \left[(R_1C_1) \cdot \sqrt{\sqrt{2} - 1}\right]$.

The circuit shown in Fig. 4.6.25c represents a band-pass filter. The transfer function of this circuit is

$$T(s) = \frac{V_2}{V_1} = \frac{sC_2R_2}{\frac{R_1+R_2}{R_3} + s\tau + s^2C_1C_2R_1R_2}, \tag{4.6.85}$$

where $\tau = \left[R_1 R_2 (C_1 + C_2) / R_3 + C_2 R_2\right]$.

It is easy to see that the gain at the origin and at infinite frequency is equal to zero (Fig. 4.6.26c). At the frequency

$$\omega_0 = \sqrt{(R_1 + R_2) / (C_1 C_2 R_1 R_2 R_3)} \tag{4.6.86}$$

the gain modulus is maximum and amounts

$$|T(j\omega_0)| = 1 / \left[1 + R_1 (1 + C_1 / C_2) / R_3\right]. \tag{4.6.87}$$

Finally, Fig. 4.6.25 shows a band-stop (notch) filter. The transfer function of this circuit is

$$T(s) = \frac{V_2}{V_1} = \frac{1 + s^2 C^2 R^2}{2\left(1 + sRC + s^2 C^2 R^2\right)}. \tag{4.6.88}$$

This function has a pair of zeros on the imaginary axis of the complex frequency plane, i.e., on the axis of real frequencies. This means that when the signal frequency is

$$\omega_0 = 1 / (CR) \tag{4.6.89}$$

the gain will be equal to zero. At the same time, the gain in the origin and at infinite frequency is equal to 1/2. Therefore, this circuit does not pass the frequency range in the neighborhood of the notch frequency (Fig. 4.6.26d), so it is said to act as a band-stop filter.

Now, by cascading the active cells shown in Fig. 4.6.25a–d, higher order functions can be synthesized, and circuit functions can be obtained by simply multiplying the circuit functions of individual cells. One example of generating a complex fourth-order low-pass filter circuit function is shown in Fig. 4.6.27. One low-pass cell and one band-stop cell are coupled. This coupling is preferable to the coupling containing two low-pass cells because the resulting function is characterized by a much higher slope of the amplitude characteristic in the vicinity of the cut-off frequency what is easily observed from Fig. 4.6.28. The resulting frequency characteristic has a nominal gain of six decibels less than the gain of the low-pass cell, but this is not a particular problem since adding one inverting or non-inverting amplifier to the cascade of Fig. 4.6.27 the gain may be compensated or increased at will.

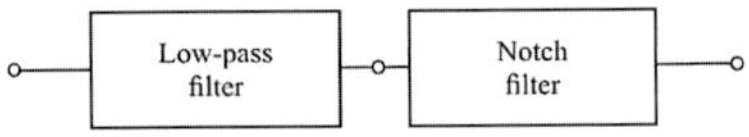

Fig. 4.6.27 Cascade coupling of two active filter cells. The low-pass filter is shown in Fig. 4.6.25a, and the notch in Fig. 4.6.25d

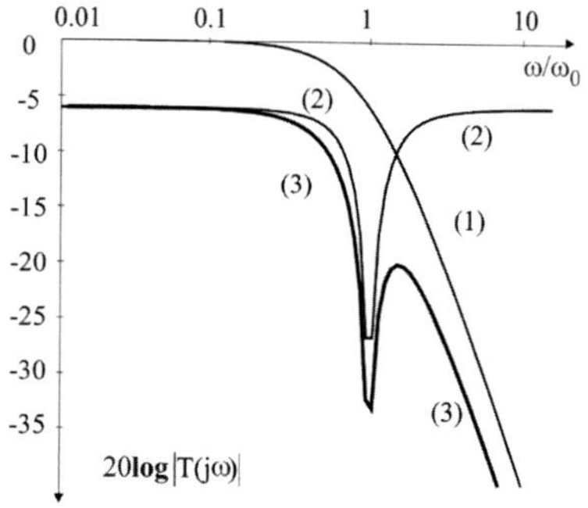

Fig. 4.6.28 Frequency characteristic of a fourth-order low-pass filter. (1) Amplitude characteristic of the low-pass stage, (2) amplitude characteristic of the notch stage and (3) the overall amplitude characteristic

4.6.3.8 Impedance Converters

A significant additional advantage of active over passive RC circuits stems from the ability to use positive feedback simultaneously with negative. In that, despite the fact that, for stability reasons, the negative feedback signal must always be dominant, very useful effects can be achieved by introducing a positive feedback signal.

One of such effects refers to the synthesis of elements that are not present in the circuit. For example, when it came to monolithic integrated circuits, it was found that inductance could be integrated but not below the RF spectrum. However, since the inductance is a very important component, the question remains how to synthesize it at low frequencies. For this purpose, the so-called impedance converters are used.

One of the possible impedance converters is shown in Fig. 4.6.29. The first stage in this circuit has two feedback loops. A negative feedback signal is supplied via the divider R_3–R_3, and a positive feedback signal is supplied via the branch R_1–J_1. This enables signal competition and a special effect that we will show soon.

For the circuit with Fig. 4.6.29 the following node equations for nodes to which inverting terminals of OAs are connected may be written

$$(V_1 - V_g)/R_3 + V_g/R_3 = 0, \tag{4.6.90}$$

and

$$-V_1/(2R_2) - V_2/Z = 0. \tag{4.6.91}$$

Now, we easily find

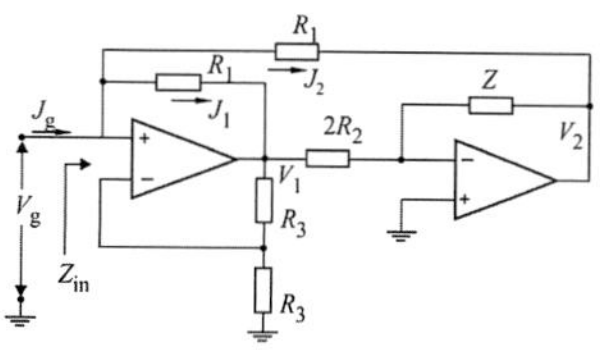

Fig. 4.6.29 Electronic circuit that performs the function of a gyrator

$$J_g = J_1 + J_2, \tag{4.6.92}$$

$$J_1 = \left(V_g - V_1\right) / R_1 \tag{4.6.93}$$

$$J_2 = \left(V_g - V_2\right) / R_1, \tag{4.6.94}$$

By solving of this system of equations we get

$$Z_{in} = \frac{V_g}{J_g} = R_1 R_2 \cdot \frac{1}{Z}. \tag{4.6.95}$$

The input impedance of this circuit is equal to a constant divided by the impedance Z. An increase in Z causes a decrease in Z_{in}. If Z is of capacitive character [Z = 1/(sC)], the input impedance will be inductive: $Z_{in} = s{\cdot}(L_{in}) = s(R_1R_2C)$. So, the circuit converted the capacitive load into an inductive input. Thus, for example, for $R_1 = R_2 = 1\ \text{k}\Omega$ an $C = 1\ \mu\text{F}$, the equivalent inductance is $L_{in} = R_1R_2C = 1$ H.

By choosing the value of the circuit elements, the value of the inductance can be controlled on a large scale. It should be noted, however, that due to the frequency dependence of the gain of OAs the value of this inductance is not constant; i.e., (4.6.95) is valid in a limited frequency range.

The circuit of Fig. 4.6.29 was used to illustrate the impedance conversion in electronics. A general version of the immittance (immittance is a common name for impedance and conductance) converter is shown in Fig. 4.6.30. This circuit was first suggested by Antonio. The input impedance of this circuit is given by

$$Z_{in} = \frac{Z_1 Z_3}{Z_2 Z_4} \cdot Z_L. \tag{4.6.96}$$

Proper choice of the impedances Z_1, Z_2, Z_3 and Z_4 will allow different conversions. For example, if $Z_1 = R$, $Z_3 = R$, $Z_2 = 1/(j\omega C)$, $Z_4 = 1/(j\omega C)$, and $Z_L = 1/(j\omega C)$, we get

$$Z_{in} = -\omega^2 R^2 C^2 Z_L = j\omega\left(R^2 C\right) \tag{4.6.97}$$

which means that the capacitive load is converted into frequency-independent inductance (of the value $L = R^2C$). This is true in the frequency range that can be considered as the one where the gains of the OAs are very large.

Fig. 4.6.30 General immittance converter

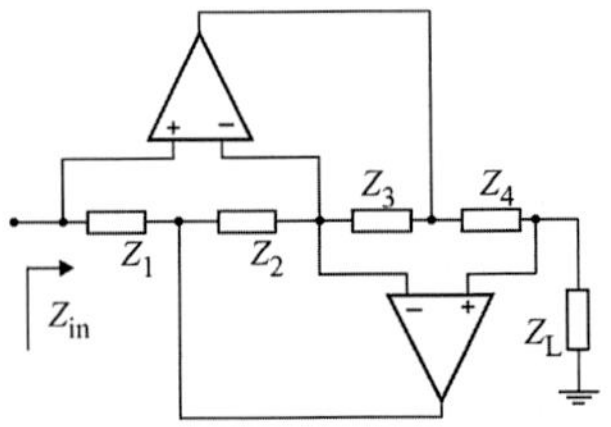

Appendix 4.A
A Short Review of Microelectronic Technology Processes

In order to set conditions for better understanding the functionality of the electronic devices discussed in this book we will here discuss briefly and partly the processes which are used in the semiconductor industry.

Cristal growth and refinement. Crystals, mostly silicon, are grown usually from a melted base with special care on alignment of the crystal axis with the crystallographic ones. Here a special refinement technology is applied in order to achieve the purity mentioned in the preamble of this book. As we could see until now it is of fundamental importance. Crystals are then cut (following the crystallographic axes) and polished in order to produce wafers needed for production. The consequent processes are taking place usually on one of the wafer's surfaces.

Diffusion. This is a process that is used to change the type of the semiconductor. Namely, through a specially worked window in the protective layer of the wafer, impurity atoms are allowed to diffuse into the semiconductor. This is a high-temperature process performed in high vacuum. The depth of penetration, the concentration, and the profile of the impurities will depend on the temperature, the time of diffusion, and the amount of available material to diffuse. While this gives opportunities to shape the newly formed semiconductor within the wafer, it asks for special care in control of all these parameters. It is important to note that, since high-temperature process, the present diffusion may strongly affect the previously performed diffusions on the same wafer.

Implantation. By this process highly controlled shallow layers of a semiconductor of a given type may be created. To achieve that the semiconductor surface is bombarded by highly accelerated ions of impurity atoms, so they become implanted into the crystal taking the places of the semiconductor atoms. To regenerate the crystal lattice, after bombardment, heating of the wafer is needed but the temperatures for that are largely lower than the ones needed for diffusion and so the implantation is considered as low-temperature process.

Epitaxial growth. This is a process which allows layers of a new crystal to grow over selected parts of the wafer's surface. If, at the same time, impurities are added,

V. B. Litovski, *Lecture Notes in Analog Electronics*, Lecture Notes in Electrical Engineering 958, https://doi.org/10.1007/978-981-19-6528-9

one may get all sorts of junctions as mentioned above. That however stands not only for grow of the original semiconductor material, e.g., silicon on silicon, but also for grow of fully new material (with different energy gap) over the given wafer. In this case the crystal becomes "hetero" and so are the potential junctions created in this way. There are several different processes allowing for epitaxial growth and, to be short, we will here mention the so-called molecular beam epitaxy (MBE) which allows layers as thin as the thickness of one molecule to be created.

Appendix 4.B
Solved Problems

PART 1: Power amplifiers

Problem 4.B.1
For the Class A amplifier depicted in Fig. 4.B.1, it is given: $V_{CC} = 10$ V; $R_E = 0.1$ kΩ; $R_L = 1$ kΩ; $R_1 = 9$ kΩ; $R_2 = 1$ kΩ; $V_{BE} = 0.7$ V; $h_{iE} = 1$ kΩ; $h_{rE} = 0$; $h_{fE} = \beta = 100$; $h_{oE} = 0$ S; $C \rightarrow \infty$.

(a) Find the value of R_C so that at the output maximal undistorted symmetrical voltage is obtained.
(b) For that value of R_C find the maximum useful power P_L.
(c) Also find the small-signal voltage gain ($A = V_{out}/V_g$).

Solution to Problem 4.B.1

(a) To solve this task, it is necessary to first determine the values of the collector current and the voltage between the collector and the emitter, which define the static and dynamic load line, as well as the quiescent operating point. To this end, we should first consider the DC circuit, which is shown in Fig. 4.B.2.

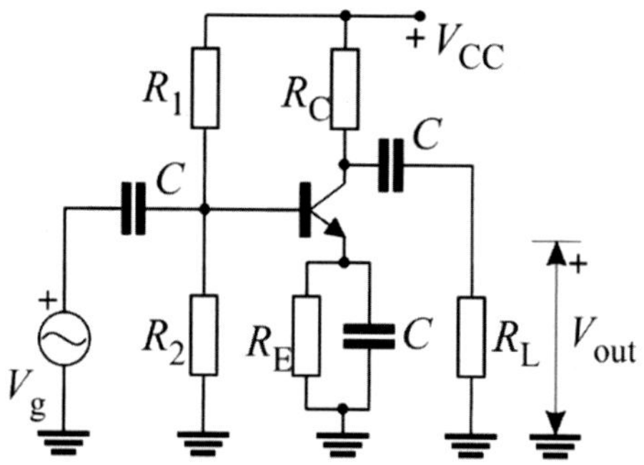

Fig. 4.B.1 Class A amplifier with BJT

V. B. Litovski, *Lecture Notes in Analog Electronics*, Lecture Notes in Electrical Engineering 958, https://doi.org/10.1007/978-981-19-6528-9

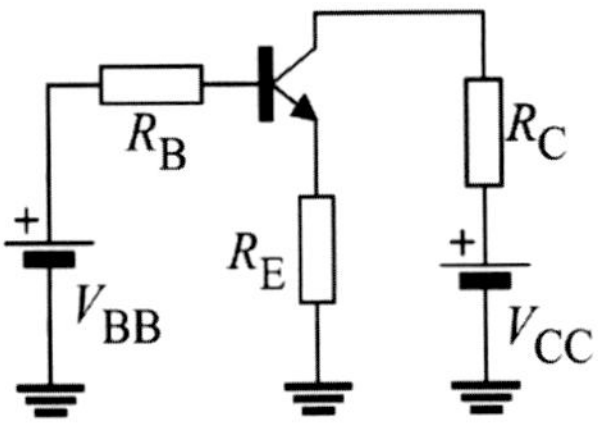

Fig. 4.B.2 DC equivalent circuit of the one of Fig. 4.B.1

Based on Fig. 4.B.1 one can claim that the resistance and voltage of the equivalent Thevenin source are given by:

$$R_B = R_1 || R_2 = 0.9\ \text{k}\Omega, \tag{4.B.1.1}$$

and

$$V_{BB} = \frac{R_2}{R_1 + R_2} \cdot V_{CC} = 1\ \text{V}. \tag{4.B.1.2}$$

As can be seen from Fig. 4.B.2 the base current of the BJT at the quiescent operating point is given by the following expression:

$$I_{BQ} = \frac{V_{BB} - V_{BE}}{R_B + (1 + \beta) \cdot R_E} = 27.3\ \mu\text{A}. \tag{4.B.1.3}$$

Now the collector current in the quiescent operating point is:

$$I_{CQ} = \beta \cdot I_{BQ} = 2.73\ \text{mA}. \tag{4.B.1.4}$$

Due to the fact that the value of the resistance R_C is unknown, it is impossible to determine the value of the voltage between the collector and the emitter at a quiescent operating point. For this purpose, it is necessary to write equations that define the DC:

$$I_C = \frac{V_{CC} - V_{CE}}{R_C + R_E}, \tag{4.B.1.5}$$

and the AC:

$$I_C - I_{CQ} = -(V_{CE} - V_{CEQ})/(R_C || R_L) \tag{4.B.1.6}$$

load line.

These two lines are depicted in Fig. 4.B.3. Based on the AC load line we can find V'_{CC} from:

$$V'_{CC} = V_{CE}|_{I_C=0} = V_{CEQ} + I_{CQ} \cdot (R_C || R_L) \tag{4.B.1.7}$$

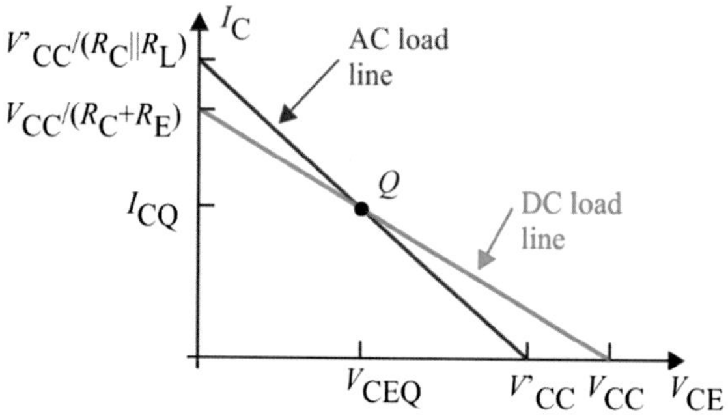

Fig. 4.B.3 Load lines for graphic analysis

On the other hand, in order to obtain maximal undistorted symmetrical voltage at the amplifier's output, which means that the amplitude of the voltage between the collector and emitter must not be greater than V_{CEQ} (so as to avoid clumping), it is necessary V'_{CC} to be:

$$V'_{CC} = 2 \cdot V_{CEQ}. \tag{4.B.1.8}$$

By equating (4.B.1.8) and (4.B.1.7) one gets:

$$V_{CEQ} = I_{CQ} \cdot (R_C \| R_L). \tag{4.B.1.9}$$

By substituting this expression into the equation of the DC load line (4.B.1.5), where V_{CE} and I_C are taken at quiescent operating point, we obtain:

$$V_{CC} - (R_C + R_E) \cdot I_{CQ} = (R_C \| R_L) \cdot I_{CQ}. \tag{4.B.1.10}$$

By proper arranging (4.B.1.10) one gets the following quadratic equation for the unknown R_C:

$$R_C^2 + R_C \cdot \left(2R_L + R_E - V_{CC}/I_{CQ}\right) + R_L \cdot \left(R_E - V_{CC}/I_{CQ}\right) = 0. \tag{4.B.1.11}$$

After solution we get:

$$R_C = 2.82\ \text{k}\Omega \tag{4.B.1.12}$$

as the only positive solution.

Now for the emitter to collector voltage at the quiescent operating point we can write:

$$V_{CEQ} = (R_C \| R_L) \cdot I_{CQ} = 2\,\text{V}.$$

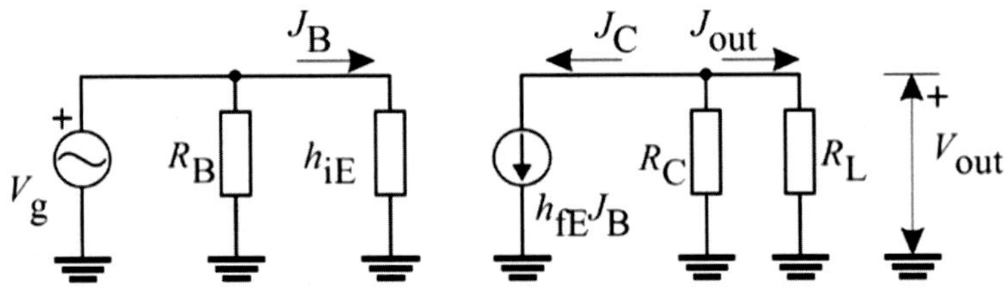

Fig. 4.B.4 AC equivalent circuit of the one of Fig. 4.B.1

(b) Due to the fact that the AC voltage on the load is practically equal to the voltage between the collector and emitter ($1/(\omega C) \approx 0$), the maximum value of voltage on the load is

$$V_{outm} = V_{CEQ}, \tag{4.B.1.13}$$

while the maximum value of the current, based on the current divider (where the maximum amplitude of the collector current is taken as the collector current at a quiescent operating point) is equal to:

$$J_{out\,max} = I_{CM} R_C/(R_C + R_L). \tag{4.B.1.14}$$

Now, for the maximum load power one may write:

$$P_{Lmax} = 0.5 \cdot V_{outmax} J_{outmax} = 2\,\text{mW}. \tag{4.B.1.15}$$

(c) In order to obtain the value of the voltage gain for small signals, it is necessary to start from the incremental equivalent circuit shown in Fig. 4.B.4. Based on this figure, we can write:

$$\begin{aligned} A_v = \frac{v_{out}}{v_g} &= \frac{v_{out}}{J_{out}} \cdot \frac{J_{out}}{J_C} \cdot \frac{J_C}{J_B} \cdot \frac{J_B}{v_g} \\ &= R_L \cdot \left(-\frac{R_C}{R_C + R_L} \right) \cdot h_{fE} \cdot \frac{1}{h_{iE}}. \end{aligned} \tag{4.B.1.16}$$

After substitution of the numerical values one gets: $A_v = -73$.

Fig. 4.B.5 Class A power amplifier using transformer

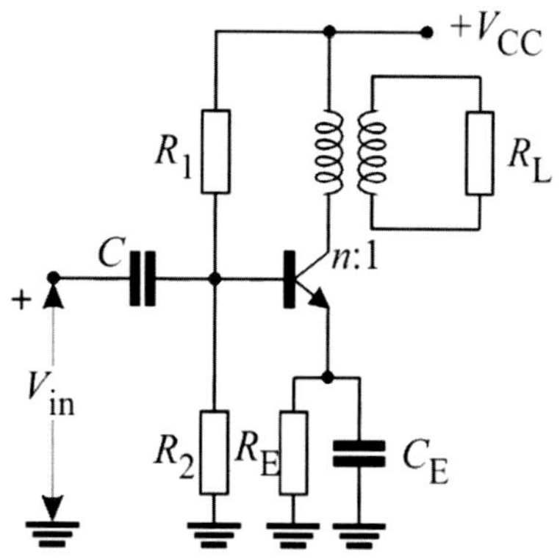

Problem 4.B.2

For the Class A power amplifier, shown in Fig. 4.B.5, determine:

(a) the quiescent operating point (V_{CEM}, I_{CM});
(b) the resistance value of the resistor R_L for which the maximum undistorted symmetrical voltage is obtained at the output and sketch the DC and AC load lines of the amplifier;
(c) maximum amplitude of the voltage and the current of the load;
(d) the efficiency of the amplifier.

It is known: $V_{CC} = 24$ V; $n = 4$; $R_E = 10\ \Omega$; C, $C_E \to \infty$; $R_1 = 0.9\ \text{k}\Omega$; $R_2 = 0.1\ \text{k}\Omega$. The BJT is characterized by: $\beta = 100$; $V_{BE} = 0.7$ V; $V_{CEmin} = 0$ V. The transformer is ideal.

Solution to Problem 4.B.2

(a) To calculate the coordinates of the quiescent operating point of the BJT, it is necessary to pre-transform the circuit of Fig. 4.B.5 for DC operation. The transformed circuit is shown in Fig. 4.B.6, wherein the elements of the Thevenin source in the base circuit are:

$$R_B = (R_1 \| R_2) = 90\ \Omega, \tag{4.B.2.1}$$

and

$$V_{BB} = \frac{R_2}{R_1 + R_2} \cdot V_{CC} = 2.4\ \text{V}. \tag{4.B.2.2}$$

Assuming that $\beta \gg 1$, i.e., that the collector current at the quiescent operating point is approximately equal to the emitter current, based on the equation for the base node and the transistor model it can be written that the collector current is equal to.

Fig. B.6 DC equivalent circuit of the one of Fig. 4.B.5

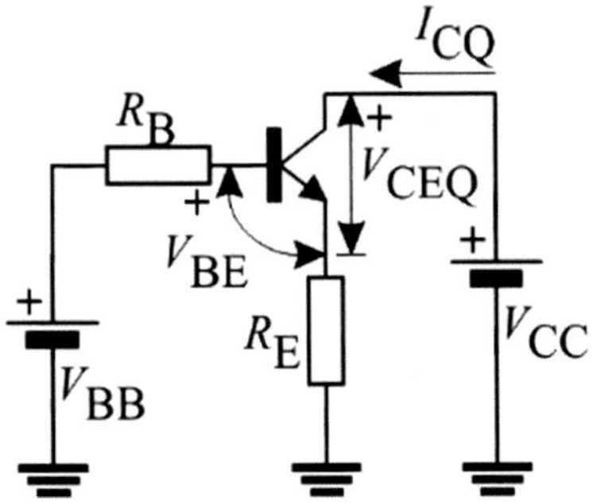

Fig. 4.B.7 AC equivalent circuit of the one of Fig. 4.B.5

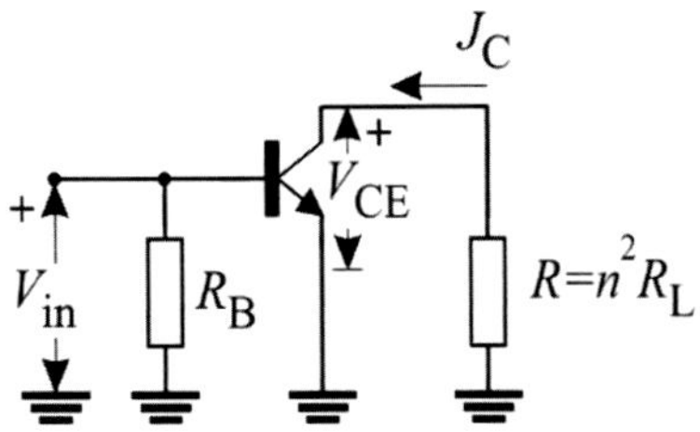

$$I_{CQ} = \beta \cdot \frac{V_{BB} - V_{BE}}{R_B + \beta \cdot R_E} = 0.15\,\text{A}. \tag{4.B.2.3}$$

Then, by analysis of the collector's part of the circuit we get:

$$V_{CEQ} = V_{CC} - R_E \cdot I_{CQ} = 22.5\ \text{V}. \tag{4.B.2.4}$$

(b) The resistance of the R_L can be determined on the basis of the AC load line written so that the maximum undistorted symmetric voltage is obtained at the output. As can be seen from Fig. 4.B.6 the DC load line is defined by:

$$I_C = (V_{CC} - V_{CE})/R_E, \tag{4.B.2.5}$$

while from Fig. 4.B.1.7 we see that the AC load line is defined by R_L:

$$I_C = \left(V'_{CC} - V_{CE}\right)/R = \left(V'_{CC} - V_{CE}\right)/\left(n^2 \cdot R_L\right), \tag{4.B.2.6}$$

where V'_{CC} is obtained from the condition for maximal non-distorted output voltage:

$$V'_{CC} = 2 \cdot V_{CEQ}. \tag{4.B.2.7}$$

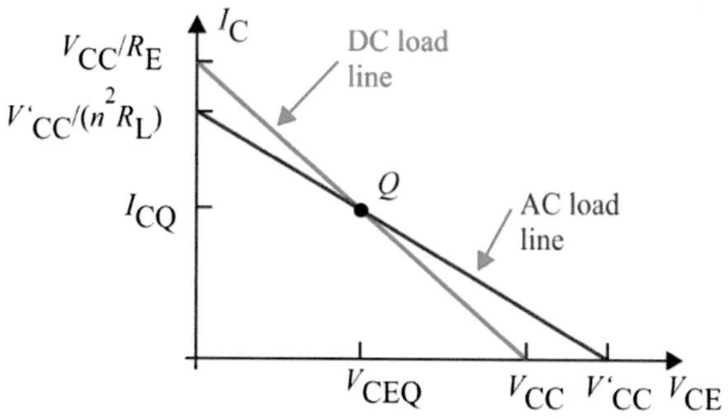

Fig. 4.B.8 Load lines for graphic analysis of the circuit of Fig. 4.B.5

The DC and AC load lines are depicted in Fig. 4.B.8. After substitution of (4.B.2.7) into (4.B.2.6), where the values for the I_C and V_{CE} should be taken at the quiescent operating point, it is possible to determine the R_L, as follows

$$R_L = V_{CEQ}/\left(n^2 \cdot I_{CQ}\right) = 9.37\ \Omega. \tag{4.B.2.8}$$

(c) The maximum amplitude of the collector current is given by:

$$J_{Cmax} = I_{CQ}, \tag{4.B.2.9}$$

so that the maximum amplitude of the load current is:

$$J_{Lmax} = n \cdot J_{Cmax} = n \cdot I_{CQ} = 0.6\ \text{A} \tag{4.B.2.10}$$

Similarly, the maximum value of the collector to emitter voltage is:

$$V_{CEmax} = V_{CEQ}, \tag{4.B.2.11}$$

which leads to the maximum amplitude of the load voltage:

$$V_{Lmax} = \frac{V_{CEmax}}{n} = \frac{V_{CEQ}}{n} = 5.62\ \text{V}. \tag{4.B.2.12}$$

(d) The efficiency is defined as the ratio of the useful power dissipated to the load and the power delivered by the battery to the amplifier. Therefore, in order to determine it, it is necessary to determine these two quantities. The useful power on the load is:

$$P_L = 0.5 \cdot V_{Lmax} L_{Lmax} = 1.68\ \text{W}, \tag{4.B.2.13}$$

while the power delivered to the amplifier by the battery is:

$$P = V_{\mathrm{CC}} \cdot I_{\mathrm{CQ}} = 3.6\ \mathrm{W}. \tag{4.B.2.14}$$

Accordingly the efficiency is

$$\eta = 100 \cdot P_{\mathrm{L}}/P = 46\%. \tag{4.B.2.15}$$

Problem 4.B.3

Figure 4.B.9 depicts a power amplifier operating in Class A. It is known: $V_{\mathrm{CC}} = 12$ V; $R_1 = 3.8$ kΩ; $R_2 = 2.2$ kΩ; $R_{\mathrm{E}} = 750$ Ω; $C, C_{\mathrm{E}} \to \infty$. The primary winding of the transformer is considered partly real, i.e., its resistive component is R_{I}. The number of windings in the primary and secondary is $N_1 = 75$ and $N_2 = 15$, respectively. The transistor parameters are: $V_{\mathrm{BE}} = 0.65$ V; $V_{\mathrm{CEmin}} = 0$ V and $\beta = 100$. Determine:

(a) the unknown resistance of the primary winding R_{I} so that the DC load line traverses the point $V_{\mathrm{CEmin}} = 0$ V, $I_{\mathrm{C}} = 15$ mA;
(b) the quiescent working point of the transistor ($V_{\mathrm{CEQ}}, I_{\mathrm{CQ}}$);
(c) the value of R_{L}, so that the maximum change in voltage between the collector and the emitter is $\Delta V_{\mathrm{CE}} = 8$ V;
(d) maximum amplitude of the current and voltage of the load;
(e) the efficiency of the amplifier.

Sketch DC and AC load lines and assign critical values.

Solution to Problem 4.B.3

(a) The circuit valid for the DC regime is depicted on Fig. 4.B.10. Since $\beta \gg 1$, the collector current is approximately equal to the emitter current, so that, based on Fig. 4.B.10, the DC load line is defined by:

$$I_{\mathrm{C}} = \frac{V_{\mathrm{CC}} - V_{\mathrm{CE}}}{R_{\mathrm{I}} + R_{\mathrm{E}}}. \tag{4.B.3.1}$$

Substitution of $I_{\mathrm{C}} = 15$ mA and $V_{\mathrm{CE}} = 0$ V leads to:

$$R_{\mathrm{I}} = \frac{V_{\mathrm{CC}}}{I_{\mathrm{C}}} - R_{\mathrm{E}} = 50\ \Omega. \tag{4.B.3.2}$$

(b) The Thevenin source in the base of the transistor circuit of Fig. 4.B.10 is described by

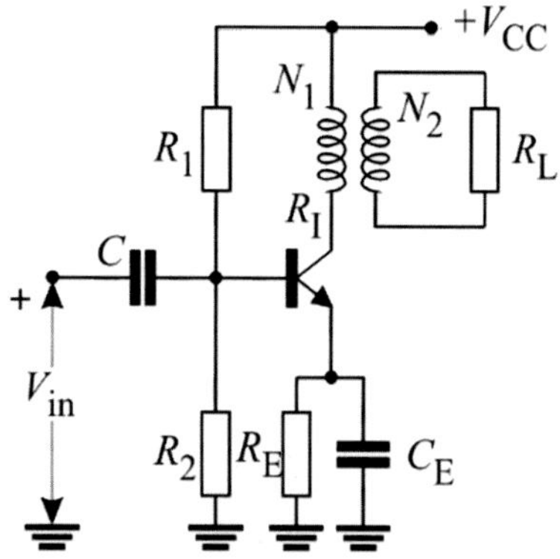

Fig. 4.B.9 Class A power amplifier using transformer

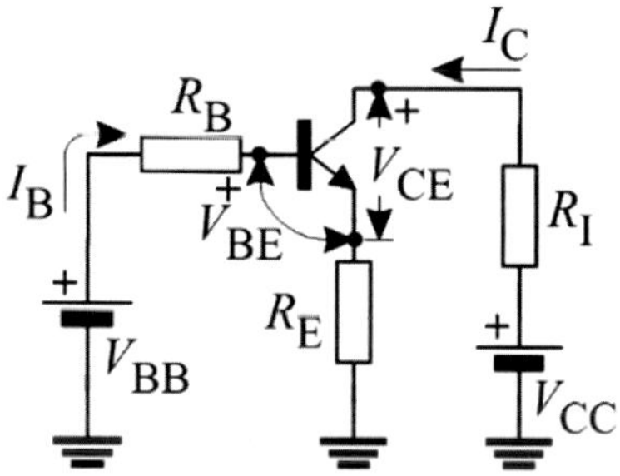

Fig. 4.B.10 DC equivalent circuit of the one of Fig. 4.B.9

$$V_{BB} = V_B = \frac{R_2}{R_1 + R_2} \cdot V_{CC} = 4.4 \text{ V}; \tag{4.B.3.3}$$

$$R_B = R_1 \| R_2 = 1.4 \text{ k}\Omega. \tag{4.B.3.4}$$

Having in mind the transistor model we have:

$$I_B = \frac{V_{BB} - V_{BE}}{R_B + (1+\beta) \cdot R_E} \approx 50 \ \mu\text{A}, \tag{4.B.3.5}$$

so that the quiescent collector current is:

$$I_{CQ} \cong I_{EQ} = \frac{V_E}{R_E} = 5 \text{ mA}. \tag{4.B.3.6}$$

Now, the voltage between the collector and the emitter at the quiescent operating point based on the DC load line given by (4.B.3.1) is equal to:

$$V_{CEQ} = V_{CC} - I_{CQ} \cdot (R_I + R_E) = 8 \text{ V}. \tag{4.B.3.7}$$

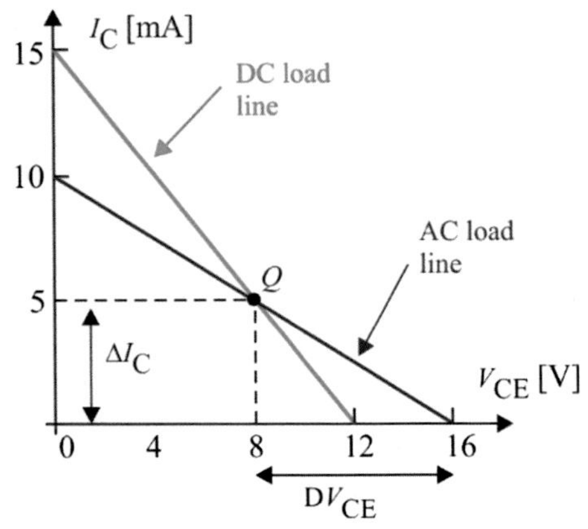

Fig. 4.B.11 Load lines for graphic analysis of the circuit of Fig. 4.B.9

(c) As can be seen from Fig. 4.B.11, the maximum change of the V_{CE} voltage, $\Delta V_{CE} = 8$ V, corresponds to a change in the current of $\Delta I_C = 5$ mA. These two quantities are related by the equation:

$$\Delta V_{CE} = R''_L \cdot \Delta I_C, \tag{4.B.3.8}$$

Where

$$R''_L = R_I + R'_L. \tag{4.B.3.9}$$

The resistance R'_L represents the mapped load resistant into the primary of the transformer:

$$R'_L = (N_1/N_2)^2 \cdot R_L. \tag{4.B.3.10}$$

By combining (4.B.3.10), (4.B.3.9), and (4.B.3.8) one can develop an expression for the unknown load resistance as

$$R_L = (N_2/N_1)^2 \cdot (\Delta V_{CE}/\Delta I_C - R_I), \tag{4.B.3.11}$$

which, after substitution of the numerical values, gives: $R_L = 62\ \Omega$.

(d) The maximum amplitude of the voltage at the load is:

$$V_{outmax} = \frac{V_{CEQ}}{N_1/N_2} = 1.6\,\mathrm{V} \tag{4.B.3.12}$$

while the maximum amplitude of the load current is:

$$J_{\text{outmax}} = \frac{I_{\text{CQ}} N_1}{N_2} = 25\,\text{mA}. \tag{4.B.3.13}$$

(e) Knowing the values of the maximum amplitudes of voltage on the load, and its current, as well as the value of the collector current at the operating point, one can find the useful power at the load to be

$$P_{\text{L}} = \frac{1}{2} \cdot V_{\text{Lm}} \cdot I_{\text{Lm}} = 20\ \text{mW}, \tag{4.B.3.14}$$

while the power delivered by the battery to be:

$$P_{\text{B}} = V_{\text{CC}} \cdot I_{\text{CQ}} = 60\ \text{mW}. \tag{4.B.3.15}$$

Hence, the efficiency is

$$\eta = 100 \cdot \frac{P_{\text{L}}}{P_{\text{B}}} = 33.3\%. \tag{4.B.3.16}$$

Problem 4.B.4

Figure 4.B.12 depicts a push–pull Class A amplifier. The excitation signal is $v_{\text{in}} = V_{\text{inmax}} \cdot \sin(\omega t)$, with $V_{\text{inmax}} = 0.5$ V.

(a) Determine the value of resistance R_1, so that the amplifier operates in Class A, without distortion and with the V_{inmax} signal, while the dissipation on the transistors at a quiescent operating point is minimal. The input characteristic of both transistors can be approximated by the graph given in Fig. 4.B.13.
(b) Determine the dissipated power on the transistors at the quiescent operating point.
(c) Determine the maximum useful power at the load (when the input is excited by the voltage of maximum amplitude) as well as the efficiency of the amplifier.

It is known that: $V_{\text{CC}} = 12$ V; $R_2 = 20$ kΩ; $R_{\text{L}} = 4$ Ω; $n = 5$; $C \to \propto$. The transistors have identical parameters: $V_\gamma = 0.5$ V; $h_{\text{inE}} = r_{\text{B}} = 1$ kΩ; $h_{\text{rE}} = 0$; $h_{\text{fE}} = \beta = 100$ and $h_{\text{oE}} = 0$ S.

Solution to Problem 4.B.4

(a) The push–pull circuit is formed of two active elements with identical characteristics, and it enables obtaining twice the power with a significant reduction of nonlinear distortions as compared to the amplifier with single element. This circuit is achieved by coupling two identical stages, so that the battery circuit is common, which means that the quiescent operating points of these two active elements are the same:

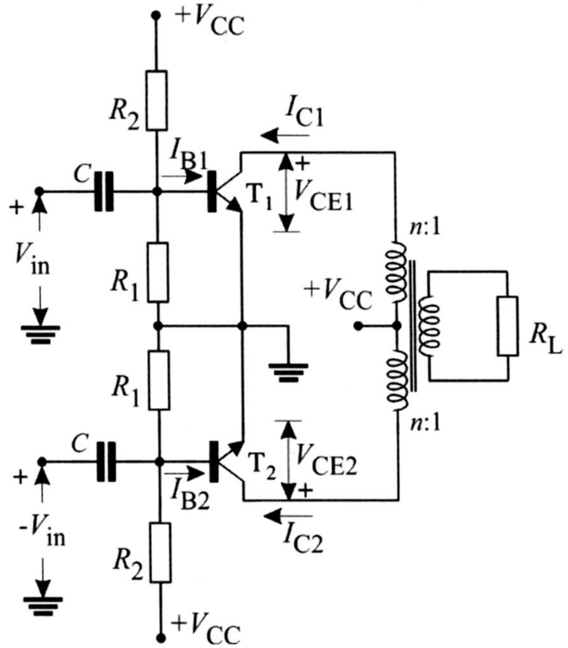

Fig. 4.B.12 Class A push-pull amplifier

$$I_{BQ1} = I_{BQ2} = I_{BQ},$$
$$I_{CQ1} = I_{CQ2} = I_{CQ},$$
$$V_{CEQ1} = V_{CEQ2} = V_{CEQ}. \tag{4.B.4.1}$$

To determine the value of resistor R_1, which provides the maximum undistorted input signal of amplitude $V_{inmax} = 0.5$ V, it should be ensured that the quiescent operating point on the input characteristic satisfies the condition given in Fig. 4.B.14. For this purpose, it is necessary to perform DC analysis of the equivalent input circuit of the transistor shown in Fig. 4.B.15, for which it is possible to write the voltage of the Thevenin source as:

$$V_{BB} = \frac{R_1}{R_1 + R_2} \cdot V_{CC} = k \cdot V_{CC}. \tag{4.B.4.2}$$

Its internal resistance is:

$$R_{BB} = \frac{R_1 \cdot R_2}{R_1 + R_2} = k \cdot R_2, \tag{4.B.4.3}$$

where

$$k = \frac{R_1}{R_1 + R_2}. \tag{4.B.4.4}$$

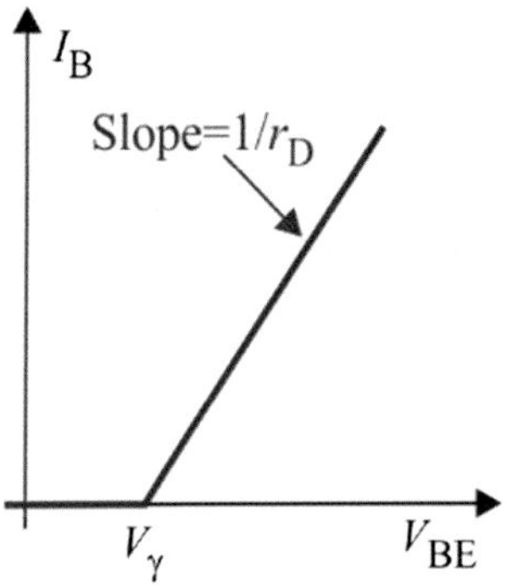

Fig. 4.B.13 Approximation of the input characteristic

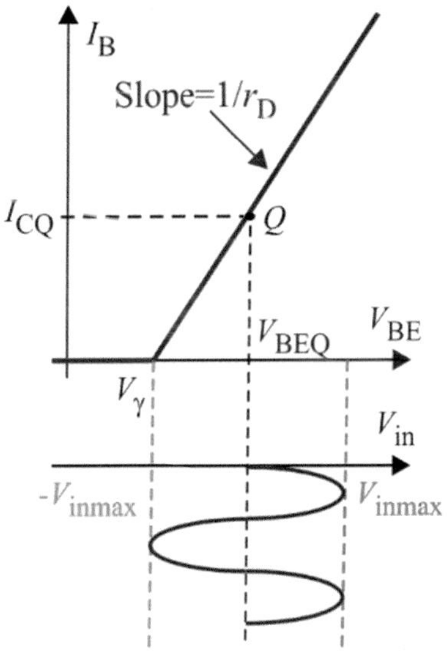

Fig. 4.B.14 Evaluation of the maximum amplitude of the input signal

Based on Fig. 4.B.15 it is possible to find the voltage between the base and the emitter of the transistor at the operating point, V_{BEQ}, as

$$V_{BEQ} = \frac{r_B}{r_B + R_B} \cdot V_{BB} + \frac{R_B}{R_B + r_B} \cdot V_\gamma. \tag{4.B.4.5}$$

On the other side to satisfy the condition given in the problem it is necessary that

$$V_{BEQ} = V_\gamma + V_{inmax}. \tag{4.B.4.6}$$

By substitution of (4.B.4.2) and (4.B.4.3) in (4.B.4.5) one gets

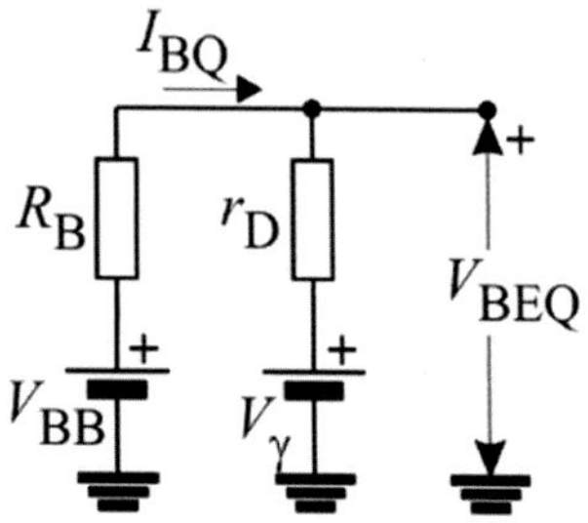

Fig. 4.B.15 Circuit for finding I_{BEQ}

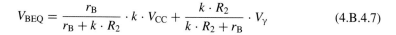

$$V_{BEQ} = \frac{r_B}{r_B + k \cdot R_2} \cdot k \cdot V_{CC} + \frac{k \cdot R_2}{k \cdot R_2 + r_B} \cdot V_\gamma \tag{4.B.4.7}$$

Then, by equating the expressions (4.B.4.6) and (4.B.4.7), one gets the value of k as

$$k = \frac{r_B \cdot (V_{inmax} + V_\gamma)}{r_B \cdot V_{CC} - R_2 \cdot V_{inmax}} = 0.5. \tag{4.B.4.8}$$

Now it easy to extract the value of R_1 as

$$R_1 = R_2 \cdot \frac{k}{1-k} = R_2 = 20\ \mathrm{k}\Omega.$$

(b) To determine the DC dissipation on the transistors, it is necessary to previously determine the collector currents and voltages between the collectors and the emitters at a quiescent operating point for both transistors:

$$P_D = I_{CQ1} \cdot V_{CEQ1} + I_{CQ2} \cdot V_{CEQ2}. \tag{4.B.4.9}$$

Since the configuration is symmetrical, the collector currents can be considered equal, as well as the voltages between the collectors and emitters of these two transistors, so that for the dissipation power we have:

$$P_D = 2 \cdot I_{CQ} \cdot V_{CEQ}. \tag{4.B.4.10}$$

This means that one needs to find I_{CQ} and V_{CEQ} for a single transistor. Based on the circuit of Fig. 4.B.15 it can be found that the base current at the quiescent operating point is

$$I_{BQ} = \frac{V_{BB} - V\gamma}{R_B + r_B} = 0.5\ \mathrm{mA}. \tag{4.B.4.11}$$

where

$$R_B = 0.5 \cdot R_2 = 10\ \mathrm{k}\Omega, \tag{4.B.4.12}$$

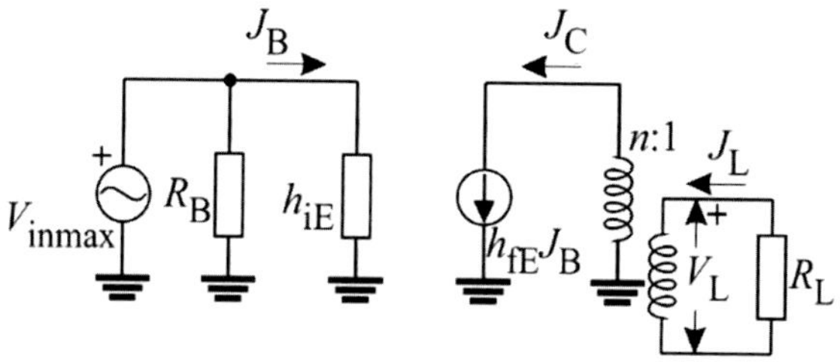

Fig. 4.B.16 AC equivalent circuit of the one of Fig. 4.B.12

and

$$V_{BB} = 0.5 \cdot V_{CC} = 6 \text{ V}. \tag{4.B.4.13}$$

The collector current is now:

$$I_{CQ} = \beta \cdot I_{BQ} = 50 \text{ mA}, \tag{4.B.4.14}$$

while based on Fig. 4.B.12 it is obvious that the voltage between the collector and the emitter is

$$V_{CEQ} = V_{CC} = 12 \text{ V}. \tag{4.B.4.15}$$

Finally, for the dissipation power on the transistors we get

$$P_D = 1.2 \text{ W}. \tag{4.B.4.16}$$

(c) To determine the useful power on the load, it is necessary to analyze the circuit shown in Fig. 4.B.16 under the presumption that $J_{B1} = -J_{B2} = J_B$ and $J_{C1} = -J_{C2} = J_C$. For the circuit with Fig. 4.B.16 it is possible to claim that the base current is

$$J_B = \frac{V_{inmax}}{h_{iE}} = 0.5 \text{ mA}. \tag{4.B.4.17}$$

Now, the collector current is

$$J_C = h_{fE} \cdot J_B = 50 \text{ mA}, \tag{4.B.4.18}$$

so that the load current is

$$J_{out} = n \cdot (J_{C1} - J_{C2}) = 2nJ_C = 0.5 \text{ A}. \tag{4.B.4.19}$$

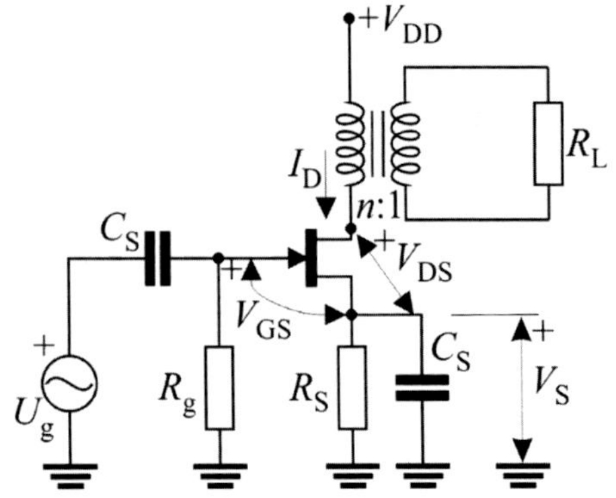

Fig. 4.B.17 Class A amplifier using transformer and JFET

The useful power dissipated on the load is now

$$P_L = \frac{1}{2} \cdot R_L \cdot J_{out}^2 = 0.5 \text{ W}. \quad (4.B.4.20)$$

Finally, the efficiency is:

$$\eta = 100 \cdot \frac{P_L}{P_D} = 41\%. \quad (4.B.4.21)$$

Problem 4.B.5

Design the large signal amplifier with a JFET, shown in Fig. 4.B.17, so that it operates in Class A, and that on the load resistance $R_L = 10\ \Omega$ delivers maximum useful power, with minimal distortion of the input signal, when powered by a battery voltage $V_{DD} = 24$ V. The circuit uses JFET with $V_P = -3$ V and $P_{dmax} = 6$ W. Assume that the minimum drain current is $I_{Dmin} = 10$ mA.

Solution to Problem 4.B.5

The design of the amplifier means the determination of unknown circuit elements, namely the values of the R_S resistance and the transformation ratio n, based on the data given by the task conditions:

- In order for the amplifier to operate in Class A, the operating point must not migrate out of the area of voltage saturation.
- In order to obtain maximum power, the dynamic load line must touch the dissipation hyperbola defined by P_{dmax}.
- In order to obtain the maximum undistorted signal, it is necessary that the quiescent operating point is in the middle between I_{Dmax} and I_{Dmin}.

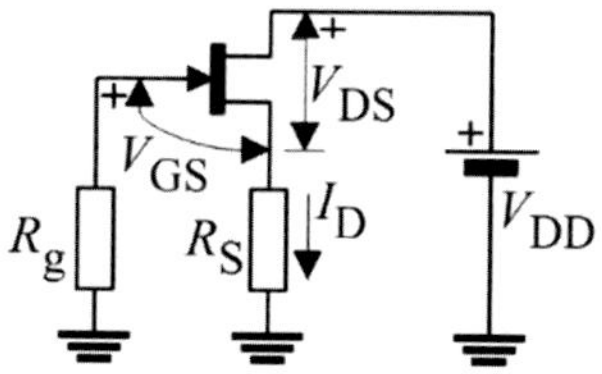

Fig. 4.B.18 DC equivalent circuit of the one of Fig. 4.B.17

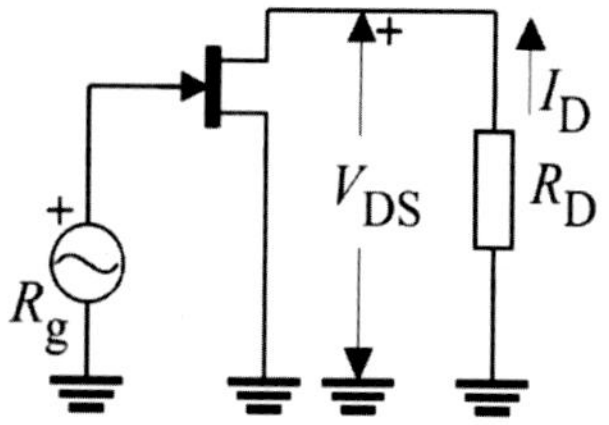

Fig. 4.B.19 AC equivalent circuit of the one of Fig. 4.B.17

It is obvious, from Fig. 4.B.18, that the DC operating point is on a DC load line defined by:

$$I_D = (V_{DD} - V_{DS})/R_s, \tag{4.B.5.1}$$

At the same time, since there is no gate current, one may write

$$I_D = -V_{GS}/R_S. \tag{4.B.5.2}$$

For the transistor to operate in saturation it is necessary that $V_{DS} > V_{DSsat}$, where

$$V_{DSsat} = V_{GS} - V_P. \tag{4.B.5.3}$$

After substitution of V_{GS} from (4.B.5.2), and R_S from (4.B.5.1), one gets

$$V_{DSsat} = -R_S \cdot I_D - V_P = -\frac{V_{DD} - V_{DSQ}}{I_{DQ}} \cdot I_D - V_P. \tag{4.B.5.4}$$

The other two conditions refer to the AC mode of operation of the amplifier, for which the equivalent circuit shown in Fig. 4.B.19 is valid.

The drain circuit is loaded with the mapped resistance

$$R_D = n^2 \cdot R_L, \tag{4.B.5.5}$$

where n is the transformer's turn ratio. The AC load line is determined by this resistance in the following way

$$I_D = -V_{DS}/R_D. \tag{4.B.5.6}$$

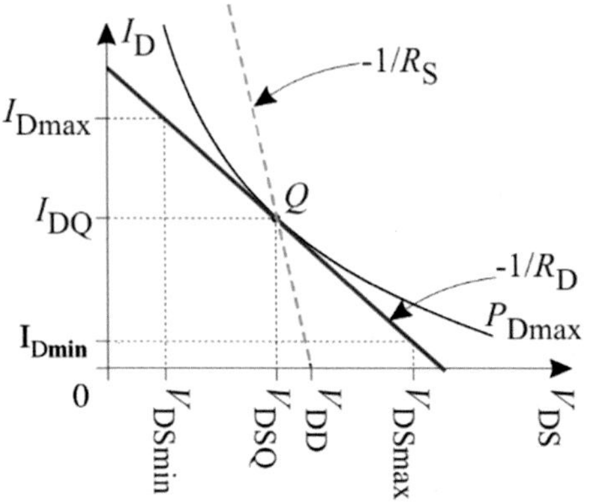

Fig. 4.B.20 Graphic analysis of the Class A amplifier using transformer and JFET

This line must touch the power hyperbola P_{dmax} which, in the quiescent point, is defined by

$$I_D = P_{Dmax}/V_{DS} \tag{4.B.5.7}$$

This means that the slope of the line (4.B.5.6) must be equal to the derivative of (4.B.5.7) at $I_D = I_{DQ}$, or:

$$-\frac{1}{R_D} = \left.\frac{dI_D}{dV_{DS}}\right|_{V_{DS}=V_{DSQ}} = -\frac{P_{Dmax}}{V_{DSQ}^2}. \tag{4.B.5.8}$$

So, we obtain

$$R_D = \frac{V_{DSQ}^2}{P_{Dmax}} = \frac{P_{Dmax}}{I_{DQ}^2}. \tag{4.B.5.9}$$

Now, since we know the slope of the AC load line and the quiescent operating point defined as (I_{DQ}, V_{DSQ}), we can write:

$$I_D - I_{DQ} = -\frac{1}{R_D}\left(V_{DS} - V_{DSQ}\right) = -\frac{I_{DQ}^2}{P_{Dmax}}\left(V_{DS} - \frac{P_{Dmax}}{I_{DQ}}\right). \tag{4.B.5.10}$$

To obtain minimal distortion of the output signal, as can be seen from Fig. 4.B.20, it is necessary to position the quiescent operating point in the middle of the two limiting values of drain current, i.e.,

$$I_{DQ} = \frac{1}{2}(I_{Dmax} - I_{Dmin}). \tag{4.B.5.11}$$

The unknown value of I_{Dmax} is found in the interception of the AC load line (4.B.5.10) and the line defining the voltage V_{DSsat} (4.B.5.4), or:

$$I_{\mathrm{Dmax}} - I_{\mathrm{DQ}} = -\frac{I_{\mathrm{DQ}}^2}{P_{\mathrm{Dmax}}}\left(V_{\mathrm{DSmin}} - \frac{P_{\mathrm{Dmax}}}{I_{\mathrm{DQ}}}\right), \tag{4.B.5.12}$$

$$V_{\mathrm{DSmin}} = -\frac{V_{\mathrm{DD}} - V_{\mathrm{DSQ}}}{I_{\mathrm{DQ}}} \cdot I_{\mathrm{Dmax}} - V_{\mathrm{P}}. \tag{4.B.5.13}$$

By elimination of V_{DSmin} from the previous expressions and by substitution of I_{Dmax} via I_{DQ} and I_{Dmin} from (4.B.5.11) we obtain a quadratic equation in I_{DQ}:

$$I_{\mathrm{DQ}}^2(2V_{\mathrm{DD}} + V_{\mathrm{P}}) - I_{\mathrm{DQ}}(2P_{\mathrm{Dmax}} + I_{\mathrm{Dmin}}V_{\mathrm{DD}}) + 2P_{\mathrm{Dmax}} \cdot I_{\mathrm{Dmin}} = 0, \tag{4.B.5.14}$$

whose solution is

$$I_{\mathrm{DQ}} = \frac{2P_{\mathrm{Dmax}} + I_{\mathrm{Dmin}} \cdot V_{\mathrm{DD}}}{2(2V_{\mathrm{DD}} + V_{\mathrm{P}})} \cdot \left[1 + \sqrt{1 - \frac{8I_{\mathrm{Dmin}} \cdot P_{\mathrm{Dmax}} \cdot (2V_{\mathrm{DD}} + V_{\mathrm{P}})}{(2P_{\mathrm{Dmax}} + I_{\mathrm{Dmin}} \cdot V_{\mathrm{DD}})^2}}\right]. \tag{4.B.5.15}$$

By substitution of the numerical values, we get

$$I_{\mathrm{DQ}} = 0.26 \text{ A}. \tag{4.B.5.16}$$

Then, using this current for (4.B.5.9) one gets

$$R_{\mathrm{D}} = 88.76\ \Omega. \tag{4.B.5.17}$$

Based on (4.B.5.5) it is easy to find that a turn ratio required is

$$n = \sqrt{R_{\mathrm{D}}/R_{\mathrm{L}}} \approx 3. \tag{4.B.5.18}$$

To find R_{S} from (4.B.5.1) we first need to calculate V_{DSQ} from (4.B.5.7). The result is

$$R_{\mathrm{S}} = V_{\mathrm{DD}}/I_{\mathrm{DQ}} - P_{\mathrm{Dmax}}/I_{\mathrm{DQ}}^2 = 3.46\ \Omega. \tag{4.B.5.19}$$

Problem 4.B.6

Figure 4.B.21 shows a power amplifier with a complementary pair, operating in Class B.

- Draw the transfer characteristic of the amplifier: $V_{\mathrm{out}} = f(V_{\mathrm{in}})$.
- Determine the optimal value of the resistance R_{L} at which maximum load power is obtained.
- Determine the waveforms of the voltage on the load and the current through the transistors if the amplifier is excited by a signal of the form $v_{\mathrm{in}} = V_{\mathrm{inm}}\sin(\omega t)$.

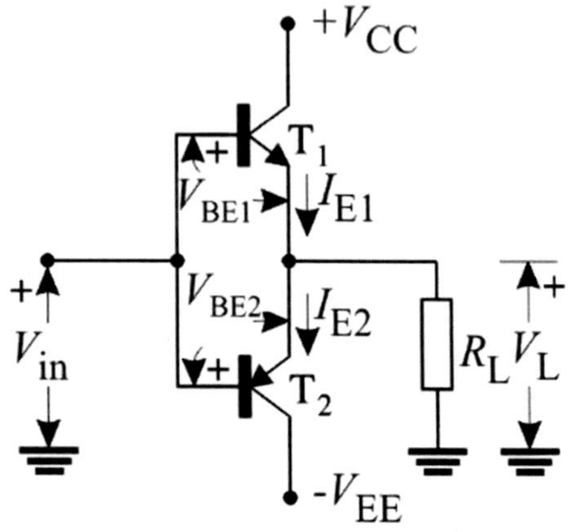

Fig. 4.B.21 A simple amplifier with complementary pair of BJTs

The supply voltages are $V_{CC} = V_{EE} = 12$ V, while the maximum dissipation power for each transistor is $P_{Dmax} = 3.6$ W. It is known: $V_{BE1} = -V_{BE2} = 0.7$ V; $V_{\gamma 1} = -V_{\gamma 2} = 0.4$ V; $V_{BES1} = -V_{BES2} = 0.8$ V; $V_{CES1} = -V_{CES2} = 0.2$ V.

Solution to Problem 4.B.6

(a) To start the analysis of this circuit, it is most convenient to say that the voltage at the input is negative and by modulus higher than the supply voltage V_{CC} (assume that $V_{in} = -15$ V). Also, suppose both transistors do not conduct. In this case the voltage on the load is $V_L = 0$ V. Now the voltages between the base and the emitter of both transistors are:

$$V_{BE1} = V_{in} - R_L \cdot I_{E1} = V_{in} < 0; \tag{4.B.6.1}$$

$$V_{BE2} = V_{in} - V_L = V_{in} < V_{\gamma 2}. \tag{4.B.6.2}$$

Given the obtained values, it is obvious that the base-emitter junction of transistor T_1 is backward biased, so this transistor does not conduct, while the base-emitter junction of transistor T_2 is forward biased, which means that our assumption was incorrect, i.e., transistor T_2 conducts. To determine the operating region of transistor T_2, let us start from the assumption that it conducts in the active region. In this case, it turns out that the voltage on the load is

$$V_L = V_{in} - V_{BE2} = -14.3 \text{ V}. \tag{4.B.6.3}$$

Now for the voltage between the collector and emitter of the second transistor we get that:

$$V_{CE2} = -V_{CC} - V_L = 2.3 \text{ V}, \tag{4.B.6.4}$$

that is, it turns out that the collector is at a higher potential than the emitter, which means that the assumption that the transistor works in the active area is not correct. So, transistor T_2 works in the (current) saturation region, i.e., $V_{CE2} = V_{CEs2} = -0.2$ V, so that the voltage on the load is $V_L = -11.8$ V.

Increasing the input voltage will change the operating region of T_2, which will switch to the active area at certain input voltage. That voltage is:

$$V_{inmin} = -V_{EE} - V_{CEs2} + V_{BEs2} = -12.6\text{V}, \tag{4.B.6.5}$$

so that, at this input voltage, the voltage at the load is:

$$V_{Lmin} = -V_{EE} - V_{CEs2} = -11.8\text{V}. \tag{4.B.6.6}$$

By further increasing the input voltage, transistor T_1 still does not conduct, while T_2 goes into the active area. It will operate in the active area, until the input voltage drops to the value $V_{\gamma 2}$, when the transistor stops conducting. So, for input voltages $-V_{EE} - V_{CEs2} + V_{BEs2} \leq V_{in} < V_{\gamma 2}$ the voltage on the load will change according to

$$V_L = V_{in} - V_{BE2}. \tag{4.B.6.7}$$

For input voltages higher than $V_{\gamma 2}$ neither transistor conducts, nor the voltage on the load is $V_L = 0$ V. This situation is valid until the input voltage reaches the value $V_{\gamma 1}$, when the transistor T_1 starts conducting and that in the active area. The load voltage is now

$$V_L = V_{in} - V_{BE1}. \tag{4.B.6.8}$$

This will be valid until transistor T_1 goes into (current) saturation, which will happen at the input voltage

$$V_{inmax} = V_{CC} - V_{CEs1} + V_{BEs1} = 12.6 \text{ V}. \tag{4.B.6.9}$$

Then we will have

$$V_{Lmax} = V_{CC} - V_{CEs1} = 11.8 \text{ V}. \tag{4.B.6.10}$$

By further increasing the input voltage, the situation in the circuit remains unchanged. Finally, the complete expression for the transfer characteristic of the circuit is:

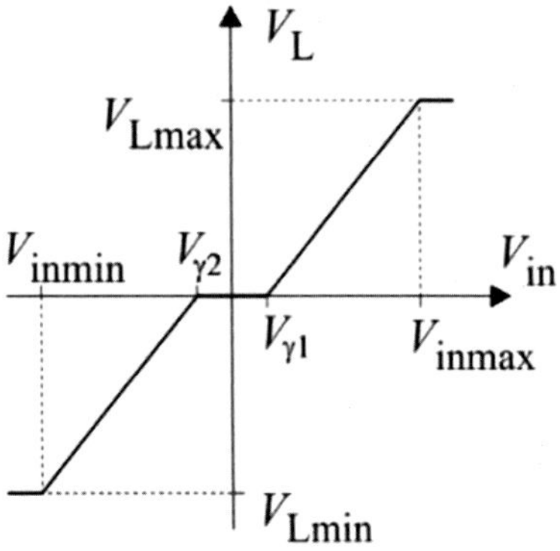

Fig. 4.B.22 Transfer characteristic of the simple amplifier with complementary pair of BJTs

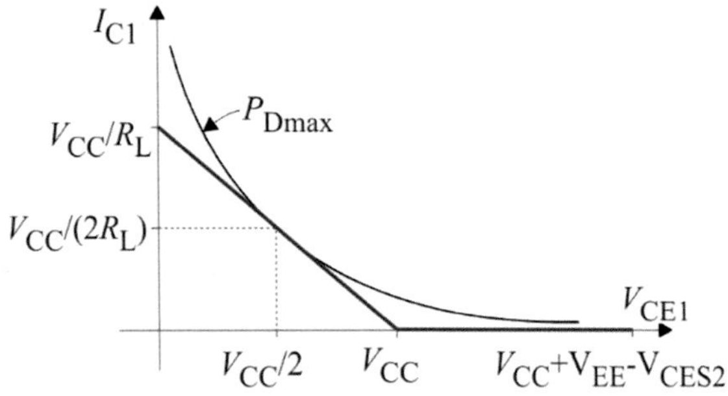

Fig. 4.B.23 Graphic analysis of the simple amplifier with complementary pair of BJTs

$$V_L = \begin{cases} -V_{EE} - V_{CES2}, & V_{in} < -V_{EE} - V_{CES2} + V_{BES2} \\ V_{in} - V_{BE2}, & -V_{EE} - V_{CES2} + V_{BES2} \le V_{in} \le V_{\gamma 2} \\ 0, & V_{\gamma 2} \le V_{in} \le V_{\gamma 1} \\ V_{in} - V_{BE1}, & V_{\gamma 1} \le V_{in} \le V_{CC} - V_{CES1} + V_{BES1} \\ V_{CC} - V_{CES1}, & V_{in} > V_{CC} - V_{CES1} + V_{BES1} \end{cases} \tag{4.B.6.11}$$

This transfer characteristic is graphically represented in Fig. 4.B.22.

(b) To determine the optimal value of the resistor R_L at which the maximum power is obtained on the load, it must be considered that the load line touches the power hyperbola, as shown in Fig. 4.B.23. It is obvious from Fig. 4.B.21 that the AC and DC load lines coincide. Figure 4.B.23 shows the operation of the circuit in the positive half-cycle of the input signal. The load lines drawn are described by the equation

$$I_{C1} = (V_{CC} - V_{C1})/R_L.$$

It was written with the presumption that the lower transistor is cut-off. It was assumed that R_L was chosen so as to obtain maximum useful power.

The instantaneous value of the power which is dissipated on the load in the positive half-cycle of the input signal is

$$p_L = v_{CE} \cdot i_C = (V_{CC} - R_L \cdot i_C) \cdot i_C. \tag{4.B.6.12}$$

To determine the optimal value of R_L, the value of the current i_C, at which the maximum power is dissipated at the load, must be determined first. This value is determined by equalizing the first derivative of the instantaneous power p_L with respect to i_C with zero, i.e.,

$$\mathrm{d}p_L/\mathrm{d}i_C = V_{CC} - 2R_L \cdot i_C = 0. \tag{4.B.6.13}$$

Here, the current J_{Cm} producing maximum power is

$$J_{Cm} = V_{CC}/(2R_L). \tag{4.B.6.14}$$

Accordingly, the maximum load power is

$$P_{Lmax} - (V_{CC} - R_L J_{Cm}) \cdot J_{Cm} = V_{CC}^2/(4R_L). \tag{4.B.6.15}$$

Hence, the optimum value of the load resistance R_L is

$$R_{Lopt} = V_{CC}^2/(4P_{Lmax}) = 10\,\Omega. \tag{4.B.6.16}$$

(c) The voltage and current waveforms in the circuit are shown in Fig. 4.B.24. The mean value of the current flowing through the battery V_{CC} can be determined as:

$$I_{CC} = \frac{1}{T} \cdot \int_0^T i_{C1}(t) \cdot \mathrm{d}t = \frac{1}{T} \cdot \int_0^{T/2} J_{C1m} \sin(\omega t) \cdot \mathrm{d}t, \tag{4.B.6.17}$$

where $J_{C1m} = V_{CC}/(2R_L)$ is the maximum value of the collector current of T_1. By calculating this integral, we obtain the current I_{CC} as

$$I_{CC} = J_{C1m}/\pi. \tag{4.B.6.18}$$

Similarly, the mean value of the current flowing through the battery V_{EE} is

$$I_{EE} = -J_{C2m}/\pi, \tag{4.B.6.19}$$

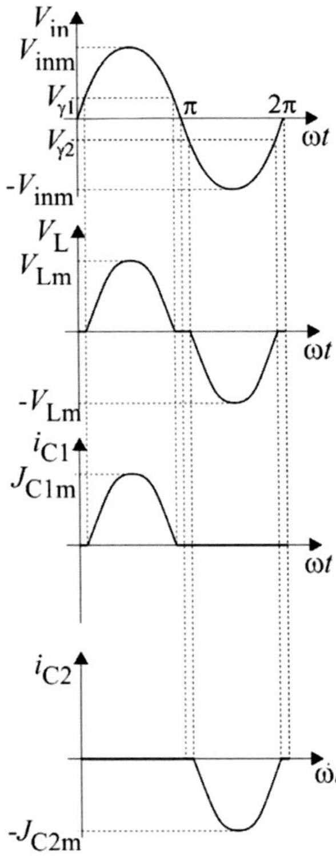

Fig. 4.B.24 Waveforms in the circuit of Fig. 4.B.21

where $J_{C2m} = V_{EE}/(2R_L)$ is the modulus of the maximum value of the collector current of T_2. Since the voltages V_{EE} and V_{CC} are equal the following is valid: $J_{C2m} = J_{C1m}$. Therefrom we conclude that $I_{CC} = -I_{EE}$ is also valid.

(d) The power delivered by the power sources can be calculated as

$$P = V_{CC} \cdot I_{CC} - V_{EE} \cdot I_{EE}, \tag{4.B.6.20}$$

which after substitution for I_{CC} and I_{EE} becomes

$$P = 2V_{CC} \cdot J_{C1m}/\pi. \tag{4.B.6.21}$$

The useful power delivered to the load can be defined as

$$P_L = 0.5 \cdot V_{Lm} \cdot J_{C1m}, \tag{4.B.6.22}$$

so that the efficiency is

Fig. 4.B.25 A simple Class B amplifier

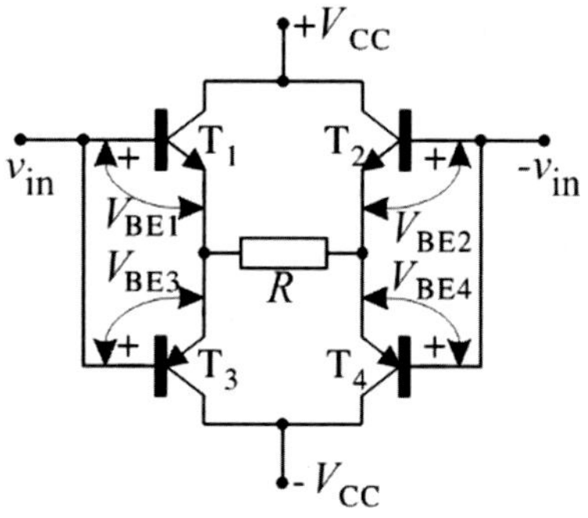

$$\eta = 100\frac{P_L}{P} = 100\frac{\pi}{4}\frac{V_{Lm}}{V_{CC}}. \tag{4.B.6.23}$$

Having in mind that $V_{Lm} \approx V_{CC}$ we get

$$\eta = 78.6\%. \tag{4.B.6.24}$$

Problem 4.B.7

Figure 4.B.25 shows a power amplifier operating in Class B. It is known: $V_{CC} = 40$ V; $R = 25\ \Omega$; $v_{in} = V_m \cdot \sin(\omega t)$.

Determine:

- maximum instantaneous power delivered to R, if $V_{CEmin} = 0$ V;
- maximum average undistorted power on resistor R;
- maximum mean power dissipation on any transistor;
- the output resistance of the amplifier seen by the resistor R, under the assumption that the circuit is excited by a signal of small amplitude, if: $h_{rE} = 0$ and $h_{oE} = 0$ S;
- the efficiency η of the power amplifier.

Solution to Problem 4.B.7

(a) In order to determine the maximum power, it is necessary to first pay attention to the principle of operation of this circuit. It is obvious that the transistors in the circuit are brought into the conducting state by means of the input voltage v_{in}. It is also obvious that transistors T_1 and T_4 conduct in the positive half-cycle of v_{in} (base-emitter junctions are forward biased, i.e., $V_{BE1} > 0$ and $V_{BE4} < 0$), while in the negative half-cycle conduct the transistors T_2 and T_3 ($V_{BE2} > 0$ and $V_{BE3} < 0$). It should be noted that the voltages between the collector and emitter of the transistors conducting in one half-cycle are the same at all times due to the identical characteristics of the transistors. Observed in one half-cycle, it is possible to notice that when between the collector and emitter of the two transistors leading in that half-cycle the voltage is minimal, $V_{CEmin} = 0$ V, the

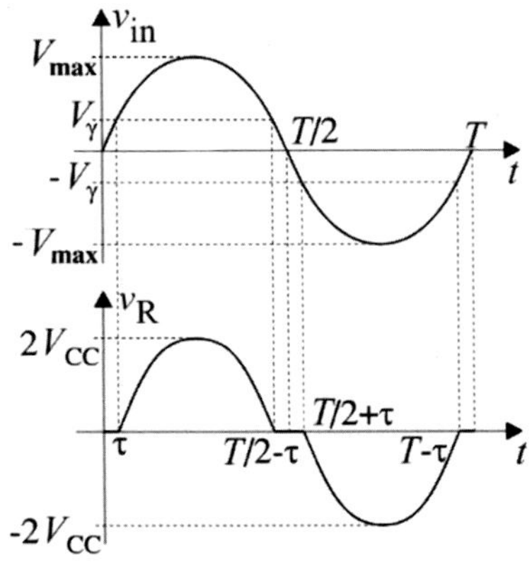

Fig. 4.B.26 Waveforms in the circuit of Fig. 4.B.22

voltage across the resistor will be maximum: $V_{Rmax} = 2V_{CC}$. So the maximum instantaneous power dissipated on resistor R is now equal to

$$P_{Rmax} = (2V_{CC})^2/R = \ 256 \text{ W}. \tag{4.B.7.1}$$

(b) To determine the maximum of the average undistorted power on the resistor, it is necessary to analyze the voltage waveform on the resistor, which is shown in Fig. 4.B.26. As can be seen from this figure, transistors T_1 and T_4 will conduct in the interval from τ to $(T/2 - \tau)$, while transistors T_2 and T_3 will lead in the interval from $(T/2 + \tau)$ do $(T - \tau)$. The amplitude of this distorted sinusoid is obviously equal to $2V_{CC}$, as shown under a). Given the voltage values in the circuit, it can be considered that τ is negligibly small, i.e., that $\tau \to 0$. Taking this into account, we get that the time dependence of the voltage on the resistor as

$$v_R(t) = 2V_{CC} \cdot \sin(\omega t). \tag{4.B.7.2}$$

It is now possible to write that the maximum average undistorted power on resistor R as

$$P_R = \frac{1}{T}\int_0^T \frac{v_R^2(t)}{R} \cdot dt = 128 \text{ W}. \tag{4.B.7.3}$$

(c) When determining the maximum average power on one transistor, it is necessary to first determine the instantaneous value of this power. This can be done by first determining the instantaneous value of the voltage between the collector

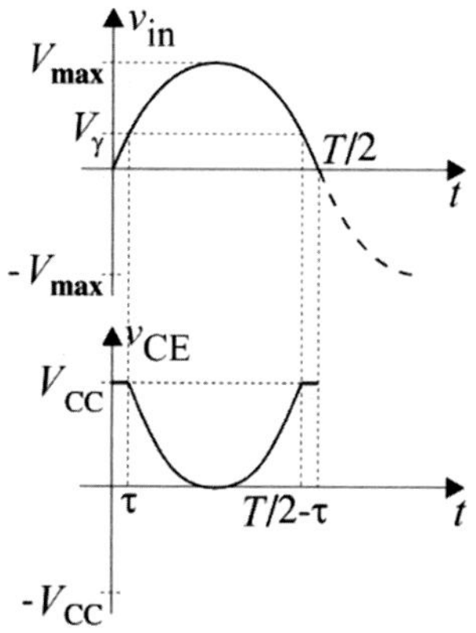

Fig. 4.B.27 Voltage waveforms in the circuit of Fig. 4.B.25

and the emitter, as well as the instantaneous value of the collector current. It is clear from Fig. 4.B.26 that the voltage between the collector and the emitter in one half-cycle is opposite to the voltage across the resistor (by this is meant that the voltage between the collector and the emitter is maximum per modulus when the voltage across the resistor is 0 V or equal to zero when the voltage across the resistor is maximum per modulus). The voltage waveform v_{CE} in the first half-cycle for transistor T_1 is defined by the expression

$$v_{CE}(t) = V_{CC} \cdot [1 - \sin(\omega t)] \tag{4.B.7.4}$$

and depicted in Fig. 4.B.27.

The collector current of any conducting transistor is equal to the current through the resistor, i.e., it is equal to the quotient of the voltage across the resistor and the resistance R so as

$$i_C(t) = i_R(t) = \frac{2V_{CC}}{R} \cdot \sin(\omega t). \tag{4.B.7.5}$$

The instantaneous value of the dissipation power on the transistor is now

$$p_T(t) = v_{CE} \cdot i_C = \frac{2V_{CC}^2}{R} \cdot [1 - \sin(\omega t)] \cdot \sin(\omega t), \tag{4.B.7.6}$$

so that the maximum average power on single transistor can be calculated as

$$P_T = \frac{1}{T} \int_0^{T/2} p_T(t) \cdot \mathrm{d}t. \tag{4.B.7.7}$$

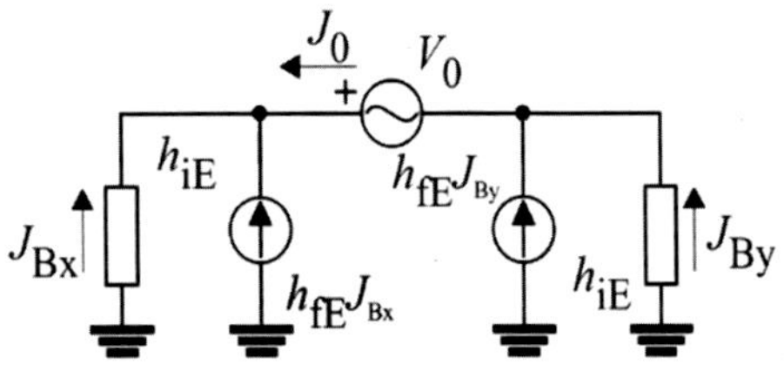

Fig. 4.B.28 AC equivalent circuit of the circuit of Fig. 4.B.25

After substitution of (4.B.7.6) and calculating the integral, we get

$$P_T = \frac{2V_{CC}^2}{R} \cdot \left(\frac{1}{\pi} - \frac{1}{4}\right) = 8.744\ \text{W}. \quad (4.B.7.8)$$

(d) Since the transistors are identical and two transistors are conducting in one half-cycle, the output resistance of the amplifier seen by the resistor R can be determined by analyzing the AC circuit in one half-cycle, shown in Fig. 4.B.28. According to the notation used in the figure, it is possible to get J_0 as

$$J_0 = -(1 + h_{fE}) \cdot J_{Bx} = (1 + h_{fE}) \cdot J_{By}, \quad (4.B.7.9)$$

where x ∈ (1, 2) and y ∈ (3, 4). From (4.B.7.9) we conclude that $J_{By} = -J_{Bx}$, and, based on the figure, we can write

$$J_{By} = \frac{V_0}{2h_{iE}}. \quad (4.B.7.10)$$

After substitution of (4.B.7.10) into (4.B.7.9) for the output resistance of the amplifier we get

$$R_{out} = \frac{V_0}{J_0} = \frac{2h_{iE}}{1 + h_{fE}}. \quad (4.B.7.11)$$

(e) The power that the battery delivers to the circuit is

$$P_B = 4P_T + P_R, \quad (4.B.7.12)$$

so that the efficiency is

$$\eta = \frac{P_R}{P_B} \cdot 100 = 78.5\%. \quad (4.B.7.13)$$

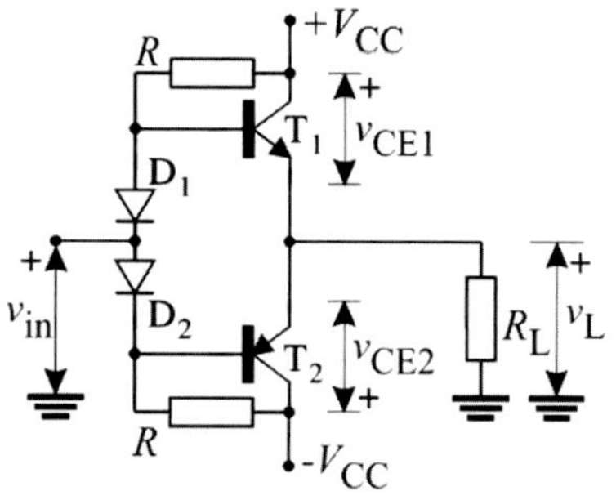

Fig. 4.B.29 Class B power amplifier using two diodes

Problem 4.B.8

Figure 4.B.29 shows a power amplifier operating in Class B. The power amplifier can deliver 50 W of the maximum average power to the load resistor R_L. If the excitation voltage $v_{in}(t) = V_m \cdot \sin(\omega t)$, of sufficiently large amplitude, V_m is applied to the circuit input, determine:

- the minimum value of the supply voltage V_{CC}, which can be used;
- the amplitude of the sinusoidal voltage at the output of the circuit for which transistors T_1 and T_2 dissipate maximum average collector power ($P_C = I_C \cdot V_{CE}$), as well as P_{Cmax};
- the maximum average power that can be delivered by the power supply V_{CC} and $-V_{CC}$;
- the efficiency of the power amplifier.

Assume that $V_{CEmin} = 0$ V, the diodes are ideal, and that the input characteristic of the transistor is given in Fig. 4.B.13. It is known $R_L = 10\ \Omega$.

Solution to Problem 4.B.8

(a) By analysis of the circuit depicted in Fig. 4.B.29 it can be observed that in the positive half-cycle the transistor T_1 is conducting, while in the negative half-cycle conducts transistor T_2. During the positive half-cycle of the input voltage, the voltage on the load is equal to $v_L = V_{CC} - v_{CE1}$, while during the negative half-cycle the voltage on the load is equal to $v_L = -V_{CC} - v_{CE2}$. If the transistors are symmetrical, a sinusoidal voltage is obtained at the output as shown in Fig. 4.B.30. By definition, the load power can be calculated as

$$P_L = 0.5 \cdot V_{Lm} \cdot J_{Lm}. \quad (4.B.8.1)$$

To determine the maximum value of the voltage on the load, as a function of V_{CC}, it is sufficient to analyze the gain of the circuit during one half-cycle of the input signal. Considering that the voltage on the load during the positive half-cycle

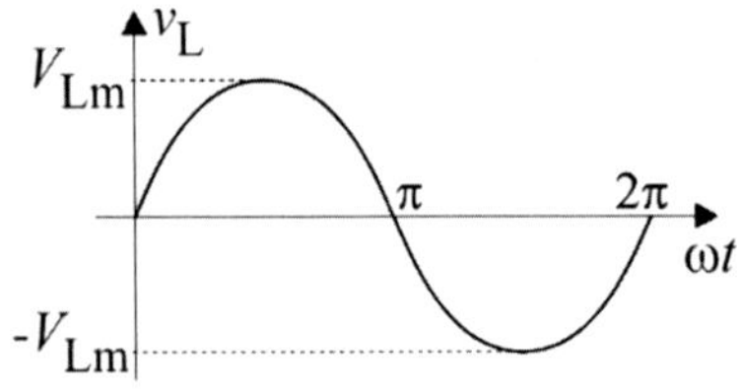

Fig. 4.B.30 Output voltage Waveform of the circuit of Fig. 4.B.29

is $v_L = V_{CC} - v_{CE1}$, it is obvious that the voltage v_L will be maximum when v_{CE1} is minimum, i.e.,

$$V_{Lm} = V_{CC} - V_{CEmin} = V_{CC}. \tag{4.B.8.2}$$

If so, the following is valid

$$J_{Lm} = \frac{V_{Lm}}{R_L} = \frac{V_{CC}}{R_L}. \tag{4.B.8.3}$$

By substituting the expressions (4.B.8.2) and (4.B.8.3) into the expression for the maximum power on the resistor R_L, (4.B.8.1), and rearranging it, for the minimum value of supply voltage we can write

$$V_{CCmin} = \sqrt{2R_L \cdot P_L} = 31.62 \text{ V}. \tag{4.B.8.4}$$

(b) To determine the maximum value of the sinusoidal voltage at the output of the circuit for which the transistors dissipate maximum average collector power, it is necessary to determine the expression for the average collector power. Due to the symmetry of the circuit, it is again sufficient to analyze the operation of the circuit in one half-cycle only.

To determine the average collector dissipation of one of the transistors, it is necessary to determine the waveforms of the collector current and voltage between the collector and the emitter, as well as the expressions for these two quantities.

The transistor T_1 conducts in the positive half-cycle, and during that time interval the collector current of this transistor is equal to the current through the resistor R_L. This further means that the collector current of transistor T_1 changes according to the sinusoidal law and that its amplitude is equal to the amplitude of the current through R_L, i.e., that the collector current of transistor T_1 in this half-period can be described by

$$i_{C1}(t) = (V_{Lm}/R_L) \cdot \sin(\omega t), \tag{4.B.8.5}$$

considering that it is equal to zero in the negative half-cycle.

As for the voltage between the collector and the emitter of the transistor T_1, it can be seen that at the moment when the transistor begins to conduct it is at its maximum

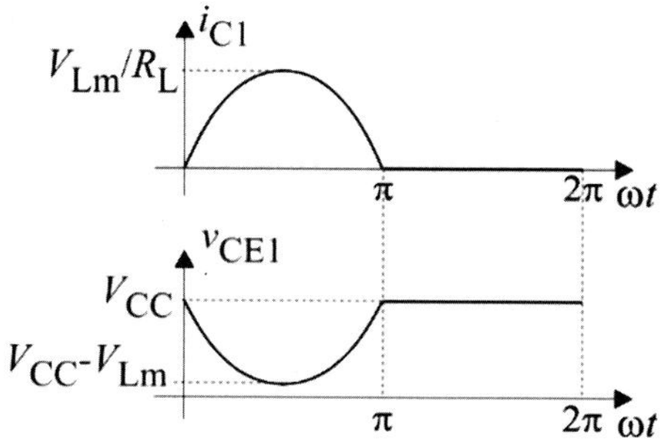

Fig. 4.B.31 Transistor related waveforms in the circuit of Fig. 4.B.29

and is equal to V_{CC}. As the input voltage increases further, the voltage between the collector and the emitter decreases according to the sinusoidal law, and at the moment when the input voltage reaches its maximum value, this voltage is equal to the supply voltage V_{CC} reduced by the voltage amplitude on the resistor R_L, i.e.,

$$v_{CE1}(t) = V_{CC} - V_{Lm} \cdot \sin(\omega t). \tag{4.B.8.6}$$

The waveforms of the collector current and voltage between the collector and the emitter of the transistor T_1 are shown in Fig. 4.B.31. The average collector dissipation of T_1 (where this power is equal to that corresponding to transistor T_2, but during the negative half-cycle) is now

$$P_{C1} = \frac{1}{T}\int_0^{T/2} i_{C1}(t) \cdot v_{CE1}(t) \cdot \mathrm{d}t. \tag{4.B.8.7}$$

By substituting the expressions for i_{C1} and v_{CE1} into the expression (4.B.8.7) and solving the integral, one gets the average collector dissipation power as a function of the amplitude of the voltage on R_L.

$$P_{C1} = \frac{V_{Lm} \cdot V_{CC}}{\pi \cdot R_L} - \frac{V_{Lm}^2}{4R_L}. \tag{4.B.8.8}$$

To determine the required value of V_{Lm}, it is necessary to differentiate the previous expression with respect to V_{Lm}, and to equate the obtained expression with zero. From this it is obtained that the value of the amplitude of the sinusoidal voltage on R_L for which the transistors dissipate the maximum power is given by

$$V_{Lm} = (2V_{CC})/\pi = 20.13 \text{ V}. \tag{4.B.8.9}$$

Substituting this numerical value in (4.B.8.8) gives the maximum average dissipation on the collector of one transistor, and it is:

$$P_{\mathrm{Cmax}} = 10.13\ \mathrm{W}. \tag{4.B.8.10}$$

(c) When determining the maximum average power that can be delivered from the power sources, V_{CC} and $-V_{\mathrm{CC}}$, it is necessary to determine the average value of the current flowing through these sources for which we use

$$I_{\mathrm{CC}} = \frac{1}{T}\int_{0}^{T/2} \frac{V_{\mathrm{CC}}}{R_{\mathrm{L}}} \cdot \sin(\omega t) \cdot \mathrm{d}t = \frac{V_{\mathrm{CC}}}{\pi \cdot R_{\mathrm{L}}}. \tag{4.B.8.11}$$

The maximum average power delivered by one power supply source is now

$$P_{\mathrm{CC}} = V_{\mathrm{CC}} \cdot I_{\mathrm{CC}} = V_{\mathrm{CC}}^2/(\pi \cdot R_{\mathrm{L}}) = 31.83\ \mathrm{W}. \tag{4.B.8.12}$$

(d) Since there are two power sources, the efficiency in this case is:

$$\eta = P_{\mathrm{L}}/(2P_{\mathrm{CC}}), \tag{4.B.8.13}$$

By substituting numerical values and using $P_{\mathrm{L}} = 50$ W, we get $\eta = 78.5\%$.

Problem 4.B.9

Figure 4.B.32 shows a large signal amplifier with a JFET.

- Determine the position of the quiescent operating point of the transistor, $Q_0(V_{\mathrm{DSQ}}, I_{\mathrm{DQ}})$.
- If the amplifier is excited by a voltage of the form $v_{\mathrm{g}} = V_{\mathrm{gm}} \cdot \cos(\omega t)$, $V_{\mathrm{gm}} = 0.5$ V, determine the change in the position of the operating point Q_0 and the THD at the output.

It is known that $V_{\mathrm{DD}} = 25$ V; $V_{\mathrm{P}} = -4$ V; $I_{\mathrm{DSS}} = 8$ mA; $R_{\mathrm{g}} = 1$ MΩ; $R_{\mathrm{D}} = 4$ kΩ; $R_{\mathrm{S}} = 0.5$ kΩ; $C \rightarrow \propto$.

Solution to Problem 4.B.9

(a) The DC circuit, based on which the position of the quiescent operating point is determined, is shown in Fig. 4.B.33. Here, the following is valid

$$V_{\mathrm{GS}} = -R_{\mathrm{S}} \cdot I_{\mathrm{D}}. \tag{4.B.9.1}$$

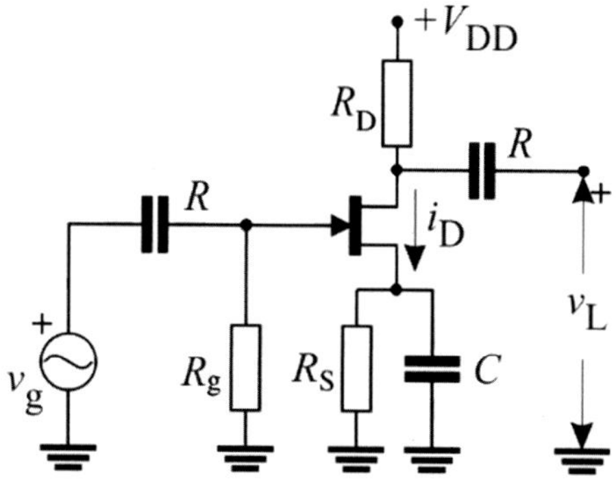

Fig. 4.B.32 Large signal amplifier using JFET

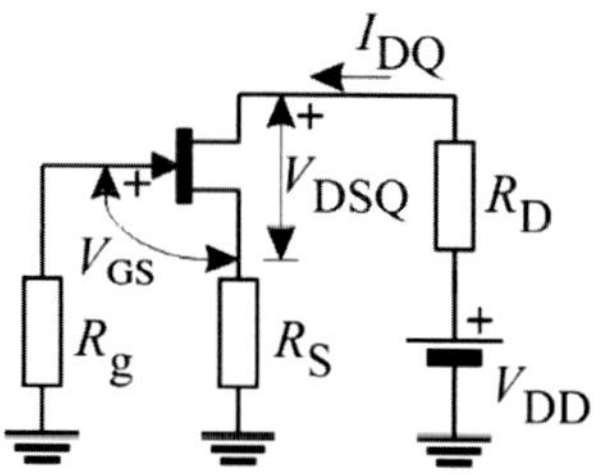

Fig. 4.B.33 DC equivalent circuit of the one of Fig. 4.B.32

By substituting the model for the drain current given by (4.2.32) we get

$$V_{GS} = -R_S \cdot I_{DSS} \cdot \left(1 - \frac{V_{GS}}{V_P}\right)^2. \tag{4.B.9.2}$$

Dividing the left and right sides of this equation by $-V_P$ and adding one, leads to the following quadratic equation:

$$1 - \frac{V_{GS}}{V_P} = 1 + \frac{R_S \cdot I_{DSS}}{V_P} \cdot \left(1 - \frac{V_{GS}}{V_P}\right)^2. \tag{4.B.9.3}$$

Now we introduce the substitution $a = (1 - V_{GS}/V_P)$ so that the previous equation becomes

$$\frac{R_S \cdot I_{DSS}}{V_P} \cdot a^2 - a + 1 = 0. \tag{4.B.9.4}$$

By solving this quadratic equation, we obtain that the voltage between the gate and the source of the transistor at the operating point to be

$$V_{GSQ} = V_P \cdot (1 - a) = \ -1.5\ \text{V}. \tag{4.B.9.5}$$

By substituting this voltage into the drain current we get

$$I_{DQ} = I_{DSS} \cdot \left(1 - \frac{V_{GSQ}}{V_P}\right)^2 = 3 \text{ mA}. \tag{4.B.9.6}$$

Finally for the voltage between the drain and the source in the quiescent operating point we have

$$V_{DSQ} = V_{DD} - (R_D + R_S) \cdot I_{DQ} = 11.5 \text{ V}. \tag{4.B.9.7}$$

(b) To find the THD we start with the following definition

$$THD(\%) = \frac{\sqrt{J_{2m}^2 + J_{3m}^2 + \cdots}}{J_{1m}} \cdot 100\%, \tag{4.B.9.8}$$

where J_{2m}, J_{3m}, ... are the amplitudes of the higher harmonics while J_{1m} is the amplitude of the fundamental component of the current signal.

As the output voltage is equal to the AC component on the resistor R_D, in order to determine the *THD*, it is necessary to determine the instantaneous drain current in advance as

$$i_D = I_{DSS} \cdot (1 - v_{GS}/V_P)^2, \tag{4.B.9.9}$$

which, when one considers that the voltage between the gate and the source consists of a DC and AC component, leads to the following expression:

$$i_D = I_{DSS} \cdot \left(1 - \frac{\overline{V_{GS}} + v_g}{V_P}\right)^2, \tag{4.B.9.10}$$

where

$$\overline{V_{GS}} = -R_S \cdot I_{D0}. \tag{4.B.9.11}$$

In expression (4.B.9.11), the average value of the drain current is denoted with I_{D0}, which, in fact, represents a DC component different from I_{DQ}. This inequality with the I_{DQ} current stems from the nonlinearity of the JFET characteristic. Based on (4.B.9.10) we get

$$i_D = I_{DSS} \cdot \left[\left(1 - \frac{\overline{V_{GS}}}{V_P}\right)^2 - 2\left(1 - \frac{\overline{V_{GS}}}{V_P}\right) \cdot \frac{v_g}{V_P} + \frac{v_g^2}{V_P^2}\right]. \tag{4.B.9.12}$$

By substitution of the expression for v_g, this equation becomes

$$i_D = I_{DSS} \cdot \left(1 - \frac{\overline{V_{GSQ}}}{V_P}\right)^2 - \frac{2I_{DSS}}{V_P} \cdot \left(1 - \frac{\overline{V_{GSQ}}}{V_P}\right) \cdot V_{gm} \cdot \cos(\omega t)$$
$$+ \frac{I_{DSS}}{V_P^2} \cdot V_{gm}^2 \cdot \cos^2(\omega t), \tag{4.B.9.13}$$

while $\cos^2(\omega t)$ may be expressed as

$$\cos^2(\omega t) = \frac{1 + \cos(2\omega t)}{2}. \tag{4.B.9.14}$$

After substitution of (4.B.9.14) into (4.B.9.13) one gets

$$i_D = I_{DSS} \cdot \left(1 - \frac{\overline{V_{GSQ}}}{V_P}\right)^2 + \frac{I_{DSS}}{V_P^2} \cdot \frac{V_{gm}^2}{2}$$
$$- \frac{2I_{DSS}}{V_P} \cdot \left(1 - \frac{\overline{V_{GSQ}}}{V_P}\right) \cdot V_{gm} \cdot \cos(\omega t) + \frac{I_{DSS}}{V_P^2} \cdot \frac{V_{gm}^2}{2} \cdot \cos(2\omega t). \tag{4.B.9.15}$$

Based on this expression, it is possible to write that the instantaneous value of the drain current as

$$i_D = I_{D0} + I_{D1} \cdot \cos(\omega t) + I_{D2} \cdot \cos(2\omega t). \tag{4.B.9.16}$$

The DC component of the drain current I_{D0} may be now extracted as

$$I_{D0} = \frac{1}{T}\int_0^T i(t) \cdot dt = I_{DSS} \cdot \left(1 + \frac{V_{GSQ}}{V_P}\right)^2 + \frac{I_{DSS}}{V_P^2} \cdot \frac{V_{gm}^2}{2}. \tag{4.B.9.17}$$

Substituting the expression (4.B.9.11) into (4.B.9.17) produces a quadratic equation with I_{D0} as the unknown. After solving we get $I_{D0} = 3.038$ mA. Based on (4.B.9.11) we also get $V_{GSQ} = -1.541$ V.

Now, we find that the amplitudes of the fundamental and second harmonics of the drain current are equal, respectively

$$I_{D1} = -\frac{2I_{DSS}}{V_P} \cdot \left(1 - \frac{V_{GSQ}}{V_P}\right) \cdot V_{gm} = 1.23 \text{ mA} \tag{4.B.9.18}$$

$$I_{D2} = \frac{I_{DSS}}{V_P^2} \cdot \frac{V_{gm}^2}{2} = 0.062 \text{ V}. \tag{4.B.9.19}$$

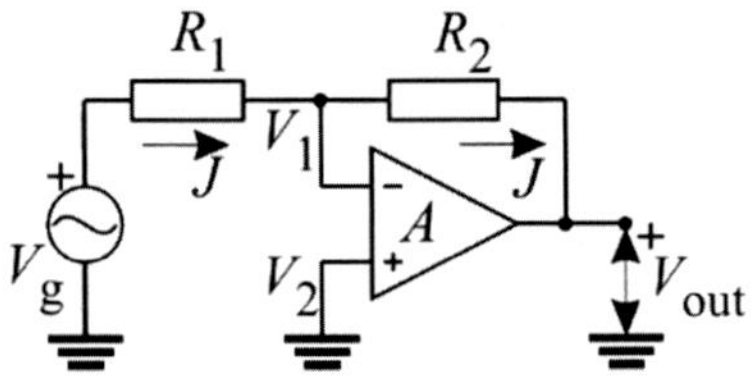

Fig. 4.B.34 Inverting amplifier

On the drain of the transistor, we have

$$\begin{aligned} \upsilon_{\mathrm{D}} &= V_{\mathrm{DD}} - R_{\mathrm{D}} i_{\mathrm{D}} \\ &= V_{\mathrm{DD}} - R_{\mathrm{D}} I_{\mathrm{D0}} - R_{\mathrm{D}} J_{\mathrm{D1m}} \cos(\omega t) - R_{\mathrm{D}} J_{\mathrm{D2m}} \cos(2\omega t), \end{aligned} \tag{4.B.9.20}$$

where from for the output voltage we get

$$\upsilon_{\mathrm{L}} = -R_D J_{\mathrm{D1m}} \cos(\omega t) - R_{\mathrm{D}} J_{\mathrm{D2m}} \cos(2\omega t). \tag{4.B.9.21}$$

Based on this expression it is possible to find that the amplitudes of the fundamental and second harmonics of the output voltage are, respectively:

$$V_{\mathrm{L1m}} = R_{\mathrm{D}} J_{\mathrm{D1m}}; \tag{4.B.9.22}$$

$$V_{\mathrm{L2m}} = R_{\mathrm{D}} J_{\mathrm{D2m}} \tag{4.B.9.23}$$

Since there is only one harmonic, after substitution of numerical values, we get

$$THD = HD_2 = \frac{J_{\mathrm{D2m}}}{J_{\mathrm{D1m}}} \cdot 100 \approx 5\%. \tag{4.B.9.24}$$

PART 2: Analog computation.

Problem 4.B.10

(a) Show the voltage gain of the inverting (from Fig. 4.B.34) and non-inverting amplifier (from Fig. 4.B.35), realized using operational amplifier (OA) with finite differential gain, *A*, and ideal the rest of its characteristics, in the form

$$A_{\mathrm{v}} = \frac{V_{\mathrm{out}}}{V_{\mathrm{g}}} = \frac{A_{\mathrm{v0}}}{1 + (1 + R_2/R_1)/A},$$

where A_{v0} represents the gain of the same amplifier realized using an ideal OA with infinite gain.

(b) Determine the sensitivity of the gain of the inverting and non-inverting amplifier to changes in the gain of the OA, if the resistance ratio is $R_2/R_1 = 100$, the gain of the OA $A = 10^4$, and the relative change of the gain of the OA $100{\bullet}\Delta A/A = 50\%$.

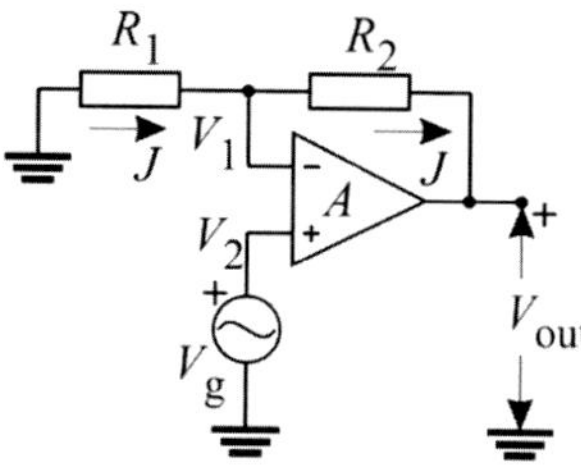

Fig. 4.B.35 Non-inverting amplifier

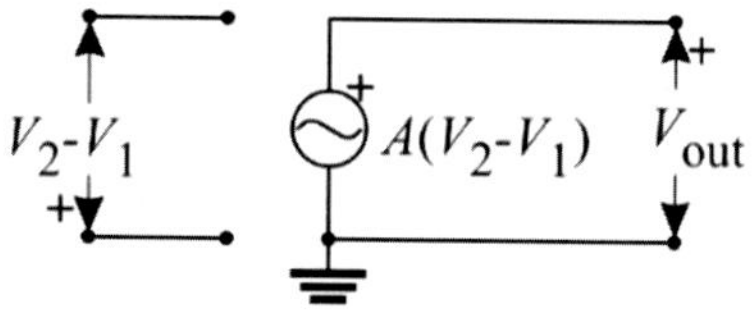

Fig. 4.B.36 Model of a differential amplifier with a finite gain

Solution to Problem B.10

The equivalent circuit (model) of an OA with final gain and ideal rest of its characteristics is shown in Fig. 4.B.36.

(a) In the inverting amplifier, the non-inverting input of the OA is connected to ground, i.e., $V_2 = 0$. The resistors R_1 and R_2 form a negative feedback circuit. If it is assumed that the input impedance of the operational amplifier is infinite, then the input current of the OA will be equal to zero, and the current, J, through resistors R_1 and R_2 (Fig. 4.B.34) will be the same:

$$J = \frac{V_g - V_1}{R_1} = \frac{V_1 - V_{out}}{R_2}. \tag{4.B.10.1}$$

Since the non-inverting input of the OA is connected to ground ($V_2 = 0$) and since, according to the model of the OA shown in Fig. 4.B.36, it is

$$V_{out} = A \cdot (V_2 - V_1), \tag{4.B.10.2}$$

then

$$V_{out} = -A \cdot V_1. \tag{4.B.10.3}$$

By substitution of $V_1 = -V_{out}/A$ from (4.B.10.3) into (4.B.10.1) the following equation is formed

$$V_g + \frac{V_{out}}{A} = \frac{R_1}{R_2}\left(-\frac{V_{out}}{A} - V_{out}\right)$$

where from the voltage gain may be found as

$$A_\mathrm{v} = \frac{V_\mathrm{out}}{V_\mathrm{g}} = \frac{-R_2/R_1}{1 + (1 + R_2/R_1)/A}. \tag{4.B.10.4}$$

In the case of very large gain of the OA, A, and having in mind realistic values of R_1 and R_2, the following is valid

$$(1 + R_2/R_1)/A \ll 1. \tag{4.B.10.5}$$

If the condition (4.B.10.5) is met, the expression for the voltage gain of the inverting amplifier for the case of ideal OA is obtained as

$$A_\mathrm{v0} = -R_2/R_1. \tag{4.B.10.6}$$

In the non-inverting amplifier, the non-inverting input is connected to the excitation voltage V_g ($V_2 = V_\mathrm{g}$). Considering the expression (4.B.10.2), the voltage at the inverting input of OA, V_1, is:

$$V_1 = V_\mathrm{g} - V_\mathrm{out}/A. \tag{4.B.10.7}$$

Due to the infinitely large input impedance of the OA, the currents through resistors R_1 and R_2 (Fig. 4.B.35) must be the equal:

$$J = \frac{-V_1}{R_1} = \frac{V_1 - V_\mathrm{out}}{R_2}. \tag{4.B.10.8}$$

By substituting (4.B.10.7) into (4.B.10.8) and after arranging, the voltage gain of the non-inverting amplifier is obtained as

$$A_v = \frac{V_\mathrm{out}}{V_\mathrm{g}} = \frac{1 + R_2/R_1}{1 + (1 + R_2/R_1)/A}. \tag{4.B.10.9}$$

If the condition (4.B.10.5) is met, which is certainly met when the gain of the OA is infinite, the voltage gain of the non-inverting amplifier using ideal OA is obtained as

$$A_\mathrm{v0} = 1 + R_2/R_1. \tag{4.B.10.10}$$

(b) The sensitivity of the gain of inverting and non-inverting amplifiers can be obtained from the common form:

$$A_\mathrm{v} = \frac{A_\mathrm{v0}}{1 + (1 + R_2/R_1)/A}. \tag{4.B.10.11}$$

If the gain of the OA, changes from A to $A + \Delta A$, the gain of amplifier A_v changes to and becomes

$$A_v + \Delta A_v = \frac{A_{v0}}{1 + \beta/(A + \Delta A)}. \quad (4.B.10.12)$$

where $\beta = 1 + R_2/R_1$.

The combination of the previous expressions gives the expression for the sensitivity of the gain of the amplifier as

$$\frac{\Delta A_v}{A_{v0}} = \frac{\beta \frac{\Delta A}{A(A+\Delta A)}}{(1 + \beta/A)[1 + 1 + \beta/(A + \Delta A)]}. \quad (4.B.10.13)$$

This expression can be simplified if the following conditions are met

$$\frac{1 + R_2/R_1}{A} \ll 1 \quad \text{and} \quad \frac{1 + R_2/R_1}{A + \Delta A} \ll 1, \quad (4.B.10.14)$$

which is most often fulfilled, which gives

$$\frac{\Delta A_v}{A_{v0}} = \frac{\Delta A}{A} \cdot \frac{1 + R_2/R_1}{A(1 + \Delta A/A)}. \quad (4.B.10.15)$$

Based on this expression, the sensitivity of the gain of the inverting and non-inverting amplifiers can be calculated. If A is the largest possible value of the gain of the OA, then $\Delta A/A$ must be a negative number, and it is $\Delta A/A = -0.5$. By substituting numerical values, the following is obtained:

$$\frac{\Delta A_v}{A_{v0}} = -0.5\frac{1 + 100}{10{,}000 \cdot (1 - 0.5)} = -\frac{101}{10{,}000} = -0.01 \quad (1\%). \quad (4.B.10.16)$$

Problem 4.B.11

Determine the voltage gain and the input impedance of the inverting (Fig. 4.B.34) and non-inverting amplifier (Fig. 4.B.35), realized by an OA with finite differential gain A and finite input resistance R_{in}. Using the derived expressions, calculate the voltage gain and output impedance of both amplifiers for $R_1 = 1\ \text{k}\Omega$ and $R_2 = 100\ \text{k}\Omega$, if $R_{out} = 0$ and:

(a) $A = 10{,}000$, $R_{in} = 100\ \text{k}\Omega$;
(b) $A = 10{,}000$, $R_{in} \to \infty$;
(c) $A \to \infty$, $R_{in} = 100\ \text{k}\Omega$;
(d) $A \to \infty$, $R_{in} \to \infty$.

Solution to the Problem 4.B.11

The inverting amplifier can be represented by a circuit of Fig. 4.B.37.

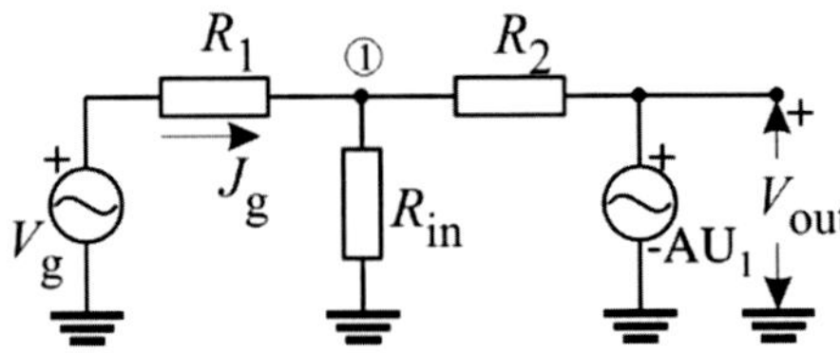

Fig. 4.B.37 Equivalent circuit of the inverting amplifier

For node (1) of Fig. 4.B.37 the following equation may be written

$$(V_1 - V_g)/R_1 + V_1/R_{in} + (V_1 + A \cdot V_1)/R_2 = 0. \quad (4.B.11.1)$$

By solving for V_1 we get

$$V_1 = \frac{V_g}{1 + R_1/R_{in} + (1 + A) \cdot R_1/R_2}. \quad (4.B.11.2)$$

The voltage gain of the inverting amplifier is now

$$A_v = \frac{V_{out}}{V_g} = \frac{-A \cdot V_1}{V_g} = \frac{-R_2/R_1}{1 + (1 + R_2/R_1 + R_2/R_{in})/A}. \quad (4.B.11.3)$$

The input impedance of the inverting amplifier is

$$Z_{in} = V_g/J_g, \quad (4.B.11.4)$$

where

$$J_g = (V_g - V_1)/R_1. \quad (4.B.11.5)$$

By substitution of (4.B.11.2) and (4.B.11.5) into (4.B.11.4) the expression for the input impedance becomes

$$Z_{in} = R_1 + \frac{1}{1/R_{in} + (1 + A)/R_2}. \quad (4.B.11.6)$$

Using (B.11.3) for the voltage gain and (B.11.6) for inverting amplifier's input impedance we get:

(a) $A = 10{,}000$ and $R_{in} = 100$ kΩ: $A_v = -98.99$ and $Z_{in} = 1009.998$ Ω;
(b) $A = 10{,}000$ and $R_{in} \to \infty$: $A_v = -99$ and $Z_{in} = 1009.999$ Ω;
(c) $A \to \infty$ and $R_{in} = 100$ kΩ: $A_v = -100$ and $Z_{in} = 1000$ Ω;
(d) $A \to \infty$ *i* $R_{in} \to \infty$: $A_v = -100$ and $Z_{in} = 1000$ Ω.

The equivalent circuit of the non-inverting amplifier is shown in Fig. 4.B.38.

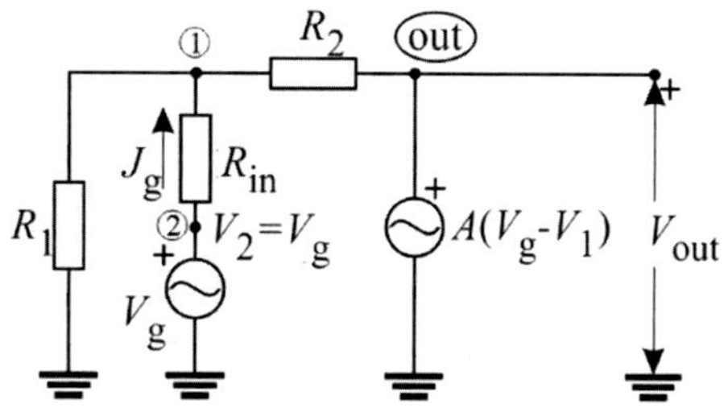

Fig. 4.B.38 Equivalent circuit of the non-inverting amplifier

For node (1) we may write the following

$$V_1/R_1 + (V_1 - V_g)/R_{in} + [V_1 - A \cdot (V_g - V_1)]/R_2 = 0. \qquad (4.B.11.7)$$

After solving for V_1 we get

$$V_1 = V_g \frac{1/R_{in} + 1/R_2}{1/R_1 + 1/R_{in} + (1 + A)/R_2}. \qquad (4.B.11.8)$$

The voltage gain of the non-inverting amplifier is now

$$A_v = \frac{V_{out}}{V_g} = \frac{A \cdot (V_g - V_1)}{V_g} = \frac{1 + R_2/R_1 + R_2/R_{in}}{1 + (1 + R_2/R_1 + R_2/R_{in})/A}. \qquad (4.B.11.9)$$

The input impedance of a non-inverting amplifier is determined by

$$Z_{in} = V_g/J_g, \qquad (4.B.11.10)$$

where

$$J_g = (V_g - V_1)/R_{in}. \qquad (4.B.11.11)$$

After substituting of (4.B.11.8) and (4.B.11.11) into (4.B.11.10) the expression for the input impedance obtains the following form

$$Z_{in} = R_{in}\left(1 + \frac{1 + R_2/R_{in}}{A + R_2/R_1}\right). \qquad (4.B.11.12)$$

For the given numerical values of A and R_{in}, using (4.B.11.9) for the voltage gain and (4.B.11.12) for the input impedance we get

(a) A= 10,000 and R_{in}= 100 kΩ: A_v= 100.97 and Z_{in}= 100.02 kΩ;
(b) A= 10,000 and $R_{in} \to \infty$: A_v= 99.99 and $Z_{in} \to \infty$;
(c) $A \to \infty$ and R_{in}= 100 kΩ: A_v= 102 and Z_{in}= 100 kΩ;
(d) $A \to \infty$ and $R_{in} \to \infty$: A_v= 100 and $Z_{in} \to \infty$.

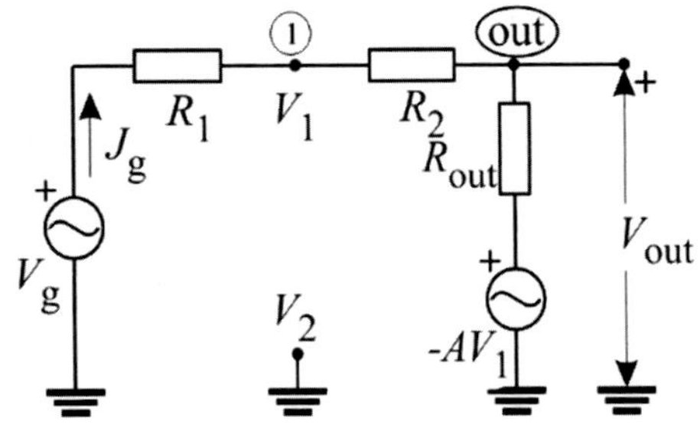

Fig. 4.B.39 Equivalent circuit of the inverting amplifier

Problem 4.B.12

Determine the voltage gain and output impedance of the inverting (Fig. 4.B.34) and non-inverting amplifier (Fig. 4.B.35), realized by an OA with finite differential gain A and output resistance R_{out}. Using the derived expressions, calculate the voltage gain and output impedance of both amplifiers for $R_1 = 1\ \text{k}\Omega$ and $R_2 = 100\ \text{k}\Omega$, if $R_{in} \to \infty$ and:

(a) $A = 10{,}000$, $R_{out} = 500\ \Omega$;
(b) $A = 10{,}000$, $R_{out} = 0$;
(c) $A \to \infty$, $R_{out} = 500\ \Omega$;
(d) $A \to \infty$, $R_{out} = 0$.

Solution to the Problem 4.B.12

Figure 4.B.39 represents the equivalent circuit of an inverting amplifier when an OA having infinite input and finite output resistance is used.

For the nodes (1) and (out) the following is valid

$$(V_1 - V_g)/R_1 + (V_1 - V_{out})/R_2 = 0 \tag{4.B.12.1}$$

$$(V_{out} - V_1)/R_2 + (V_{out} + A \cdot V_1)/R_{out} = 0. \tag{4.B.12.2}$$

By solving this system of equations, the voltage gain of the inverting amplifier is obtained as

$$A_v = \frac{V_{out}}{V_g} = \frac{-R_2/R_1}{1 + \frac{R_1+R_2}{R_1}\,\frac{R_2+R_{out}}{A \cdot R_2 - R_{out}}}. \tag{4.B.12.3}$$

The output resistance of the inverter amplifier is determined by analysis of the circuit shown in Fig. 4.B.40 and by finding the ratio of the voltage V_0 and the current J_0. Of course, on this occasion the excitation source V_g is short-circuited, i.e., we use $V_g = 0$.

For the nodes (1) and (0) the following pair of equations is valid

$$V_1/R_1 + (V_1 - V_0)/R_2 = 0 \tag{4.B.12.4}$$

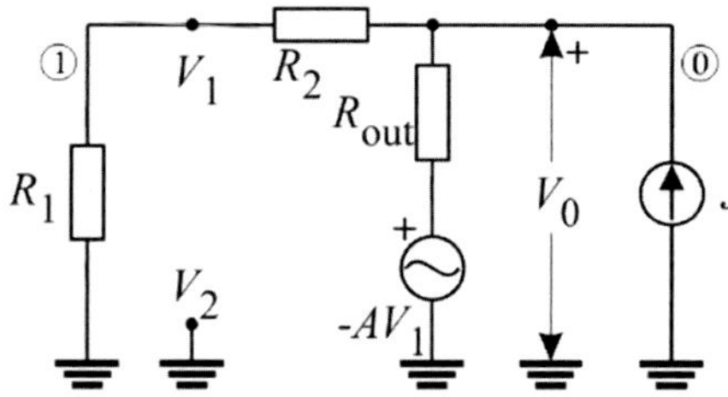

Fig. 4.B.40 Equivalent circuit of the inverting amplifier used for extraction of the output impedance

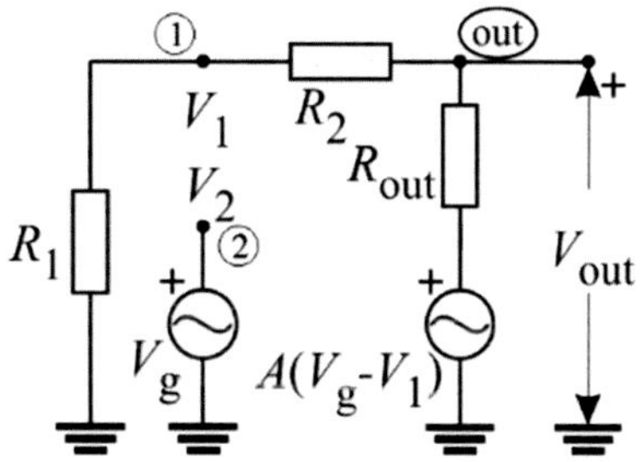

Fig. 4.B.41 Equivalent circuit of the non-inverting amplifier

$$(V_0 - V_1)/R_2 + (V_0 + \Delta V_1)/R_{out} = J_0. \tag{4.B.12.5}$$

After solving this system of equations, the output impedance is obtained as

$$Z_{out} = \frac{V_0}{J_0} = \frac{R_{out}}{1 + (A \cdot R_1 + R_{out})/(R_1 + R_2)}. \tag{4.B.12.6}$$

By substitution, the numerical values for A and R_{out} into (4.B.12.3) and (4.B.12.6) we get.

(a) $A = 10{,}000$ and $R_{out} = 500\ \Omega$: $A_v = -98.995$ and $Z_{out} = 4.999\ \Omega$;
(b) $A = 10{,}000$ and $R_{out} = 0$: $A_v = -99$ and $Z_{out} = 0$;
(c) $A \to \infty$ and $R_{out} = 500\ \Omega$: $A_v = 100$ and $Z_{out} = 0$;
(d) $A \to \infty$ and $R_{out} = 0$: $A_v = -100$ and $Z_{out} = 0$.

The analysis of the non-inverting amplifier of Fig. 4.B.41 starts with creation of the node equations for nodes (1) and (out):

$$V_1/R_1 + (V_1 - V_{out})/R_2 = 0 \tag{4.B.12.7}$$

$$(V_{out} - V_1)/R_2 + \left[V_{out} - A \cdot \left(V_g - V_1\right)\right]/R_{out} = 0 \tag{4.B.12.8}$$

After elimination of V_1 the voltage gain of the non-inverter amplifier is obtained as

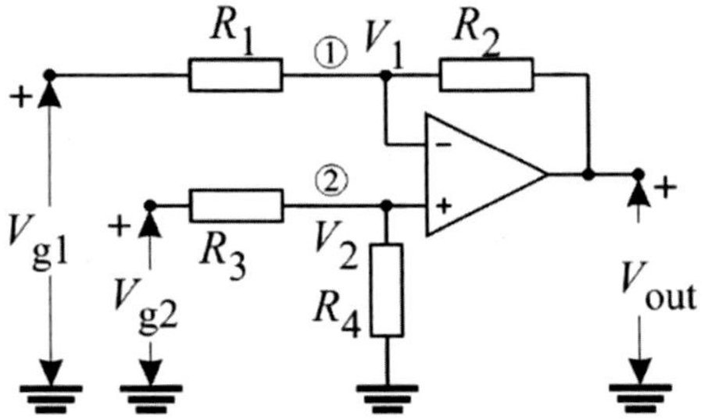

Fig. 4.B.42 Differential balanced amplifier

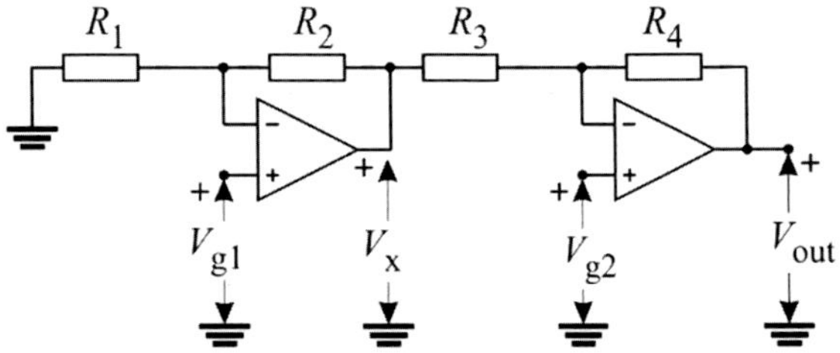

Fig. 4.B.43 Differential amplifier using two OAs

$$A_v = \frac{V_{out}}{V_g} = \frac{1 + R_2/R_1}{1 + (1 + R_2/R_1 + R_{out}/R_1)/A}. \tag{4.B.12.9}$$

The output impedance of the non-inverter amplifier is equal to the output impedance of the inverting amplifier since both amplifiers can be represented by the equivalent circuit of Fig. 4.B.40 ($V_g = 0$, when determining Z_{out}). Therefore, in this case, only the voltage gain of the non-inverting amplifier given by expression (4.B.12.9) will be determined for the prescribed values of the gain A and the output resistance R_{out} of the OA. These are

(a) $A = 10{,}000$ and $R_{out} = 500\ \Omega$: $A_v = 87.06$;
(b) $A = 10{,}000$ and $R_{out} = 0$: $A_v = 99.99$;
(c) $A \to \infty$ and $R_{out} = 500\ \Omega$: $A_v = 101$;
(d) $A \to \infty$ and $R_{out} = 0$: $A_v = 101$.

Problem 4.B.13
For the realization of a differential amplifier (subtraction circuit), based on single, two, or three OAs, the schematics shown in Figs. 4.B.42, 4.B.43, and 4.B.44 will be used. Determine the conditions that must be met by the elements of these circuits in order for the output voltage to be proportional to the difference of the input voltages. The OAs used are ideal with infinite gain.

Solution to the Problem 4.B.13
For the nodes denoted V_1 and V_2 of the amplifier of Fig. 4.B.42 the following system is valid

$$\left(V_1 - V_{g1}\right)/R_1 + (V_1 - V_{out})/R_2 = 0 \tag{4.B.13.1}$$

$$\left(V_2 - V_{g2}\right)/R_3 + V_2/R_4 = 0. \tag{4.B.13.2}$$

As a reminder, if the OA has infinite gain, we have

$$V_1 = V_2. \tag{4.B.13.3}$$

After solving the above system of equations using the relation (4.B.13.3) we get

$$V_{out} = (1 + p)\frac{1}{1 + q}V_{g2} - p \cdot V_{g1}, \tag{4.B.13.4}$$

or

$$V_{out} = p\left(\frac{q}{p} \cdot \frac{1 + p}{1 + q}V_{g2} \quad V_{g1}\right). \tag{4.B.13.5}$$

where $p = R_2/R_1$ and $q = R_4/R_3$.

For the circuit of Fig. 4.B.42 to represent a differential balanced amplifier, the coefficients accompanying V_{g1} and V_{g2} in this expression must be equal, which defines the required ratios of resistors in the circuit as

$$\frac{R_2}{R_1} = \frac{R_4}{R_3}. \tag{4.B.13.6}$$

If so, the voltage at the output of the differential balanced amplifier is proportional to the difference of the input voltages and is given by

$$V_{out} = \frac{R_2}{R_1}\left(V_{g2} - V_{g1}\right) = A_{db}\left(V_{g2} - V_{g1}\right), \tag{4.B.13.7}$$

where the proportionality constant (the gain of the differential balanced amplifier) is denoted by A_{db}.

For a circuit with two OAs shown in Fig. 4.B.43, based on the equation for the inverting node, we set

$$V_{out} = (1 + q) \cdot V_{g2} - q \cdot V_x, \tag{4.B.13.8}$$

where

$$V_x = (1 + p) \cdot V_{g1} \tag{4.B.13.9}$$

After substitution of (4.B.13.9) into (4.B.13.8) one gets

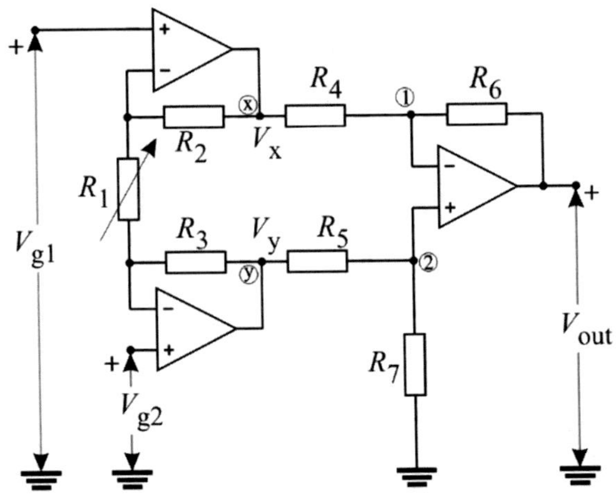

Fig. 4.B.44 Instrumentation amplifier

$$V_{\text{out}} = (1+q) \cdot V_{\text{g2}} - q \cdot (1+p) \cdot V_{\text{g1}}. \tag{4.B.13.10}$$

This circuit becomes a differential amplifier if the following condition is satisfied

$$R_4/R_3 = R_1/R_2. \tag{4.B.13.11}$$

In that case the output voltage is proportional to the difference of the input voltages:

$$V_{\text{out}} = (1 + R_1/R_2)\left(V_{\text{g2}} - V_{\text{g1}}\right) = A_{\text{dZ}}\left(V_{\text{g2}} - V_{\text{g1}}\right), \tag{4.B.13.12}$$

with a proportionality constant (differential amplifier gain) A_{dZ}.

Instrumentation amplifier (Fig. 4.B.44) is a circuit with a large input impedance that is very often used in electronic measurement instruments. The following expressions derived from node equations can be written for this circuit:

for node (x)

$$V_{\text{x}} = (1+p) \cdot V_{\text{g1}} - p \cdot V_{\text{g2}}, \tag{4.B.13.13}$$

for node (y)

$$V_{\text{y}} = (1+s) \cdot V_{\text{g2}} - s \cdot V_{\text{g1}}, \tag{4.B.13.14}$$

and for nodes (1) and (2)

$$V_{\text{out}} = -t \cdot V_{\text{x}} + (1+t)\frac{1}{1+r}V_{\text{y}}. \tag{4.B.13.15}$$

where $s = R_3/R_1$, $t = R_6/R_4$ and $r = R_5/R_7$.

After substitution of (4.B.13.13) and (4.B.13.14) into (4.B.13.15) we get

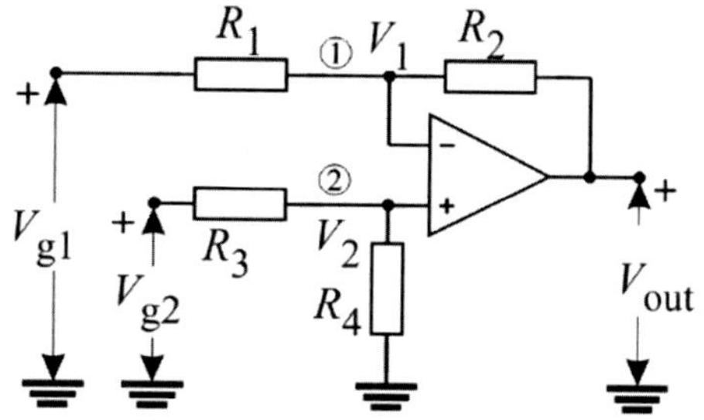

Fig. 4.B.45 Differential balanced amplifier

$$V_{out} = \left[p \cdot t + \frac{(1+s)(1+t)}{1+r}\right] \cdot V_{g2} - \left[t(1+p) + s\frac{1+t}{1+r}\right] \cdot V_{g1}. \tag{4.B.13.16}$$

If we adopt: $R_2 = R_3$, $R_4 = R_5 = R_6 = R_7$, we get a simplified expression for the output voltage which will be proportional to the difference of the input voltages. That is

$$V_{out} = (1 + 2R_2/R_1)(V_{g2} - V_{g1}) = A_{iZ}(V_{g2} - V_{g1}), \tag{4.B.13.17}$$

with a proportionality constant (differential amplifier gain) $A_{iZ.}$ It is obvious that the gain of the instrumentation amplifier can be controlled by changing only one resistor in the circuit (R_1).

Problem 4.B.14

Figure 4.B.45 shows the circuit of a differential balanced amplifier. Determine the minimum value of the common mode rejection ratio (*CMRR*) of the differential balance amplifier, if the gain of the OA is $A = 10{,}000$, its common mode gain is $A_s = 10$, and the tolerance of the resistors used is 1%. The differential gain of the differential balance amplifier when using an OA with infinite gain should be equal to one ($A_{db} = 1$).

Solution to the Problem 4.B.14

When using an infinite gain OA, the differential gain of the differential balanced amplifier should be

$$A_{db} = R_2/R_1 = R_4/R_3 = 1. \tag{4.B.14.1}$$

For a circuit in which the OA's gain is finite one may write

$$(V_2 - V_{g2})/R_3 + V_2/R_4 = 0 \tag{4.B.14.2}$$

$$(V_1 - V_{g1})/R_1 + (V_1 - V_{out})/R_2 = 0. \tag{4.B.14.3}$$

From these two equations we get the values of V_1 and V_2 as

$$V_1 = \frac{R_2}{R_1 + R_2} V_{g1} + \frac{R_1}{R_1 + R_2} V_{out} \tag{4.B.14.4}$$

$$V_2 = \frac{R_4}{R_3 + R_4} V_{g2}. \tag{4.B.14.5}$$

The output voltage can be represented as a function of these voltages, the differential gain, A, and the common mode gain of the OA, A_s, by the expression

$$V_{out} = A \cdot (V_2 - V_1) + A_s(V_1 + V_2)/2. \tag{4.B.14.6}$$

After substitution of (4.B.14.4) and (4.B.14.5) into (4.B.14.6) we get the output voltage as

$$V_{out} = \frac{\frac{R_4}{R_3+R_4}(A + A_s/2)}{1 + \frac{R_1}{R_1+R_2}(A - A_s/2)} V_{g2} - \frac{\frac{R_2}{R_1+R_2}(A - A_s/2)}{1 + \frac{R_1}{R_1+R_2}(A - A_s/2)} V_{g1}. \tag{4.B.14.7}$$

The output voltage can be expressed via the differential, A_{db}, and the common mode, A_{sb}, gain of the differential balanced amplifier in a manner similar to the expression (4.B.14.6). That is

$$V_{out} = A_{db}\left(V_{g2} - V_{g1}\right) + A_{sb}\left(V_{g1} + V_{g2}\right)/2. \tag{4.B.14.8}$$

From the pair of equations (4.B.14.7) and (4.B.14.8) one can easily get the gains needed as

$$A_{db} = \frac{1}{2}\frac{\alpha(A - A_s/2) + \beta(A + A_s/2)}{1 + \alpha(A - A_s/2)}, \tag{4.B.14.9}$$

and

$$A_{sb} = \frac{\alpha(-A + A_s/2) + \beta(A + A_s/2)}{1 + \alpha(A - A_s/2)}, \tag{4.B.14.10}$$

where je $= R_2/(R_1 + R_2)$ and $\beta = R_4/(R_3 + R_4)$.

The *CMRR* of the differential balanced amplifier is the ratio of its differential and its common mode gain: $\rho_b = A_{db}/A_{sb}$, or

$$\rho_b = \frac{1}{2}\frac{A(\alpha + \beta) - \frac{A_s}{2}(\alpha - \beta)}{-A(\alpha - \beta) + \frac{A_s}{2}(\alpha + \beta)}. \tag{4.B.14.11}$$

This expression gives simultaneously the influence of the *CMRR* of the OA ($\rho = A/A_s$) and the influence of the resistor mismatch (ρ_r) on the *CMRR* of the differential balanced amplifier.

The *CMRR* of the signal due to the mismatch of the resistors can be estimated from expression (4.B.14.11) when it is assumed that the *CMRR* of the OA is infinite, i.e., if $A_s = 0$. If so

$$\rho_r = \frac{1}{2}\frac{2R_2R_4 + R_2R_3 + R_1R_4}{R_1R_4 - R_2R_3}. \qquad (4.B.14.12)$$

The CMRR of the differential balanced amplifier can be now expressed as

$$\rho_b = (\rho_r\rho + 1/4)/(\rho_r + \rho). \qquad (4.B.14.13)$$

If the condition

$$\rho_r \cdot \rho \gg 1/4, \qquad (4.B.14.14)$$

is satisfied (this is most frequently the case), we may write

$$1/\rho_b = 1/\rho_r + 1/\rho. \qquad (4.B.14.15)$$

With the given tolerance of the resistors of 1% ($p = 0.01$) taking the worst case of resistor failure, i.e.,

$$R_4 = R_2(1+p) \quad \text{and} \quad R_3 = R_1(1-p), \qquad (4.B.14.16)$$

we get the *CMRR* due to the resistor mismatch as

$$\rho_r = \frac{1}{2}\frac{2R_2^2(1+p) + 2R_1R_2}{R_1R_2(2p)} = \frac{1 + (1+p)(R_2/R_1)}{2p}. \qquad (4.B.14.17)$$

For the given differential gain of the balanced amplifier $A_{db} = R_2/R_1 = 1$, the *CMRR* due to the mismatch of the resistors becomes

$$\rho_r = (2+p)/(2p) = 100.5 \qquad (4.B.14.18)$$

or 40.04 dB. Since the *CMMR* of the OA is

$$\rho = A/A_s = 1000 \quad (60\,\text{dB}), \qquad (4.B.14.20)$$

it is obtained that the *CMRR* of the differential balanced amplifier (since the condition (4.B.14.14) is met) according to the expression (4.B.14.15) is

$$\rho_b = 91.32 \quad (39.2\,\text{dB}). \qquad (4.B.14.21)$$

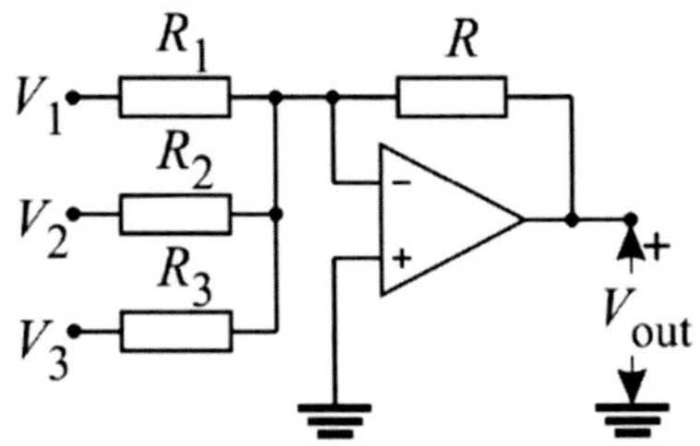

Fig. 4.B.46 Inverting adder

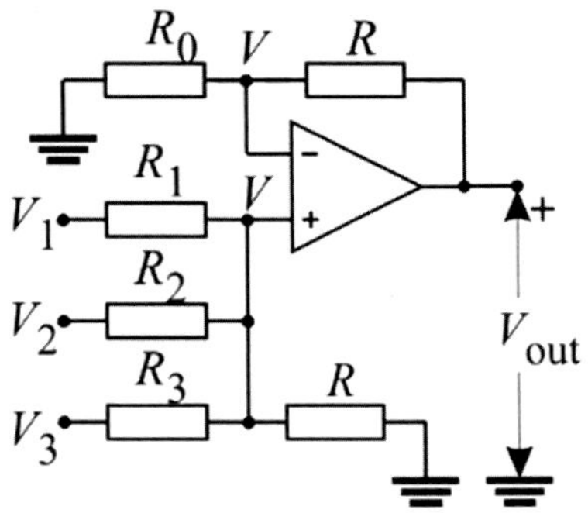

Fig. 4.B.47 Non-inverting adder

Problem 4.B.15

Figure 4.B.46 shows an inverting, and Fig. 4.B.47 a non-inverting adder circuit.

(a) For the circuit in Fig. 4.B.46 determine the dependence of the output on the input voltages.
(b) Determine the range of changes in output voltage if the input voltages change from 0 to 1 V, provided that $R_1 = R$, $R_2 = R/2$, and $R_3 = R/4$.
(c) Determine the value of resistance $R_{(i=1,2,3)0}$ in the circuit of the non-inverting adder (Fig. 4.B.47) with the values of resistance R given in case (b), so that the absolute value of the output voltage has the same dependence on the input voltages as under (b).

Solution to the Problem 4.B.15

(a) For the inverting adder of Fig. 4.B.46 the following is valid

$$V_1/R_1 + V_2/R_2 + V_3/R_3 + V_{out}/R = 0. \tag{4.B.15.1}$$

From here the output voltage is obtained as

$$V_{out} = -R\left(\frac{V_1}{R_1} + \frac{V_2}{R_2} + \frac{V_3}{R_3}\right) \tag{4.B.15.2}$$

(b) When the resistances R_i ($i = 1, 2, 3$) are assigned the proper values we have

$$V_{out} = -(V_1 + 2 \cdot V_2 + 4 \cdot V_3) = -\left(2^0 \cdot V_1 + 2^1 \cdot V_2 + 2^2 \cdot V_3\right). \qquad (4.B.15.3)$$

If the input voltage change from zero to 1 V, the output voltage will have minimum value when all voltages are $V_i = 0$ V ($i = 1, 2, 3$) and becomes $V_{out} = 0$ V. It will get maximum value for $V_i = 1$ V ($i = 1, 2, 3$) which will be $V_{out} = -7$ V.

(c) For the non-inverting adder of Fig. 4.B.47 the following equation can be written for the non-inverting input node of the OA

$$(V - V_1)/R_1 + (V - V_2)/R_2 + (V - V_3)/R_3 + V/R = 0. \qquad (4.B.15.4)$$

where, based on the node equation for the inverting input terminal of the OA,

$$V = V_{out} R_0/(R + R_0). \qquad (4.B.15.5)$$

Substituting (4.B.15.5) into (4.B.15.4) gives the expression for the output voltage:

$$V_{out} = \frac{(1 + R/R_0) \cdot R}{1 + \frac{R}{R_1} + \frac{R}{R_2} + \frac{R}{R_3}} \left(\frac{V_1}{R_1} + \frac{V_2}{R_2} + \frac{V_3}{R_3} \right). \qquad (4.B.15.6)$$

For the dependence of the output voltage of the non-inverting adder given by expression (4.B.15.6) to be (in absolute value) equal to the dependence given by (4.B.15.2) for the inverting adder, the following condition must be met:

$$1/R_0 = 1/R_1 + 1/R_2 + 1/R_3. \qquad (4.B.15.7)$$

For the values of R_i ($i = 1, 2, 3$) given under (b) we have:

$$1/R_0 = 1/R + 2/R + 4/R = 7/R, \qquad (4.B.15.8)$$

or

$$R_0 = R/7. \qquad (4.B.15.9)$$

Problem 4.B.16
Figure 4.B.48 depicts an integrator circuit.

(a) Determine the dependence of the output voltage of the integrator on the input voltage if at the time instant $t = 0$ the voltage on the capacitor is equal to zero.

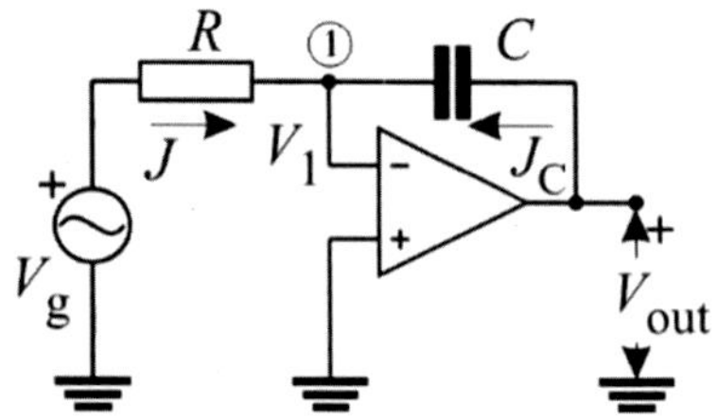

Fig. 4.B.48 The integrator circuit

(b) If the time constant $RC = 1$ s, draw a diagram and determine the elements of the circuit for solving the differential equation: $3\frac{dy}{dt} + 2y = 3 \cdot \sin(\omega t)$ [V].

(c) Determine the transfer function of the integrator itself used in the circuit for solving the differential equation given in point (b) if the operational amplifier has a finite gain $A = 105$ and a finite upper cut-off frequency $f_0 = 200$ Hz. Up to what frequency does such a circuit work as an integrator?

Solution to the Problem 4.B.16

(a) For node (1) in the integrator circuit of Fig. 4.B.48 the following is valid

$$\frac{v_g}{R} + C\frac{dv_{out}}{dt} = 0. \tag{4.B.16.1}$$

Since the initial value of the capacitor voltage (the same as the output voltage) is equal to zero the output voltage is

$$v_{out} = -\frac{1}{RC}\int_0^t v_g dt. \tag{4.B.16.2}$$

(b) The given differential equation may be rewritten into the form

$$-\frac{dy}{dt} = \frac{2}{3}y - \sin(\omega t). \tag{4.B.16.3}$$

We now introduce the analogy (v_{out}, y) for convenience. So, if the time constant $RC = 1$ s, the output voltage of the integrator, y, to whose input the signal $-dy/dt$ is applied, will be the solution of the differential equation.

The $-dy/dt$ signal can be obtained at the output of the differential balanced amplifier if the excitation sinusoidal voltage is applied to the inverting input and the output signal of the integrator is fed to the non-inverting input. Based on all this, the circuit for solving the given differential equation can be represented by the schematic depicted in Fig. 4.B.49.

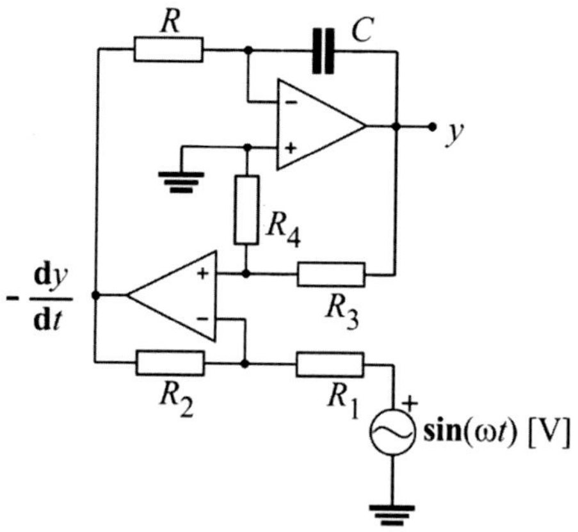

Fig. 4.B.49 The circuit for solving a differential equation with zero boundary condition

The output voltage of the differential balanced amplifier in the circuit in Fig. 4.B.49 may be represented by the expression:

$$-\frac{\mathrm{d}y}{\mathrm{d}t} = \frac{R_4(R_1 + R_2)}{R_1(R_3 + R_4)} \cdot y - \frac{R_2}{R_1} \cdot \sin(\omega t). \tag{4.B.16.4}$$

Comparing the corresponding coefficients in expressions (4.B.16.3) and (4.B.16.4) we obtain

$$R_2/R_1 = 1, \tag{4.B.16.5}$$

and

$$\frac{1 + R_2/R_1}{1 + R_3/R_4} = \frac{2}{3}, \tag{4.B.16.6}$$

or

$$R_3/R_4 = 2. \tag{4.B.16.7}$$

(c) The circuits of Figs. 4.B.48 and 4.B.34 have the same topology with the difference that instead of R_2 in the circuit of Fig. 4.B.48 has X_c, while R_1 is replaced by R. It is obvious that expression (4.B.10.4) can be used in which R_1 is replaced by R and R_2 by $X_C = 1/(s \cdot C)$. In addition, the finite gain A of the operational amplifier in (4.B.10.4) should be replaced by an expression in which the finite cut-off frequency of the amplifier f_0 is included: $A_n(s) = A/(1 + s/\omega_0)$, where $\omega_0 = 2\pi f_0$.

After the proper substitutions, the transfer function of the integrator becomes

$$A_{\text{int}} = \frac{-A\omega_0/(RC)}{s^2 + s\left(\omega_0 A + \omega_0 + \frac{1}{RC}\right) + \frac{\omega_0}{RC}}. \tag{4.B.16.8}$$

If $A \gg 1$ and $A\omega_0 \gg 1/(RC)$ the transfer function of the circuit is

$$A_{\text{int}} = \frac{-A\omega_0/(RC)}{s^2 + s \cdot (\omega_0 A) + \frac{\omega_0}{RC}}. \tag{4.B.16.9}$$

The denominator polynomial of (4.B.16.9) has two zeros, one of which (s_1) is dominant (since $s_1 \ll s_2$) so that it can be transformed as

$$(s - s_1)(s - s_2) = s^2 - (s_1 + s_2)s + s_1 s_2 \approx s^2 - s_2 s + s_1 s_2 \tag{4.B.16.10}$$

By comparison with the expression for the denominator we can find the following analogies

$$s_1 = -1/(A \cdot R \cdot C) \quad \text{and} \quad s_2 = -A \cdot \omega_0. \tag{4.B.16.11}$$

The frequency values corresponding to s_1 and s_2 are $f_1 = 1.59 \times 10^{-6}$ Hz and $f_2 = 20$ MHz.

When an infinite gain OA is used, the circuit of Fig. 4.B.48 acts as an ideal integrator whose transfer function is (put $A \to \infty$ into (4.B.16.8)): $A_{\text{id}} = -1/(s \cdot R \cdot C) = -\omega_0/s$.

So, the ideal integrator has a simple pole in the origin.

When calculating for the real circuit, if we consider the frequency range in which the frequency of the signal f (frequency of the components of the spectrum of the input signal) is significantly less than $f_2 = 20$ MHz, we can write the following approximate expression for integrator's gain:

$$A_{\text{int}} \approx \frac{-A \cdot \frac{\omega}{RC}}{s \cdot (-s_2)} = \frac{-A \cdot \frac{\omega}{RC}}{s \cdot (A \cdot \omega_0)} = -\frac{1}{sRC}.$$

Therefore, for signals whose frequency is $f_1 < f < f_2$, the circuit acts as an ideal integrator. Outside this range, the circuit does not have an integrator's function.

Problem 4.B.17

In Fig. 4.B.50 a differentiation circuit (the differentiator) is shown.

(a) Determine the frequency response of the differentiation circuit if the OA used is ideal, and $RC = 10$ μs.

(b) Determine the frequency range in which the given circuit operates as a differentiation circuit if the OA has a finite gain $A = 10^5$ and a finite upper cut-off frequency $f_0 = 200$ Hz.

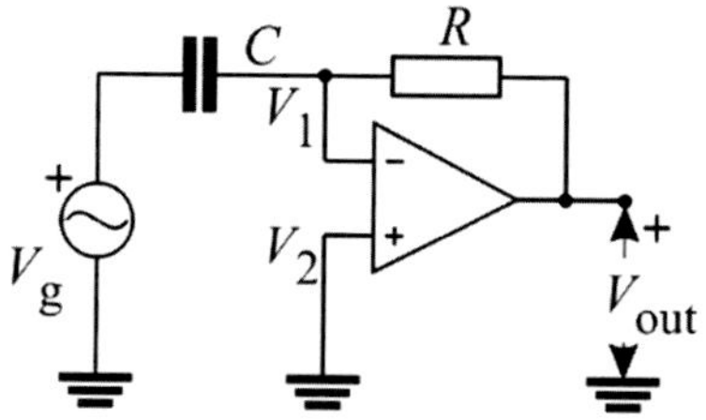

Fig. 4.B.50 Differentiation circuit

Solution to the Problem 4.B.17

(a) If the operational amplifier is ideal, for the circuit of Fig. 4.B.50, using the expression for the gain of the inverting amplifier (4.B.10.6), in which R_1 is replaced by $1/(sC)$ and R_2 by R, the expression for transfer function becomes

$$A_v = V_{out}/V_g = -sRC. \tag{4.B.17.1}$$

The transfer function has a zero at the origin.

(b) If the operational amplifier is not ideal, and its gain is given by the expression

$$A_{v1}(s) = A/(1 + s/\omega_0), \tag{4.B.17.2}$$

based on (4.B.10.4) in which R_1 is substituted with $1/(sC)$, R_2 with R, and A with (4.B.17.2) we get the transfer function of the differentiator as

$$A_v = \frac{-A\omega_0 s}{s^2 + s\left(\omega_0 + \frac{1}{RC}\right) + (1 + A)\frac{\omega_0}{RC}}, \tag{4.B.17.3}$$

whose poles are

$$s_{1/2} = -\frac{1 + \omega_0 RC}{2RC}\left[1 \pm j\sqrt{1 - \frac{4\omega_0 RC(1 + A)}{(1 + \omega_0 RC)^2}}\right]. \tag{4.B.17.4}$$

For $A \gg 1$ the poles become

$$s_{1/2} = -\frac{1 + \omega_0 RC}{2RC} \pm j\sqrt{\frac{\omega_0 A}{RC}}. \tag{4.B.17.5}$$

If $RC = 10\ \mu s$, $A = 10^5$ and $\omega_0 = 2\pi \cdot 200$ rad/s we have

$$s_{1/2} = \left(-50.63 \times 10^3 \pm j3.54 \times 10^6\right) \text{ rad/s}$$

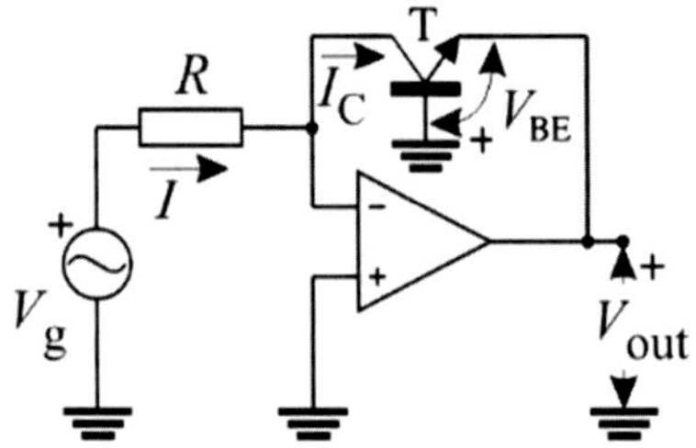

Fig. 4.B.51 Logarithmic amplifier

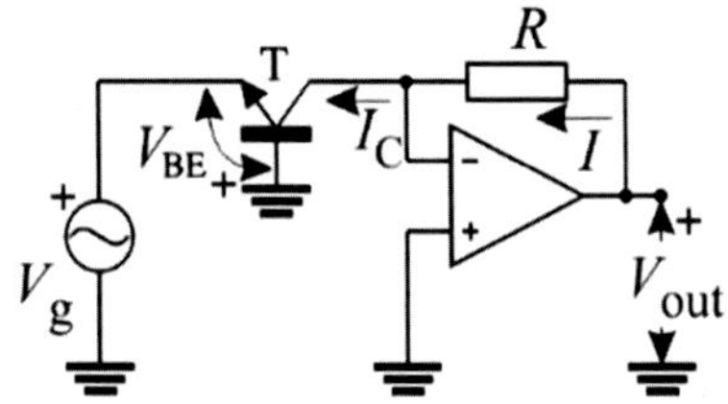

Fig. 4.B.52 Anti-logarithmic amplifier

$$\approx \pm j3.54 \times 10^6 \text{ rad/s}. \tag{4.B.17.6}$$

The corresponding frequencies are: $f_{1/2} = |s_{1/2}|/2\pi = 563.4$ kHz.

For frequencies that are significantly less than this value, the denominator acts as a constant so that the circuit acts as a differentiator. Above this frequency, both poles become active, so that the total asymptotic slope becomes −6 dB/oct, which means that the circuit acts as an integrator.

Problem 4.B.18

Figure 4.B.51 shows a logarithmic, while Fig. 4.B.52 an anti-logarithmic (exponential) amplifier. In both cases, bipolar transistors with a voltage between the base and the emitter $V_{BE} = 0.6$ V at a collector current $I_C = 1$ mA were used.

Determine the dependence of the output voltage on the input voltage for the logarithmic and anti-logarithmic amplifier, for $R = 10$ kΩ. Consider an ideal OA.

Solution to the Problem 4.B.18

During normal polarization, the collector current I_C can be represented by the approximate expression:

$$I_C \approx I_s \cdot e^{(V_{BE}/V_T)}, \tag{4.B.18.1}$$

where I_s is the inverse saturation current, V_{BE} the base-to-emitter voltage and V_T the voltage equivalent to the temperature ($V_T = kT/q$), which at room temperature has the value of 26 mV.

From this expression we extract I_s as

$$I_s = I_C \cdot e^{-(V_{BE}/V_T)} = 95 \times 10^{-15}\ A. \tag{4.B.18.2}$$

The output voltage of the logarithmic amplifier is

$$V_{out} = -V_{BE} = -V_T \cdot \ln(I_C/I_s). \tag{4.B.18.3}$$

When the operational amplifier is ideal, the collector current of the transistor is equal to the current through the resistor *R*, i.e.,

$$I_C = V_g/R. \tag{4.B.18.4}$$

Accordingly, for the output voltage we have

$$V_{out} = -V_T \cdot \ln\left[V_g/(RI_s)\right]. \tag{4.B.18.5}$$

Taking the approximation $\ln(x) \approx 2.3 \cdot \log(x)$, and by substitution of the numerical values we obtain

$$V_{out} = -0.06 \cdot \log\left(V_g/(95 \times 10^{-11})\right). \tag{4.B.18.6}$$

Of course, this circuit is functioning in the fourth quadrant (if $V_g > 95 \times 10^{-11}$ V, which means that it is necessary for the excitation voltage V_g to be positive (and $V_g > 95 \times 10^{-11}$ V, which is normally fulfilled).

The output voltage of the anti-logarithmic (exponential) amplifier from Fig. 4.B.52, in the case of an ideal infinite gain OA, may be represented by the expression:

$$V_{out} = I \cdot R. \tag{4.B.18.7}$$

Since the current *I* through the resistor *R* is equal to the collector current of the transistor I_C, which can be represented by the expression (4.B.18.1), the output voltage becomes

$$V_{out} = I_C R = I_s R \cdot e^{(V_{BE}/V_T)}. \tag{4.B.18.8}$$

On the other hand, the voltage between the emitter and the base is equal to the input voltage, i.e.,

$$V_{BE} = -V_g, \tag{4.B.18.9}$$

so, the output voltage of the anti-logarithmic amplifier is:

$$V_{out} = R \cdot I_s \cdot e^{\left(-V_g/V_T\right)}. \tag{4.B.18.10}$$

Having in mind $e^x \approx 10^{x/2.3}$ and considering the values of the elements in the circuit and the parameters of the transistor at room temperature, we obtain

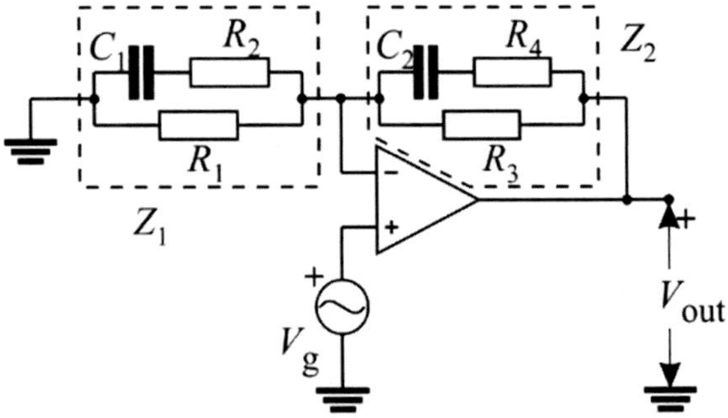

Fig. 4.B.53 Active filter

$$V_{out} = R \cdot I_s \cdot 10^{-[V_g/(2.3 \cdot V_T)]} = 95 \times 10^{-11} \times 10^{-16.72 \cdot V_g}. \tag{4.B.18.11}$$

Of course, the input voltage V_g, in this case, must be negative for the calculation to be realizable. We say that this circuit operates in the second quadrant.

Problem 4.B.19
Determine the transfer function $H(s) = V_{out}(s)/V_g(s)$ of the circuit from Fig. 4.B.53, draw an asymptotic approximation of the amplitude and phase characteristic if $R_1 = R_2 = R_3 = R_4$ and $C_1 = C_2$. The operational amplifier used is ideal with infinite gain.

Solution to the Problem 4.B.19
The impedances in the negative feedback circuit of the circuit of Fig. 4.B.53 may be represented by the expressions:

$$Z_1 = \frac{R_1[R_2 + 1/(sC_1)]}{R_1 + R_2 + 1/(sC_1)} = \frac{R_1(1 + s\tau_1)}{1 + s\tau_2}, \tag{4.B.19.1}$$

and

$$Z_2 = \frac{R_3(1 + sC_2R_4)}{1 + sC_2(R_3 + R_4)} = \frac{R_3(1 + s\,\tau_3)}{1 + s\,\tau_4}, \tag{4.B.19.2}$$

where the time constants are given by

$$\tau_1 = C_1R_2,\ \tau_2 = C_1(R_1 + R_2),\ \tau_3 = C_2R_4, \text{ and } \tau_4 = C_2(R_3 + R_4). \tag{4.B.19.3}$$

The transfer function of the non-inverting amplifier can be obtained by replacing (4.B.19.1) and (4.B.19.2) into (4.B.10.10), instead of R_1 and R_2, respectively:

$$H(s) = V_{out}/V_g = 1 + Z_2/Z_1. \tag{4.B.19.4}$$

after proper rearranging we get

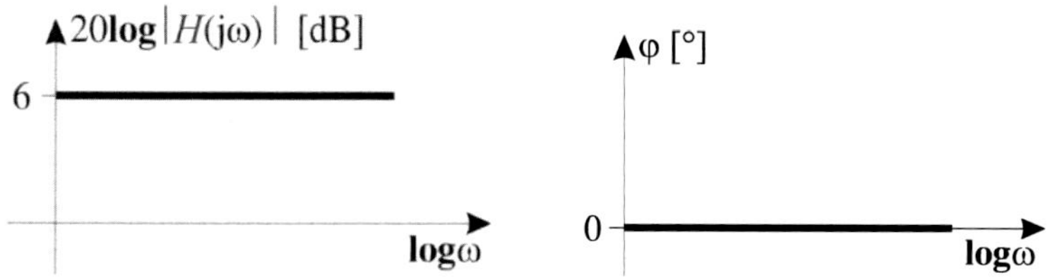

Fig. 4.B.54 Amplitude (left) and phase (right) characteristic of the circuit of Fig. 4.B.53

$$H(s) = (1 + R_3/R_1)\frac{1 + s\,\tau_5 + s^2\tau_6^2}{(1 + s\,\tau_1)(1 + s\,\tau_4)}, \tag{4.B.19.5}$$

where the time constants are

$$\tau_5 = C_1R_2 + C_2R_4 + (C_1 + C_2)\frac{R_1R_3}{R_1 + R_3} \tag{4.B.19.6}$$

and

$$\tau_6^2 = C_1C_2\left[R_2R_4 + R_1R_3\frac{R_2 + R_4}{R_1 + R_3}\right]. \tag{4.B.19.7}$$

Using the conditions $R_1 = R_2 = R_3 = R_4 = R$ and $C_1 = C_2 = C$ the time constants become

$$\tau_1 = \tau_2 = CR,\quad \tau_3 = \tau_4 = 2CR,\quad \tau_5 = 3CR,\ \text{and}\ \tau_6^2 = 2C^2R^2. \tag{4.B.19.8}$$

so that

$$H(s) = 2\frac{1 + s(3CR) + s^2\left(2C^2R^2\right)}{(1 + s\cdot CR)(1 + s\cdot 2CR)}. \tag{4.B.19.9}$$

To determine the modulus and phase of the transfer function, the complex frequency s should be replaced by $j\omega$, to produce

$$H(j\,\omega) = 2\frac{1 - 2\omega^2C^2R^2 + j3\omega CR}{(1 + j\omega CR)(1 + j2\omega CR)}. \tag{4.B.19.10}$$

Now, the modulus of the transfer function (the amplitude characteristic of the filter) is

$$|H(j\,\omega)| = 2\frac{\sqrt{(1 - 2\omega^2C^2R^2)^2 + 9\omega^2C^2R^2}}{\sqrt{\left(1 + \omega^2C^2R^2\right)\left(1 + 4\omega^2C^2R^2\right)}} = 2. \tag{4.B.19.11}$$

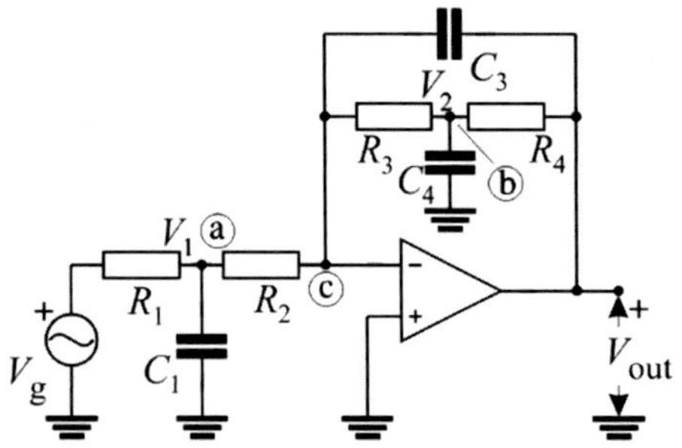

Fig. 4.B.55 A low-pass active filter

The phase characteristic is obtained as

$$\begin{aligned}\phi(\omega) &= \arg\{H(j\,\omega)\} = \phi_{\text{numerator}} - \phi_{\text{denominator}}\\ &= \text{arctg}\frac{3\omega CR}{1-2\omega^2C^2R^2} - \left[\text{arctg}(\omega\,CR) + \text{arctg}(2\omega\,CR)\right]\\ &= \text{arctg}\frac{3\omega\,CR}{1-2\omega^2C^2R^2} - \text{arctg}\frac{3\omega\,CR}{1-2\omega^2C^2R^2} = 0. \end{aligned} \quad (4.B.19.12)$$

Both the amplitude and phase characteristics are shown in Fig. 4.B.54.

Since no frequency dependence is manifested in both cases we consider this circuit is a zero delay all-pass filter which may be used as an isolation amplifier with a gain of 6 dB.

Problem 4.B.20

Determine the transfer function $H(s) = V_{out}(s)/V_g(s)$ of the circuit depicted in Fig. 4.B.55 and then write an expression for the phase characteristic of the filter. The operational amplifier used is ideal with infinite gain.

Solution to the Problem 4.B.20

For nodes a, b and c of the circuit given in Fig. 4.B.55 the nodal equations are

$$\text{a}: \quad (V_1 - V_g)/R_1 + sC_1V_1 + V_1/R_2 = 0, \quad (4.B.20.1)$$

$$\text{b}: \quad V_2/R_3 + sC_4V_2 + (V_2 - V_{out})/R_4 = 0, \quad (4.B.20.2)$$

$$\text{c}: \quad V_1/R_2 + sC_3V_{out} + V_2/R_3 = 0. \quad (4.B.20.3)$$

By solving this system of equations, the transfer function of the circuit is obtained in the form

$$H(s) = H_0\frac{1 + s/\omega_z}{\left(1+\frac{s}{\omega_{p1}}\right)\left(1+\frac{s}{\omega_{p2}}+\frac{s^2}{\omega_{p3}^2}\right)}, \quad (4.B.20.4)$$

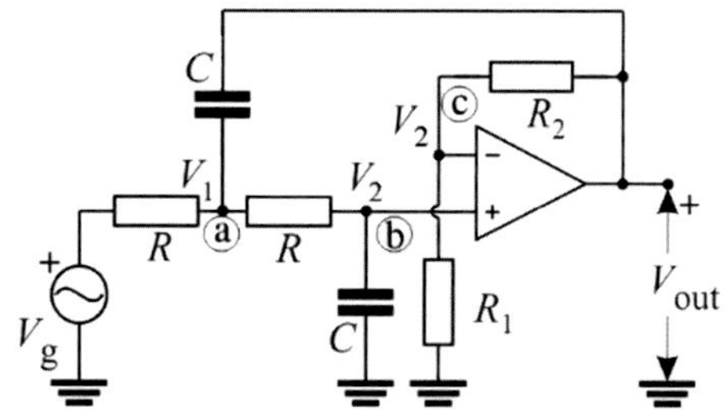

Fig. 4.B.56 Low-pass active filter

where

$$H_0 = -(R_3 + R_4)/(R_1 + R_2), \tag{4.B.20.5}$$

$$\omega_z = (R_3 + R_4)/(C_4 \cdot R_3 \cdot R_4), \tag{4.B.20.6}$$

$$\omega_{p1} = (R_1 + R_2)/(C_1 \cdot R_1 \cdot R_2), \tag{4.B.20.7}$$

$$\omega_{p2} = 1/[C_3(R_3 + R_4)], \tag{4.B.20.8}$$

$$\omega_{p3} = 1/\sqrt{C_3 \cdot C_4 \cdot R_3 \cdot R_4}. \tag{4.B.20.9}$$

When determining the phase characteristic, first, it is necessary to introduce the substitution $s = j\omega$, so that:

$$H(j\,\omega) = H_0 \frac{1 + j\,\omega/\omega_z}{\left(1 + \frac{j\,\omega}{\omega_{p1}}\right)\left(1 + \frac{j\,\omega}{\omega_{p2}} - \frac{\omega^2}{\omega_{p3}^2}\right)} \tag{4.B.20.10}$$

Now the phase characteristic is obtained as the sum of the contributions of each factor and is given by the expression:

$$\begin{aligned}\phi(\omega) &= \arg\{H(j\,\omega)\} = \arg\{H_0\} + \arg\{1 + j\,\omega/\omega_z\} \\ &\quad - \arg\left\{1 + j\,\omega/\omega_{p1}\right\} - \arg\left\{1 - \omega^2/\omega_{p3}^2 + j\,\omega/\omega_{p2}\right\} \\ &= \pi + \text{arctg}(\omega/\omega_z) - \text{arctg}\left(\omega/\omega_{p1}\right) - \text{arctg}\frac{\omega/\omega_{p2}}{1 - \left(\omega/\omega_{p3}\right)^2}.\end{aligned} \tag{4.B.20.11}$$

Problem 4.B.21

Figure 4.B.56 shows a circuit of an active low-pass filter implemented by means of an ideal operational amplifier with infinite gain.

(a) Determine the transfer function of the filter $H(s) = V_{out}(s)/V_g(s)$ if the operational amplifier is ideal with infinite gain.
(b) Determine the maximum of the amplitude characteristic and the frequency at which the maximum occurs.
(c) Draw an asymptotic approximation of the amplitude and phase characteristics.
(d) If $R_1 = R_2$ determine the amplitude characteristics of the filter at low frequencies and its upper cut-off frequency.

Solution to the Problem 4.B.21

(a) For nodes a, b, and c in the circuit of Fig. 4.B.56 the following system of equations can be written

$$\text{a}: \quad \frac{V_1 - V_g}{R} + \frac{V_1 - V_2}{R} + (V_1 - V_{out})sC = 0, \tag{4.B.21.1}$$

$$\text{b}: \quad (V_2 - V_1)/R + sCV_2 = 0, \tag{4.B.21.2}$$

$$\text{c}: \quad V_2/R_1 + (V_2 - V_{out})/R_2 = 0. \tag{4.B.21.3}$$

After solving for V_{out} the transfer function may be expressed as

$$H(s) = \frac{V_{out}(s)}{V_g(s)} = \frac{1 + R_2/R_1}{1 + sCR(2 - R_2/R_1) + s^2C^2R^2}. \tag{4.B.21.4}$$

(b) The amplitude characteristic is now:

$$|H(j\omega)| = \frac{1 + R_2/R_1}{\sqrt{(1-X)^2 + X \cdot (2 - R_2/R_1)^2}}, \tag{4.B.21.5}$$

where $X = \omega^2C^2R^2$.
For $R_1 = R_2$ the amplitude characteristic becomes

$$|H(j\omega)| = \frac{2}{\sqrt{1 - (\omega CR)^2 + (\omega CR)^4}}. \tag{4.B.21.6}$$

The transfer function of the circuit has a pair of conjugate-complex poles, so that, despite the fact that the circuit acts as a low-pass filter, the highest value of the gain does not arise at the frequency $\omega = 0$ rad/s, but at the frequency ω_{max} at which the first derivative of the modulus of the transfer function is equal to zero.

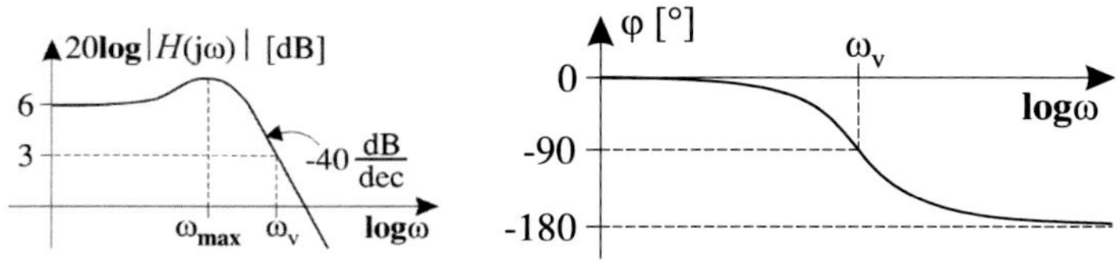

Fig. 4.B.57 Amplitude and phase characteristics of the low-pass filter depicted in Fig. 4.B.56

Therefore, the frequency at which the amplitude characteristic has its maximum was obtained from the conditions:

$$\left.\frac{\mathrm{d}|H(j\,\omega)|}{\mathrm{d}\,\omega}\right|_{\omega=\omega_{\max}}, \tag{4.B.21.7}$$

and is

$$\omega_{\max} = \frac{1}{CR\sqrt{2}}. \tag{4.B.21.8}$$

Substituting (4.B.21.8) into (4.B.21.6) gives the maximum value of the amplitude characteristic function

$$|H(j\,\omega_{\max})| = 4/\sqrt{3} = 2.31. \tag{4.B.21.9}$$

(c) The amplitude and phase characteristics are shown in Fig. 4.B.57.

(d) At low frequencies, the amplitude characteristic is

$$H_{\text{nom}} = \lim_{\omega\to 0}|H(j\,\omega)| = 2. \tag{4.B.21.10}$$

The upper cut-off frequency of the filter (frequency at which the modulus of the transfer function decreases from the nominal value $\sqrt{2}$ times) is obtained from the conditions:

$$\frac{|H(j\,\omega_{\text{h}})|}{H_{\text{nom}}} = \frac{1}{\sqrt{2}}, \tag{4.B.21.11}$$

and is

$$\omega_{\text{h}} = \frac{1}{CR}\sqrt{\frac{1+\sqrt{5}}{2}} = \frac{1.27}{CR}. \tag{4.B.21.12}$$

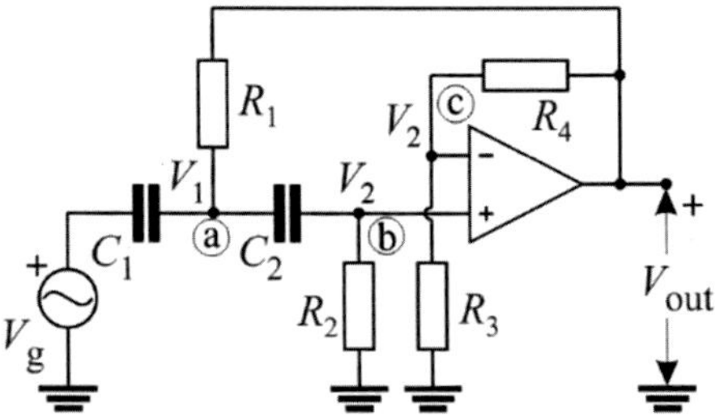

Fig. 4.B.58 A high-pass active filter

Problem 4.B.22

In Fig. 4.B.58 the circuit of the high-pass active filter is shown.

(a) Determine the transfer function of the filter $H(s) = V_{out}(s)/V_g(s)$ when the operational amplifier is ideal with infinite gain.
(b) Determine the ratio R_4/R_3 so that the nominal gain (high-frequency gain, $\omega \to \infty$) is 12 dB.
(c) Determine the lower cut-off frequency of the filter if the filter elements are: $C_1 = C_2 = 10$ nF, $R_1 = 5$ kΩ, $R_2 = 1.74$ Ω, and then draw the asymptotic approximation of the amplitude and phase characteristics.

Solution to the Problem 4.B.22

(a) For nodes a, b, and c of the circuit depicted in Fig. 4.B.58 the following equations may be written

$$\text{a}: \quad (V_1 - V_g)sC_1 + (V_1 - V_2)sC_2 + (V_1 - V_{out})/R_1 = 0, \tag{4.B.22.1}$$

$$\text{b}: \quad (V_2 - V_1)sC_2 + V_2/R_2 = 0, \tag{4.B.22.2}$$

$$\text{c}: \quad V_2/R_3 + (V_2 - V_{out})/R_4 = 0. \tag{4.B.22.3}$$

By solving this system of equations, the transfer function of the high-pass filter is obtained as

$$H(s) = \frac{s^2\tau_1\tau_2(1 + R_4/R_3)}{1 + s\left[\tau_1 + \tau_3 - \tau_2\frac{R_4}{R_3}\right] + s^2\tau_1\tau_2}. \tag{4.B.22.4}$$

where $\tau_1 = C_1R_1$, $\tau_2 = C_2R_2$, and $\tau_3 = C_2R_1$.

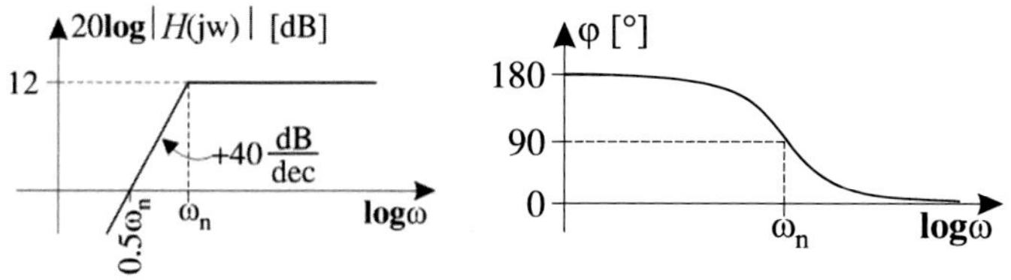

Fig. 4.B.59 Asymptotic approximation of the amplitude (left) and phase (right) characteristics of the high-pass filter

(b) The nominal gain (modulus of transfer function at high frequencies) is

$$H_0 = \lim_{\omega \to \infty} |H(j\,\omega)| = 1 + R_4/R_3. \tag{4.B.22.5}$$

or

$$H_0[\text{dB}] = 20 \cdot \log(1 + R_4/R_3). \tag{4.B.22.6}$$

From the condition that the nominal gain is $H_0[\text{dB}] = 12$ dB, the required resistance ratio R_4/R_3 is obtained as

$$R_4/R_3 = 10^{H_0[\text{dB}]/20} - 1 = 3. \tag{4.B.22.7}$$

(c) The lower cut-off frequency is obtained from the condition

$$|H(j\,\omega_l)|/H_0 = 1/\sqrt{2}, \tag{4.B.22.8}$$

where

$$|H(j\,\omega_l)| = \frac{\omega_n^2 \tau_1 \tau_2 (1 + R_4/R_3)}{\sqrt{(1 - \omega_n^2 \tau_1 \tau_2)^2 + \omega_n^2 \tau^2}} \tag{4.B.22.9}$$

with $\tau = \left[\tau_1 + \tau_3 - \tau_2 \frac{R_4}{R_3}\right]$
and is

$$f_l = \omega_l/2\pi = 6.67 \text{ kHz}. \tag{4.B.22.10}$$

The asymptotic approximation of the amplitude and phase characteristics is shown in Fig. 4.B.59.

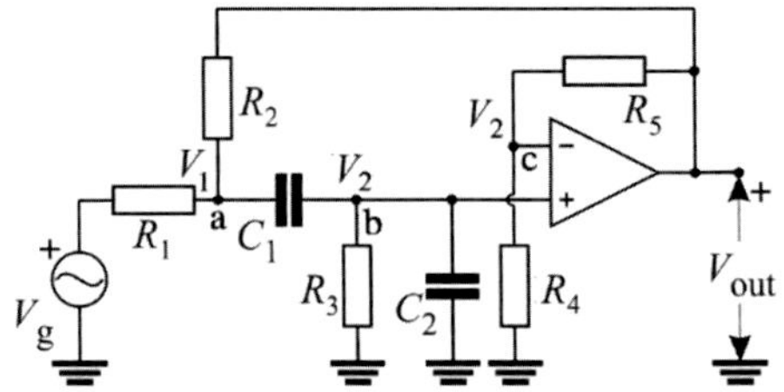

Fig. 4.B.60 An active band-pass filter

Problem 4.B.23

In Fig. 4.B.60 an active pass-band filter is given.

(a) Determine the transfer function of the filter $H(s) = V_{out}(s)/V_g(s)$ if the operational amplifier is ideal.
(b) Determine the ratio of the resistors R_5/R_4 so that the value of the modulus of the transfer function at the central frequency is 3, and then the corresponding central frequency, if $R_1 = R_2 = R_3/2 = 22\ \text{k}\Omega$ and $C_1 = 2C_2 = 2.2$ nF.
(c) Determine the upper and lower cut-off frequency of the filter and draw asymptotic approximations of the amplitude and phase characteristics.

Solution to the Problem 4.B.23

(a) For nodes a, b, and c in the circuit of Fig. 4.B.60 the following equations can be written

$$\text{a}: \quad (V_1 - V_g)/R_1 + (V_1 - V_2) \cdot sC_1 + (V_1 - V_{out})/R_2 = 0, \tag{4.B.23.1}$$

$$\text{b}: \quad (V_2 - V_1) \cdot sC_1 + V_2/R_3 + V_2 \cdot sC_2 = 0, \tag{4.B.23.2}$$

$$\text{c}: \quad V_2/R_4 + (V_2 - V_{out})/R_5 = 0. \tag{4.B.23.3}$$

By solving this system of equations, the transfer function of the band-pass filter is obtained in the form:

$$H(s) = \frac{s\tau_1}{1 + s\tau_2 + s^2\tau_3^2}, \tag{4.B.23.4}$$

where

$$\tau_1 = C_1 \cdot R_2 \cdot R_3(1 + R_5/R_4)/(R_1 + R_2), \tag{4.B.23.5}$$

$$\tau_2 = C_2R_3 + \frac{C_1R_1R_2}{R_1 + R_2}\left(1 + \frac{R_3}{R_1} - \frac{R_3R_5}{R_2R_4}\right) \tag{4.B.23.6}$$

$$\tau_3^2 = C_1 C_2 R_1 R_2 R_3/(R_1 + R_2). \tag{4.B.23.7}$$

The amplitude characteristic is

$$\left|\frac{V_{\text{out}}}{V_{\text{g}}}\right| = \frac{\omega\tau_1}{\sqrt{\left(1 - \omega^2\tau_3^2\right)^2 + \omega^2\tau_2^2}}. \tag{4.B.23.8}$$

Its maximum is at the frequency at which the following condition is met

$$\frac{\partial}{\partial\omega}\left\{\left|\frac{V_{\text{out}}}{V_{\text{g}}}\right|\right\}\Bigg|_{\omega=\omega_0} = 0, \tag{4.B.23.9}$$

which leads to

$$\omega_0 = 1/\tau_3 = 1/\sqrt{C_1 C_2 R_1 R_2 R_3/(R_1 + R_2)}. \tag{4.B.23.10}$$

The maximum value of the modulus of the transfer function is:

$$H_0 = \left|\frac{V_{\text{out}}}{V_{\text{g}}}\right|_{\omega=\omega_0} = \frac{\tau_1}{\tau_2} \tag{4.B.23.11}$$

From the conditions of the problem, $H_0 = 3$, the required ratio of resistors is

$$\frac{R_5}{R_4} = \frac{H_0 R_1 - R_2}{H_0 R_1 + R_2} + \frac{H_0 R_1 R_2}{R_3(H_0 R_1 + R_2)} + \frac{C_2}{C_1}\cdot\frac{(R_1 + R_2)H_0}{H_0 R_1 + R_2} = 1.625. \tag{4.B.23.12}$$

The cut-off frequencies of the filter are obtained from the condition

$$\frac{\omega_{\text{g}}\tau_1}{\sqrt{\left(1 - \omega_{\text{g}}^2\tau_3^2\right)^2 + \omega_{\text{g}}^2\tau_2^2}} = \frac{H_0}{\sqrt{2}} = \frac{3}{\sqrt{2}}. \tag{4.B.23.13}$$

or, as solution of

$$9\tau_3^4\omega_{\text{g}}^4 + \left(9\tau_2^2 - 18\tau_3^2 - 2\tau_1^2\right)\omega_{\text{g}}^2 + 9 = 0. \tag{4.B.23.14}$$

For the given values of the resistances and capacitances the time constants are

$$\tau_1 = 127\,\mu\text{s}, \quad \tau_2 = 42.35\,\mu\text{s}, \quad \text{and} \quad \tau_3 = 34.2\,\mu\text{s}. \tag{4.B.23.15}$$

The solutions of equation (4.B.23.14) with time constants given in (4.B.23.15) represent the cut-off frequencies which are:

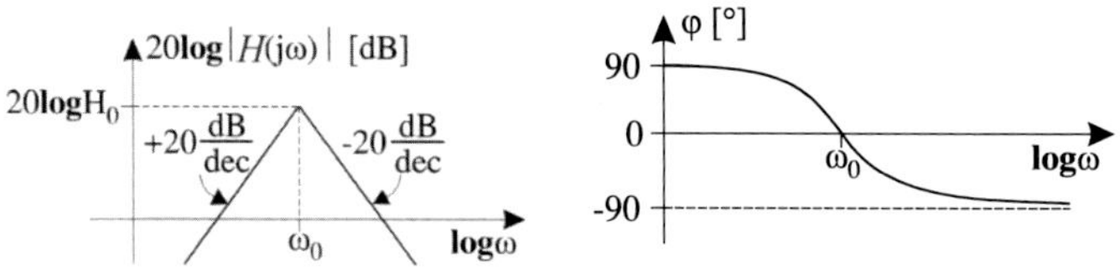

Fig. 4.B.61 Asymptotic approximation of the amplitude and phase characteristics of the band-pass filter of Fig. 4.B.60

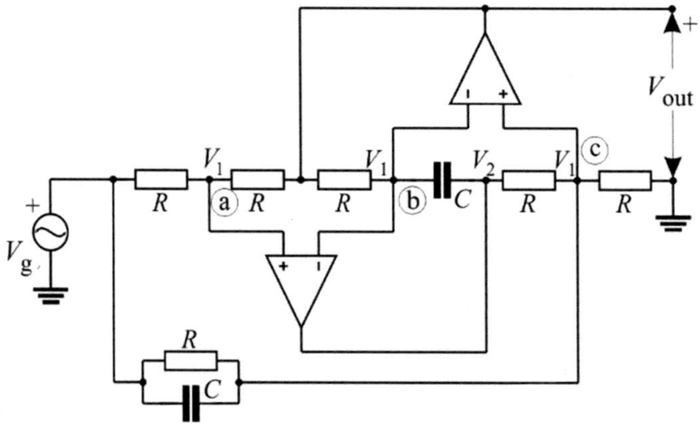

Fig. 4.B.62 A band-stop (Notch) filter

$$\omega_{c1} = 52438.3 \ \text{rad/s}, \quad \text{or} \quad f_{c1} = 8345.8 \, \text{Hz}$$

$$\omega_{c2} = 16281.3 \ \text{rad/s}, \quad \text{or} \quad f_{c2} = 2591.2 \, \text{Hz},$$

while the center frequency of the bandwidth is given by the expression (4.B.23.10) and is

$$\omega_0 = 29{,}219.3 \ \text{rad/s}, \quad \text{or} \quad f_0 = 4650.4 \ \text{Hz}.$$

The asymptotic approximation of the amplitude and phase characteristics is shown in Fig. 4.B.61.

Problem 4.B.24
For the band-stop (notch) filter of Fig. 4.B.62 determine the transfer function $H(s) = V_{\text{out}}(s)/V_{\text{g}}(s)$ and then find the minimum of the modulus of the transfer function and its frequency. The OAs used are ideal with infinite gain.

Solution to the Problem 4.B.24
Since the OAs are ideal, the potentials at points a, b, and c are equal (V_1), so the following equations can be written

$$\text{a}: \quad \left(V_1 - V_g\right)/R + (V_1 - V_{out})/R = 0, \tag{4.B.24.1}$$

$$\text{b}: \quad (V_1 - V_{out})/R + (V_1 - V_2) \cdot sC = 0, \tag{4.B.24.2}$$

$$\text{c}: \quad \frac{V_1 - V_2}{R} + \frac{V_1}{R} + \left(V_1 - V_g\right)\frac{1 + sCR}{R} = 0. \tag{4.B.24.3}$$

By solving this system of equations, the transfer function of the circuit of Fig. 4.B.62 is obtained as

$$H(s) = \frac{V_{out}}{V_g} = \frac{1 + (sCR)^2}{1 + 2sCR + (sCR)^2}. \tag{4.B.24.4}$$

Its modulus is

$$|H(j\omega)| = \frac{\left|1 - (\omega RC)^2\right|}{\sqrt{[1 - (\omega RC)^2]^2 + 4(\omega RC)^2}}. \tag{4.B.24.5}$$

The value of its minimum is obvious:

$$|H(j\omega_0)| = 0, \tag{4.B.24.6}$$

at the frequency

$$\omega_0 = 1/(CR). \tag{4.B.24.7}$$

Problem 4.B.25

(a) Determine the "y" parameters of the impedance converter from Fig. 4.B.63 if the amplifiers used are ideal with infinite gain.
(b) Determine the equivalent input impedance of the circuit if a capacitor C_2 is connected to output 2 of the circuit.
(c) Determine the resonant frequency of the circuit obtained by parallel connection of capacitors C_1 and C_2 between the input and output terminals, respectively, according to Fig. 4.B.64.

Solution to the Problem 4.B.25

(a) It is known that any linear four-terminal circuit can be described by "y" parameters in the following way:

$$\begin{aligned} J_1 &= y_{11}V_1 + y_{12}V_2 \\ J_2 &= y_{21}V_1 + y_{22}V_2 \end{aligned}. \tag{25.1}$$

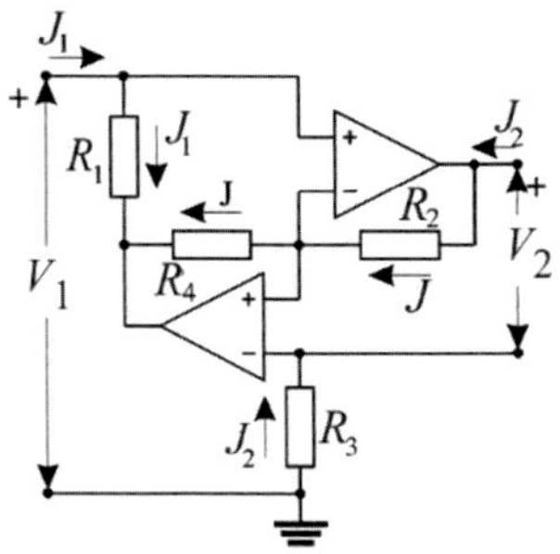

Fig. 4.B.63 Impedance converter (gyrator)

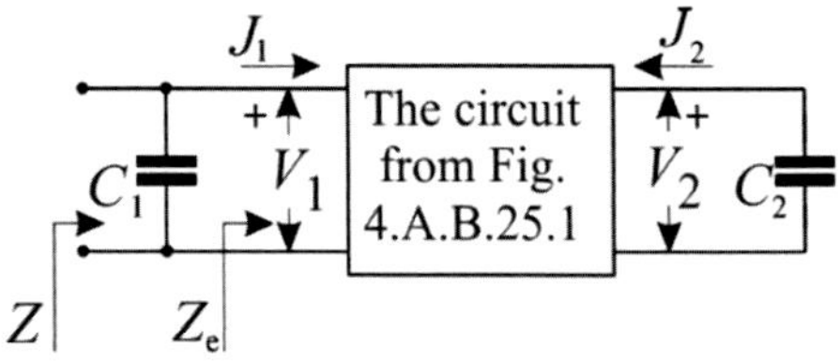

Fig. 4.B.64 Implementation of the gyrator

If OAs are ideal with infinite gain, all inputs of both operational amplifiers are at potential V_1. Therefore, for the circuit of Fig. 4.B.63 the following equations can be written

$$R_1 \cdot J_1 = R_4 \cdot J = R_4 \cdot V_2/R_2, \quad (4.B.25.2)$$

$$J_2 = -V_1/R_3, \quad (4.B.25.3)$$

or

$$J_1 = 0 \cdot V_1 + [R_4/(R_1 \cdot R_2)] \cdot V_2 \quad (4.B.25.4)$$

$$J_2 = [-1/R_3] \cdot V_1 + 0 \cdot V_2$$

By comparing the corresponding terms of (4.B.25.1) and (4.B.25.4) the "y" parameters of the circuit from Fig. 4.B.63 become

$$y_{11} = y_{22} = 0, \; y_{12} = \frac{R_4}{R_1 R_2}, \; y_{21} = -\frac{1}{R_3}. \quad (4.B.25.5)$$

(b) The equivalent input impedance Z_e of the circuit, when loaded at the output terminals by a capacitor C_2, can be determined from the expression

$$Z_e = V_1/J_1. \tag{4.B.25.6}$$

Since

$$V_2 = -J_2/(j\,\omega\,C_2), \tag{4.B.25.7}$$

we have

$$J_1 = \frac{R_4}{R_1 R_2}\left(-\frac{J_2}{j\,\omega\,C_2}\right). \tag{4.B.25.8}$$

Substituting J_2 from (4.B.25.3) leads to

$$J_1 = \frac{R_4}{R_1 R_2 R_3}\frac{1}{j\,\omega\,C_2}V_1, \tag{4.B.25.9}$$

so that the equivalent input impedance Z_e is of inductive character and is given by

$$Z_e = j\,\omega\,L_e = j\,\omega\frac{C_2 R_1 R_2 R_3}{R_4}, \tag{4.B.25.10}$$

where the equivalent input inductance is

$$L_e = C_2 R_1 R_2 R_3/R_4. \tag{4.B.25.11}$$

(c) The total input impedance of the circuit of Fig. 4.B.64 is given by

$$Z = \frac{j\,\omega\,L_e \cdot 1/(j\,\omega\,C_1)}{j\,\omega\,L_e + 1/(j\,\omega\,C_1)} = \frac{j\,\omega\,L_e}{1-\omega^2 C_1 L_e}. \tag{4.B.25.12}$$

The resonant frequency is obtained from

$$1-\omega_0^2 C_1 L_e = 0, \tag{4.B.25.13}$$

and is

$$\omega_0 = \frac{1}{\sqrt{C_1 L_e}} = \frac{1}{\sqrt{C_1 C_2 R_1 R_2 R_3/R_4}}. \tag{4.B.25.14}$$

Problem 4.B.26.

The output voltage of the circuit of Fig. 4.B.65 represents the solution of the differential equation:

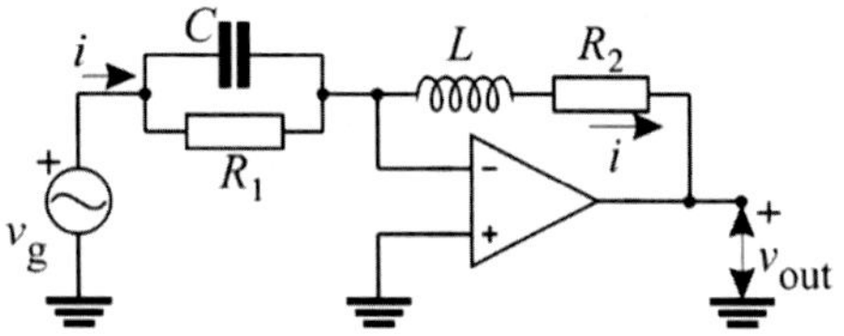

Fig. 4.B.65 Circuit solving the differential equation

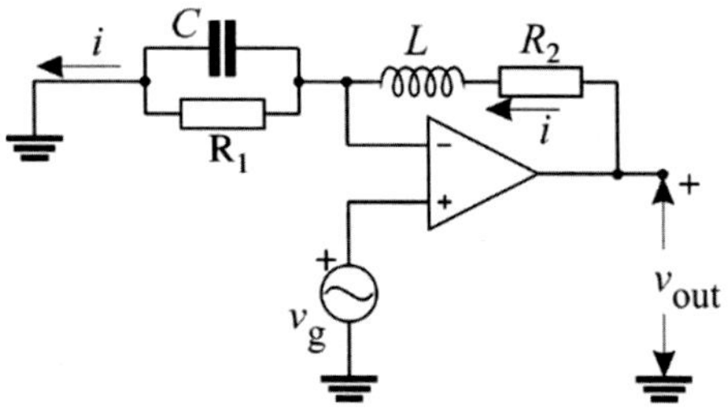

Fig. 4.B.66 A circuit complementary to the one of Fig. 4.B.65

$$v_{out}(t) = a_1 v_g(t) + a_2 \frac{dv_g(t)}{dt} + a_3 \frac{d^2 v_g(t)}{dt^2}.$$

(a) Determine the constants a_1, a_2, and a_3 as a function of the element values of the circuit. The operational amplifier is ideal with infinite gain.
(b) Draw a circuit that will have the same functional dependence of the output on the input voltage with the constants a_1, a_2, and a_3 of opposite signs as compared with the previously observed circuit of Fig. 4.B.65.

Solution to the Problem 4.B.26

(a) The inverting input of the ideal operational amplifier is at the potential of the virtual ground, so for current $i(t)$ it can be written

$$i(t) = \frac{v_g(t)}{R_1} + C\frac{dv_g(t)}{dt}. \tag{4.B.26.1}$$

The output voltage is

$$v_{out}(t) = -R_2 \cdot i(t) - L\frac{di(t)}{dt}. \tag{4.B.26.2}$$

Since by differentiating the expression (4.B.26.1) we get

$$\frac{di(t)}{dt} = \frac{1}{R_1}\frac{dv_g(t)}{dt} + C\frac{d^2 v_g(t)}{dt^2}, \tag{4.B.26.3}$$

the output voltage becomes

$$v_{\text{out}}(t) = -(R_2/R_1)\cdot v_g(t) - (R_2C + L/R_1)\frac{dv_g(t)}{dt} - LC\frac{d^2v_g(t)}{dt^2}. \quad (4.B.26.4)$$

From here the constants a_1, a_2, and a_3 may be extracted as

$$a_1 = -R_2/R_1, \quad a_2 = -(R_2C + L/R_1), \quad \text{and} \quad a_3 = -LC. \quad (4.B.26.5)$$

(b) The circuit that realizes the same functional dependence of the output on the input voltages, but with opposite signs of the constants a_1, a_2, and a_3, is shown in Fig. 4.B.66. For this circuit, the current $i(t)$ and the output voltage $v_{\text{out}}(t)$ are given by the expressions

$$i(t) = \frac{v_g(t)}{R_1} + C\frac{dv_g(t)}{dt}, \quad (4.B.26.6)$$

$$v_{\text{out}}(t) = R_2\cdot i(t) + L\frac{di(t)}{dt} + v_g(t). \quad (4.B.26.7)$$

In the same way as for the previous circuit, for the circuit of Fig. 4.B.66 one gets

$$v_{\text{out}}(t) = \left(1 + \frac{R_2}{R_1}\right)\cdot v_g(t) + \left(R_2C + \frac{L}{R_1}\right)\frac{dv_g(t)}{dt} + LC\frac{d^2v_g(t)}{dt^2}, \quad (4.B.26.8)$$

so that the constants a_1, a_2, and a_3 are

$$a_1 = 1 + R_2/R_1, \quad a_2 = R_2C + L/R_1\ , \quad \text{and} \quad a_3 = LC. \quad (4.B.26.9)$$

Problem 4.B.27

The transistors used in the circuit of Fig. 4.B.67 are identical and operate in the ohmic region in which the drain current is defined by the expression.

$$i_0 = A\cdot\left[(v_{GS} - V_T)v_{DS} - v_{DS}^2/2\right].$$

(a) Determine the dependence of the output, v_{out}, on the excitation voltages, v_x, v_y, and v_z.
(b) If $v_z = 5$ V, $v_x = V_{xm}\cdot\sin(\omega_x t)$ and $v_y = V_{ym}\cdot\sin(\omega_y t)$, where $\omega_y \gg \omega_x$, sketch the waveform of the output voltage.

The parameters A and V_T are considered known. The OAs are ideal with infinite gain.

Solution to the Problem 4.B.27

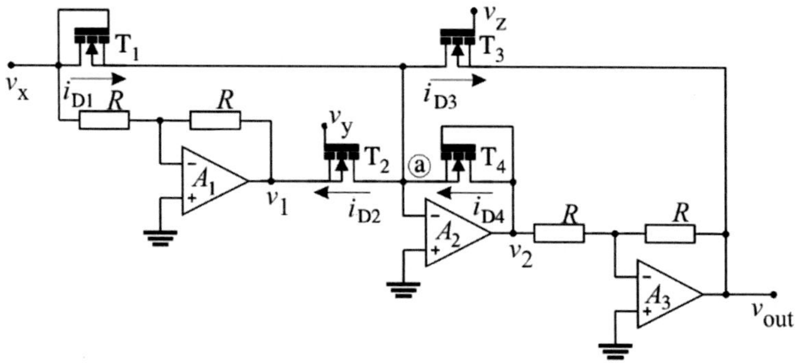

Fig. 4.B.67 Circuit of multifunctional converter

(a) The OAs A_1 and A_3 (in the configuration with equal resistors R) represent unity gain inverting amplifiers (similar to (4.B.10.6)), so that:

$$v_1 = -v_X \text{ and } v_{out} = -v_2. \tag{4.B.27.1}$$

Since the OAs are ideal with infinite gain, the inverting inputs of all OAs are at the potential of the virtual ground. Therefore, for all four transistors, the following is valid

$$v_{GS1} = v_X, \quad v_{DS1} = v_X, \tag{4.B.27.2}$$

$$v_{GS2} = v_y - v_1 = v_y + v_X, \; v_{DS2} = -v_1 = v_X, \tag{4.B.27.3}$$

$$v_{GS3} = v_Z - v_{out}, \quad v_{DS3} = -v_{out}, \tag{4.B.27.4}$$

$$v_{GS4} = v_2 = -v_{out}, \quad v_{DS4} = v_2 = -v_{out}. \tag{4.B.27.5}$$

The drain currents of transistors T_1 to T_4 in the ohmic region are given by the following expressions

$$i_{D1} = A \cdot \left[(v_X - V_T) \cdot v_X - v_X^2/2 \right], \tag{4.B.27.6}$$

$$i_{D2} = A \cdot \left[\left(v_X + v_y - V_T \right) \cdot v_X - v_X^2/2 \right] \tag{4.B.27.7}$$

$$i_{D3} = A \cdot \left[(v_Z - v_{out} - V_T) \cdot (-v_{out}) - v_{out}^2/2 \right] \tag{4.B.27.8}$$

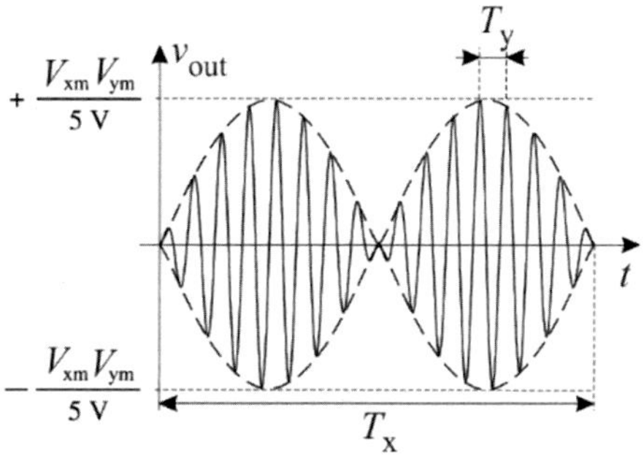

Fig. 4.B.68 The output voltage of the circuit of Fig. 4.B.67

$$i_{D4} = A \cdot \left[(-v_{out} - V_T) \cdot (-v_{out}) - v_{out}^2/2\right]. \tag{4.B.27.9}$$

The following equation can be written for node "a" (inverting input of OA A_2):

$$-i_{D1} + i_{D2} + i_{D3} - i_{D4} = 0 \tag{4.B.27.10}$$

Substituting the expressions (4.B.27.6) to (4.B.27.9) into (4.B.27.10) gives

$$-v_x v_y + v_z v_{out} = 0, \tag{4.B.27.11}$$

so that the output voltage is:

$$v_{out} = v_x v_y / v_z. \tag{4.B.27.12}$$

This expression shows that the given circuit can be used both as a multiplication circuit and as a division circuit. If $v_z = C^{te}$ is set, the output voltage is proportional to the product $v_x v_y$ and a multiplication circuit is formed. On the other hand, if v_x or v_y is chosen to be constant, the output voltage becomes proportional to the quotient of two signals: $v_{out} \sim v_y/v_z$ or $v_{out} \sim v_x/v_z$, respectively. The reader is left to determine what will happen when $v_z = 0$ occurs in the division circuit.

(b) If sinusoidal signals with significant difference in frequency are applied to inputs v_x and v_y, and constant voltage is applied to input v_z, the output signal will be of the form:

$$v_{out} = 0.2 \cdot v_{xm}\sin(\omega_x t) \cdot v_{ym}\sin(\omega_y t) \quad [V], \tag{4.B.27.13}$$

or

$$v_{out} = 0.1 \cdot V_{xm} V_{ym}\left[\cos(\omega_y - \omega_x)t - \cos(\omega_y + \omega_x)t\right] \; [V]. \tag{4.B.27.14}$$

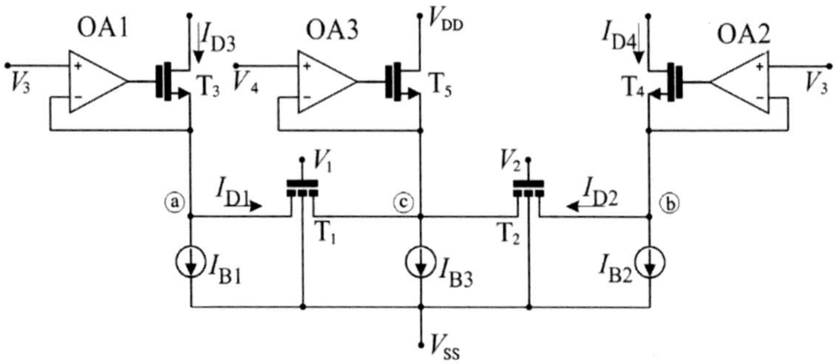

Fig. 4.B.69 Simplified schematic of the four-quadrant multiplier

The expression (4.B.27.14) represents a time-varying signal whose shape is shown in Fig. 4.B.68, where the periods T_x and T_y are:

$$T_x = 2\pi/\omega_x \text{ and } T_y = 2\pi/\omega_y. \tag{4.B.27.15}$$

Problem 4.B.28

For the circuit with Fig. 4.B.69 prove that the difference in currents ($I_{D3} - I_{D4}$) is proportional to the product of the difference in voltage $(V_1 - V_2)(V_3 - V_4)$. Using the obtained result, explain the operation of the circuit with Fig. 4.B.70.

Solution to the Problem 4.B.28

Figure 4.B.69 shows a simplified schematic of the MOS transconductance multiplier with transistors operating in the linear or ohmic region defined by the condition:

$$V_{DS} > V_{GS} - V_T, \tag{4.B.28.1}$$

where V_{DS} and V_{GS} are voltages between drain and source, or gate and source, respectively, and V_T is the threshold voltage. In the linear region, the drain current of the MOS transistor is defined by the expression:

$$I_D = A \cdot \left[(V_{GS} - V_T) \cdot V_{DS} - V_{DS}^2/2\right], \tag{4.B.28.2}$$

where the constant A depends on the technological parameters as

$$A = \mu_P C'_{ox} \cdot W/L. \tag{4.B.28.3}$$

as explained with (4.2.37).

The isolation amplifiers OA1 and OA2 ensure that the potentials at nodes "a," V_a, and "b," V_b are equal to V_3, while the isolation amplifier OA3 provides potential at points "c" as $V_c = V_4$. The voltages between the drain and the source of the transistors are equal to each other and are:

$$V_{DS1} = V_a - V_c = V_3 - V_4 \tag{4.B.28.4}$$

and

$$V_{DS2} = V_b - V_c = V_3 - V_4. \tag{4.B.28.5}$$

It is obvious that transistors T_1 and T_2 can conduct current in any direction depending on the polarity of voltage V_3 and V_4. The voltages between the gate and the source of these transistors are given as

$$V_{GS1} = V_1 - V_4 \text{ and } V_{GS2} = V_2 - V_4. \tag{4.B.28.6}$$

If both transistors operate in the linear region, the currents I_{D3} and I_{D4} are given by the following relations:

$$I_{D3} = I_{B1} + I_{D1} - I_{B1} + A_1\left[(V_{GS1} - V_{T1})V_{DS1} - V_{DS1}^2/2\right]. \tag{4.B.28.7}$$

$$I_{D4} = I_{B2} + I_{D2} = I_{B2} + A_2\left[(V_{GS2} - V_{T2})V_{DS2} - V_{DS1}^2/2\right] \tag{4.B.28.8}$$

where I_{B1} and I_{B2} are the currents of the corresponding current sources. Considering the expressions (4.B.28.4), (4.B.28.5) and (4.B.28.6) and the assumption that the transistors are identical, i.e.,

$$A_1 = A_2 = A \quad \text{and} \quad V_{T1} = V_{T2} = V_T \tag{4.B.28.9}$$

the difference of the currents I_{D3} and I_{D4} is given by

$$\Delta I = I_{D3} - I_{D4} = I_{D1} - I_{D2} = A(V_1 - V_2)(V_3 - V_4). \tag{4.B.28.10}$$

The expression (4.B.28.10) shows that the difference between the currents I_{D3} and I_{D4} is proportional to the product of the differences between the input voltages, where the differences between voltage $(V_1 - V_2)$ and $(V_3 - V_4)$ can have both positive and negative signs, i.e., this voltage multiplier works in all four quadrants; hence, it is called the four-quadrant multiplier. Of course, all this is valid only under the condition that the transistors T_1 and T_2 are paired, i.e., with identical characteristics, and that they work in the linear (ohmic) region defined by the relation (4.B.28.1).

Transistors T_{11}, T_{12}, and T_{13} in Fig. 4.B.70 represent the dynamic loads of amplifiers OA1 and OA2 composed of input transistor T_6 and transistors T_7 and T_8, respectively. Transistors T_{14} and T_{15} represent the dynamic load of amplifier OA3 consisting of input transistor T_9 and transistor T_{10}. The output transistors T_3, T_4, and T_5 have the same function as the corresponding transistors in Fig. 4.B.69. Transistors TB_1, TB_2, and TB_3 are sources of constant current, because they work in the area of saturation with constant voltage between the gate and the source $(V_8 - V_{SS})$. The output voltage between the drain of transistor T_4 and the drain of transistor T_3 is proportional to the

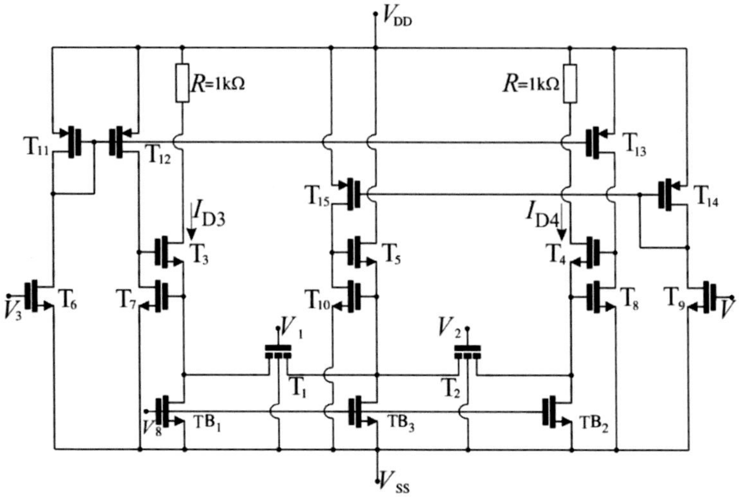

Fig. 4.B.70 Schematic of the four-quadrant multiplier

difference between the currents I_{D3} and I_{D4}, (4.B.28.10), and is

$$V_{out} = R \cdot \Delta I = R \cdot A \cdot (V_1 - V_2)(V_3 - V_4). \tag{4.B.28.11}$$

Problem 4.B.29

(a) The JFET's model given as

$$I_D = G_0\left[V_{DS} - \frac{2}{3}\left(\frac{(V_0 - V_{GS} + V_{DS})^{1.5}}{\sqrt{V_0 - V_P}} - \frac{(V_0 - V_{GS})^{1.5}}{\sqrt{V_0 - V_P}}\right)\right] \tag{4.B.29.1}$$

for small values of V_{DS} (e.g., $|V_{DS}| \leq 100$ mV), may be approximated by

$$I_D = k(1 - V_{GS}/V_P) \cdot V_{DS}. \tag{4.B.29.2}$$

The transistor parameters are $V_0 = 0.787\ V$, $G_0 = 0.32$ mA/V, and $V_P = -2.82\ V$.

Determine the difference of drain currents obtained by using the expressions (4.B.29.1) and (4.B.29.2) for $V_{GS} = 0\ V$, -1 V, and $-2\ V$ when $V_{DS} = 100$, 50, and 10 mV.

(b) Find the difference between the dynamic internal conductance $G_D = 1/r_D$ of the transistor, using the models (4.B.29.1) and (4.B.29.2), for small values of V_{DS} when $V_{GS} = 0\ V$, -1 V, and $-2\ V$.

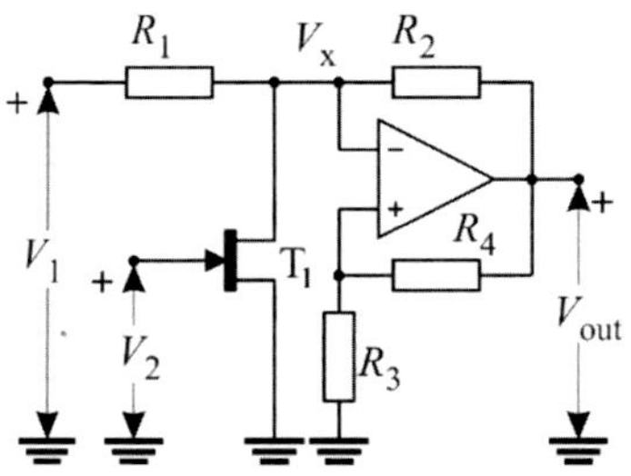

Fig. 4.B.71 A circuit performing division of voltages

(c) Using the simplified JFET model given by expression (4.B.29.2) determine the condition to be met by the resistors in the circuit of Fig. 4.B.71 that the output voltage V_{out} be proportional to the quotient of V_1 and V_2. The operational amplifier is ideal with infinite gain.

Solution to the Problem 4.B.29

Based on (4.B.29.1) it can be calculated that the drain current of the JFET, at voltages $V_{DS} = 50$ mV and $V_{GS} = -1\ V$ is

$$I_D = 4.66\,\mu A. \tag{4.B.29.3}$$

The constant k in the approximate expression (4.B.29.2) for the given values of V_{DS} and V_{GS} is:

$$k = \frac{I_D}{(1 - V_{GS}/V_P)V_{DS}} = 144.4\,\mu A/V. \tag{4.B.29.4}$$

For the comparison, the values of the drain current can be calculated using expression (4.B.29.1), which is denoted by $I_D^{(1)}$ and the approximate expression (4.B.29.2), with a constant k given in (29.4)), which is marked with $I_D^{(2)}$for voltages $V_{GS} = 0$, -1 and -2 V, when $V_{DS} = 100$, 70 and 30 mV. The calculated values are given in Table 4.B.1 where the difference between these currents $(I_D^{(2)} - I_D^{(1)})$ is substituted by ΔI_D (Table 4.B.1).

(b) It is known that a function $f(x)$ in the neighborhood of a point x for small values of the increment Δx can be approximated by

$$f(x + \Delta x) = f(x) + \frac{\partial f(x)}{\partial x}\Delta x. \tag{4.B.29.5}$$

Starting from this, it is possible to write an expression for the drain current (4.B.29.1) for small values of voltage V_{DS} (hence, the role Δx plays V_{DS}) in the form:

Table 4.B.1 Comparison od drain currents

VDS (mV)	100			50			10		
VGS (V)	0	−1	−2	0	−1	−2	0	−1	−2
$I_D^{(1)}$ (μA)	16.59	9.16	3.62	8.41	4.66	1.87	1.70	0.94	0.39
$I_D^{(2)}$ (μA)	14.44	9.32	4.20	7.22	4.66	2.1	1.44	0.93	0.42
ΔI_D (μA)	−2.15	0.16	0.52	−1.19	0	0.23	−0.26	−0.01	0.03

Table 4.B.2 The values of G_D for different values of V_{GS}

VGS (V)	0	−1	−2
$G_D^{(1)}$ (μA/V)	170.53	94.76	38.72
$G_D^{(2)}$ (μA/V)	144.40	93.19	41.99
ΔG_D (μA/V)	−26.13	−1.57	3.27

$$I_D = G_0 \cdot \left\{ V_{DS} - \frac{2}{3} \frac{(V_0 - V_{GS})^{1.5} + \frac{3}{2}(V_0 - V_{GS})^{0.5} V_{DS}}{\sqrt{V_0 - V_P}} + \frac{2}{3} \frac{(V_2 - V_{GS})^{1.5}}{\sqrt{V_0 - V_P}} \right\}, \tag{4.B.29.9}$$

or,

$$I_D = G_0 \cdot \left(1 - \sqrt{\frac{V_0 - V_{GS}}{V_0 - V_P}}\right) \cdot V_{DS}. \tag{4.B.29.7}$$

The internal conductance of a JFET (reciprocal of internal resistance) is obtained by using the expression (4.B.29.7), and it is

$$G_D^{(1)} = \frac{1}{r_D} = \frac{\partial I_D}{\partial V_{DS}} = G_0\left(1 - \sqrt{\frac{V_0 - V_{GS}}{V_0 - V_P}}\right) \tag{4.B.29.8}$$

Starting from the approximate expression for the drain current (4.B.29.2) the internal conductance is:

$$G_D^{(2)} = \partial I_D / \partial V_{GS} = k \cdot (1 - V_{GS}/V_P). \tag{4.B.29.9}$$

Using the expressions (4.B.29.8) and (4.B.29.9), the corresponding values $G_D^{(1)}$ and $G_D^{(2)}$ are obtained, as well as the difference $\Delta G_D = G_D^{(2)} - G_D^{(1)}$, for different values of V_{GS} and those are given in Table 4.B.2.

(c) For the inverting and non-inverting input of the operational amplifier in the circuit with Fig. 4.B.71 the following system of equations can be written

$$(V_x - V_1)/R_1 + V_x G_D + (V_x - V_{out})/R_2 = 0, \qquad (4.B.29.10)$$

$$(V_x - V_{out})/R_3 + V_x/R_4 = 0, \qquad (4.B.29.11)$$

where G_D denotes the internal conductance of the JFET given by (4.B.29.9), where $V_{GS} = V_2$.

From the expression (4.B.29.11) we get

$$V_x - V_{out} = -V_x R_3/R_4, \qquad (4.B.29.12)$$

or

$$V_x = V_{out} R_4/(R_3 + R_4). \qquad (4.B.29.13)$$

After substitution of (4.B.29.9) into (4.B.29.10) we get

$$V_x/R_1 - V_1/R_1 + V_x k \cdot (1 - V_2/V_P) - V_x R_3/(R_2 R_4) = 0. \qquad (4.B.29.14)$$

Small rearrangement leads to

$$V_x\left(\frac{1}{R_1} + k - \frac{R_4}{R_2 R_3}\right) - \frac{V_1}{R_1} - \frac{k \cdot R_3}{V_P(R_3 + R_4)} \cdot V_2 V_{out} = 0 \qquad (4.B.29.15)$$

For the output voltage to be proportional to the quotient of V_1 and V_2, the following condition must be met

$$1 + k \cdot R_1 = R_1 R_3/(R_2 R_4), \qquad (4.B.29.16)$$

and then we have

$$V_{out} = -\frac{V_P}{k \cdot R_1}(1 + R_4/R_3)\frac{V_1}{V_2}. \qquad (4.B.29.17)$$

It is obvious that the proportionality constant is:

$$k_P = -\frac{V_P}{k \cdot R_1}(1 + R_3/R_4). \qquad (4.B.29.18)$$

After substitution of (4.B.29.17) into (4.B.29.13) we get

$$V_x = -\frac{V_P}{k \cdot R_1}\frac{V_1}{V_2}. \qquad (4.B.29.19)$$

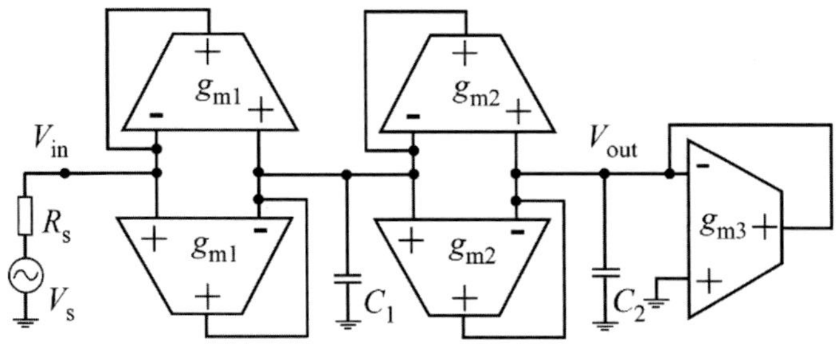

Fig. 4.B.72 Low-pass RC ladder in Gm-C technology

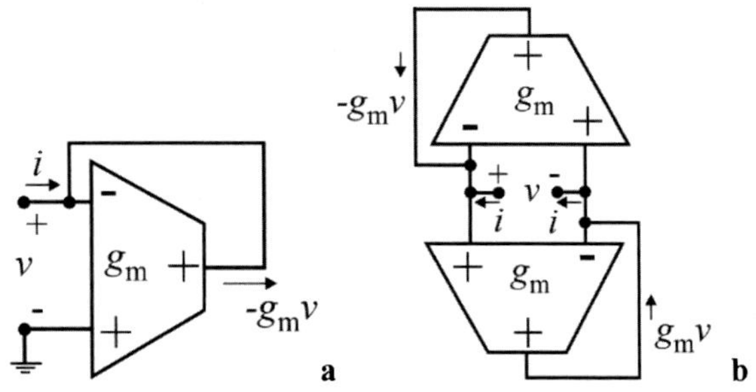

Fig. 4.B.73 Simulated resistor. **a** grounded and **b** floating

If the condition (4.B.29.16) is met by adjusting the values of resistors R_2, R_3, and/or R_4 (for fixed R_1), the voltage V_x can be used as a second output voltage proportional to the quotient the input voltages, but with a proportionality constant that does not change during adjustment for the required condition (4.B.29.16). Typically, this second output is loaded by the load via a non-inverting amplifier whose gain can be equal to or greater than one, depending on the desired proportionality constant.

Problem 4.B.30

Find the transfer function of the low-pass ladder network of Fig. 4.B.72 which is realized in Gm-C technology. The following is given: $R_s = 1\ \text{k}\Omega$, $g_{m1} = 50\ \mu\text{S}$, $g_{m2} = 100\ \mu\text{S}$, $g_{m3} = 1\ \text{mS}$, and $C_1 = C_2 = 1\ \text{nF}$. The OTAs have infinite input and output impedances (Use the model of Fig. 4.5.6).

Solution to the Problem 4.B.30

To find the requested transfer function we will first analyze the circuits depicted in Fig. 4.B.73.

For the circuit of Fig. 4.B.73a we may write the following

$$i = g_m v, \tag{4.B.30.1}$$

so that the equivalent resistance is

$$R = v/i = 1/g_m. \tag{4.B.30.2}$$

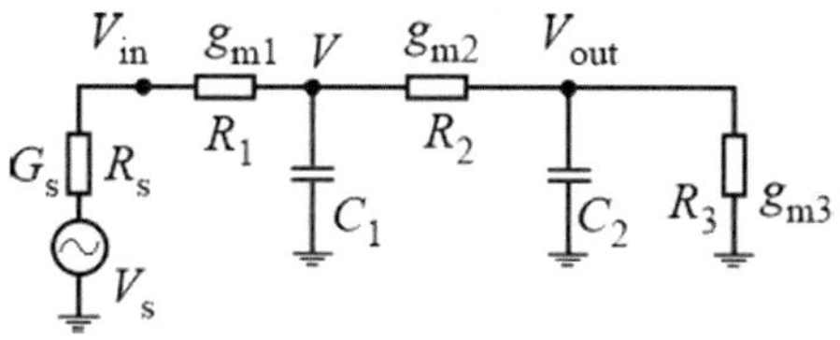

Fig. 4.B.74 The equivalent ladder circuit

Similarly, for the circuit of Fig. 4.B.73b we may write the same nodal equation (4.B.30.1) and we get the same definition for the resistance, namely (4.B.30.2).

Accordingly, for the analysis of the original circuit of Fig. 4.B.30 we may use its equivalent depicted in Fig. 4.B.74 which is a RC ladder.

In this circuit the following is valid

$$R_1 = 1/g_{m1}, \tag{4.B.30.3a}$$

$$R_2 = 1/g_{m2}, \tag{4.B.30.3b}$$

$$R_3 = \frac{1}{g_{m3}}, \tag{4.B.30.3c}$$

and

$$R_s = 1/G_s. \tag{4.B.30.3d}$$

The following system of nodal equations may be written for the unknown voltages in the circuit

$$\begin{bmatrix} G_s + g_{m1} & -g_{m1} & 0 \\ -g_{m1} & g_{m1} + g_{m2} + sC_1 & -g_{m2} \\ 0 & -g_{m2} & g_{m2} + g_{m3} + sC_2 \end{bmatrix} \times \begin{bmatrix} V_{in} \\ V \\ V_{out} \end{bmatrix} = \begin{bmatrix} G_s V_s \\ 0 \\ 0 \end{bmatrix}. \tag{4.B.30.4}$$

Its solution is

$$V_{out} = \frac{g_{m1} g_{m2} G_s V_s}{X + Ys + Zs^2} \tag{4.B.30.5a}$$

where

$$X = (G_s + g_{m1})[g_{m1}(g_{m2} + g_{m3}) + g_{m2} g_{m3}] - g_{m1}^2 (g_{m2} + g_{m3}) \tag{4.B.30.5b}$$

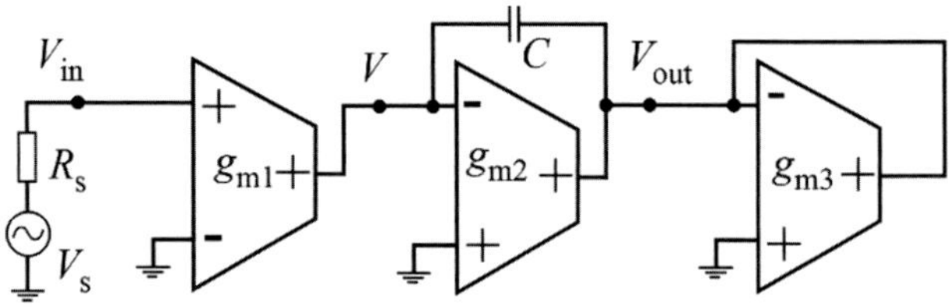

Fig. 4.B.75 Gm-C integrator circuit

Fig. 4.B.76 Equivalent circuit

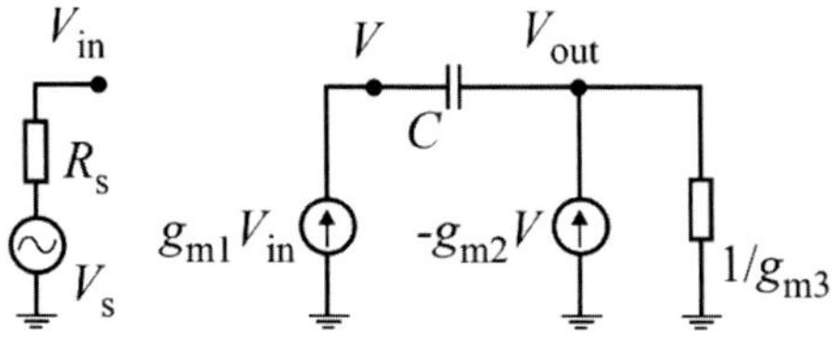

$$Y = (G_s + g_{m1})(g_{m2} + g_{m3})C_1 + \left[(G_s + g_{m1})(g_{m2} + g_{m3}) - g_{m1}^2\right]C_2 \tag{4.B.30.5c}$$

$$Z = (G_s + g_{m1})C_1C_2 \tag{4.B.30.5d}$$

We are leaving to the reader to substitute the numerical values. One is advised to consider the silicon area needed for realization of R_1 and R_2 as compared to the area of the simulated resistors used here.

Problem 4.B.31

For the circuit of Fig. 4.B.75 find the dependence of the output voltage on the input voltage symbolically. The OTAs have infinite input and output impedances (Use the model of Fig. 4.5.6a).

Solution to the Problem 4.B.31

After substitution, the model we get the equivalent circuit of Fig. 4.B.76.

The nodal equations now are:

$$g_{m1}V_{in} + C\frac{d(V_{out} - V)}{dt} = 0 \tag{4.B.31.1a}$$

$$g_{m2}V + C\frac{d(V_{out} - V)}{dt} + g_{m3}V_{out} = 0 \tag{4.B.31.1b}$$

After elimination of V, the following differential equation arises

$$C\left(1 + \frac{g_{m3}}{g_{m2}}\right)\frac{dV_{out}}{dt} = -\left(g_{m1}V_{in} + C\frac{g_{m3}}{g_{m2}}\frac{dV_{in}}{dt}\right). \tag{4.B.30.2}$$

The output signal, as can be deduced, contains two components. It may be expressed as

$$V_{\text{out}} = X \cdot \int V_{\text{in}} \mathrm{d}t + Y \cdot V_{\text{in}} \tag{4.B.30.3a}$$

where

$$X = -\frac{g_{\text{m1}}}{C\left(1 + \frac{g_{\text{m3}}}{g_{\text{m2}}}\right)} \tag{4.B.30.3b}$$

and

$$Y = \frac{-g_{\text{m3}}}{g_{\text{m2}} + g_{\text{m3}}}. \tag{4.B.30.3c}$$

The first part represents integral of the input voltage while by the second it is simply amplified. Note, in absence of the load ($g_{\text{m3}} = 0$), only the integral remains.

Appendix 4.C
Examples with SPICE Simulations

Example 4.C.1 Class A Push–Pull Amplifier
Given the amplifier circuit of Fig. 4.C.77 find the voltage, current, and the power gain developed at R_L and the efficiency for $V_g = 4$ V with the signal frequency of 1 kHz. The element values are $R_B = 1$ kΩ; $C = 100$ μF; $R_L = 8$ Ω; and $V_{CC} = 12$ V. Use proper transistor models from available library. The circuit is excited by a voltage source $V_g = 4$ V having internal resistance $R_g = 10$ Ω. Find the pulse response to a pulse characterized by: Bottom value of $V_g = -4$ V, Pulse amplitude of $V_g = 8$ V, rise time of the pulse = 1 ns, fall time of the pulse = 1 ns, duration of the pulse = 2 ms, and period of the pulse = 4 ms.

To find the requested values SPICE simulations will be performed several times. In that Fourier analysis will be requested in order to establish the amplitudes of the first harmonic of the quantities needed.

As the first result obtained by SPICE simulation for sinusoidal excitation the emitter currents of T_1 and T_2 are depicted in Fig. 4.C.78. One may observe a difference in the amplitudes which is due to the inequality of the NPN and PNP transistor characteristics. Nevertheless, as one can see from Fig. 4.C.79 the output voltage is not significantly distorted with $THD = 0.25\%$ only.

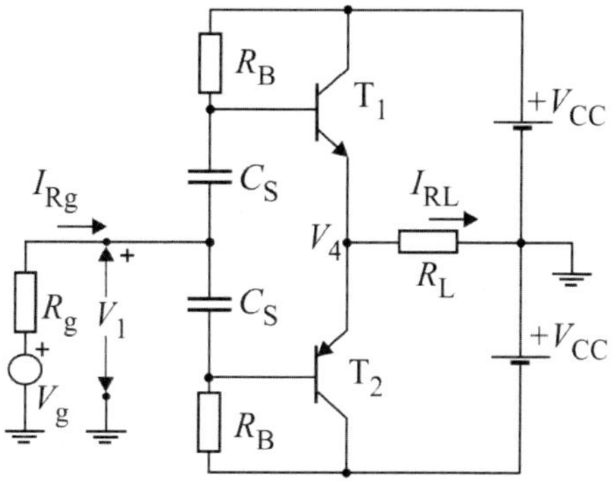

Fig. 4.C.77 Class A push–pull amplifier

V. B. Litovski, *Lecture Notes in Analog Electronics*, Lecture Notes in Electrical Engineering 958, https://doi.org/10.1007/978-981-19-6528-9

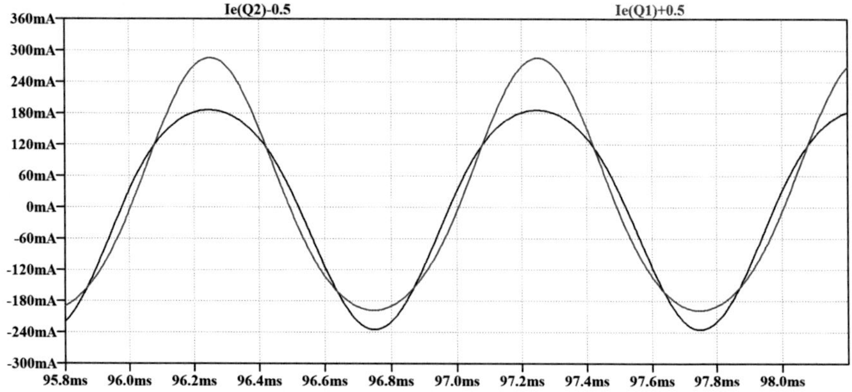

Fig. 4.C.78 Waveform of the transistors' emitter currents for sinusoidal excitation

```
Circuit: Class B amplifier circuit

Direct Newton iteration for .op point succeeded.
N-Period=1
Fourier components of V(4)
DC component:-0.164441

 Harmonic Frequency    Fourier    Normalized   Phase   Normalized
 Number      [Hz]      Component   Component  [degree] Phase [deg]
    1     1.000e+03   3.618e+00   1.000e+00    -179.71°       0.00°
    2     2.000e+03   8.668e-03   2.396e-03     -90.96°      88.74°
    3     3.000e+03   2.534e-03   7.003e-04      -4.21°     175.50°
    4     4.000e+03   2.257e-04   6.237e-05     -22.16°     157.55°
    5     5.000e+03   2.272e-04   6.279e-05     -55.04°     124.67°
Total Harmonic Distortion: 0.249739%(0.247809%)
```

Fig. 4.C.79 Fourier analysis results for the circuit of Fig. 4.C.77

In this figure the crucial values are in red. We may read that the amplitude of the first harmonic is $V_4 = 3.618$ V. Note V_g is significantly smaller than V_{CC}.

By repeating this analysis and extracting the necessary quantities we obtain the following:

$$A_C = J_{RL}/J_{Rg} = 4.008e - 01/2.320e - 02 = 17.28;$$

$$A_V = V_4/V_1 = 3.618/3.796 = 0.953;$$

$$\begin{aligned} A_P &= [V_4 * J_{RL}/2]/[V_1 * J_{Rg}/2] \\ &= [3.618 * 4.008e - 01]/[3.796 * 2.320e - 02] = 16.47; \end{aligned}$$

$$
\begin{aligned}
P_L &= V_4 * J_{RL}/2 = 3.618 * 4.008e - 01/2 = 0.725\text{W};\\
P_{CC} &= V_{CC} * I_{CC+\_DC} + V_{CC} * I_{CC-\_DC}\\
&= 12 * 0.489382 + 12 * 0.512125 = 6.15\text{W};\\
\eta &= 100 * P_L/P_{CC} = 11.8\%.
\end{aligned}
$$

I_{CC+_DC} and I_{CC-_DC} are the DC components of the power supplies $\pm V_{CC}$.

Note, the value of the efficiency is a small number due to the relatively small amplitude of the excitation voltage while the DC supply current is not signal dependent.

The three diagrams of Fig. 4.C.80 represent the pulse response of the amplifier after equilibrium. Note, sag is evident so that the top edge of the output pulse is not horizontal.

The top diagram represents a complete pulse and is intended to expose the sag.

The middle diagram depicts the rising edge of the input signal while the bottom one is related to the falling edge. In both cases delay and rising/falling time is evident.

For the falling edge of the output signal, we measured a slew rate of $\Delta V_5/\Delta t \approx 2.83$ V/ns which is for audio frequency amplification a rather acceptable value.

Example 4.C.2 Class B Push–Pull Amplifier

Given the amplifier circuit of Fig. 4.C.81 find the voltage, current, and the power gain developed at R_L and the efficiency for $V_g = 4$ V with the signal frequency of 1 kHz. The element values are $R_g = 10$ kΩ; $R_1 = 5.6$ kΩ; $R_2 = 12$ kΩ; $R_2 = 1.8$ kΩ; $C =$ 10 µF; $R_L = 50$ Ω; and $V_{CC} = 12$ V. Use proper transistor models from available library. Find the pulse response to a pulse characterized by: Bottom value of $V_g =$ -4 V, Pulse amplitude of $V_g = 8$ V, rise time of the pulse = 1 ns, fall time of the pulse = 1 ns, duration of the pulse = 2 ms, and period of the pulse = 4 ms. Finally, find the frequency response of the circuit up to $f_{max} = 10$ MHz.

To find the requested value SPICE simulation will be performed several times. In that Fourier analysis will be requested in order to establish the amplitudes of the first harmonic of the quantities needed.

As the first result obtained by SPICE simulation for sinusoidal excitation the emitter currents of T_1 and T_2 will be depicted in Fig. 4.C.82. Symmetry may be observed.

Then, Fig. 4.C.83 depicts the properties of V_5 for the same excitation.

In this figure the crucial values are in red. We may read that the amplitude of the first harmonic is $V_5 = 4.522$ V while $THD = 0.584\%$. We may say that the distortions are low even for a large amplitude of the output voltage (as compared to the maximum possible which is $V_{CC}/2 - V_{CES1} = 5.5$ V, $V_{CES1} \approx 0.5$ being the saturation voltage of T_1).

By repeating this analysis and extracting the necessary quantities we obtain the following:

$$A_C = J_{RL}/J_{C1} = 90.45e - 03/0.3979e - 03 = 227.32;$$

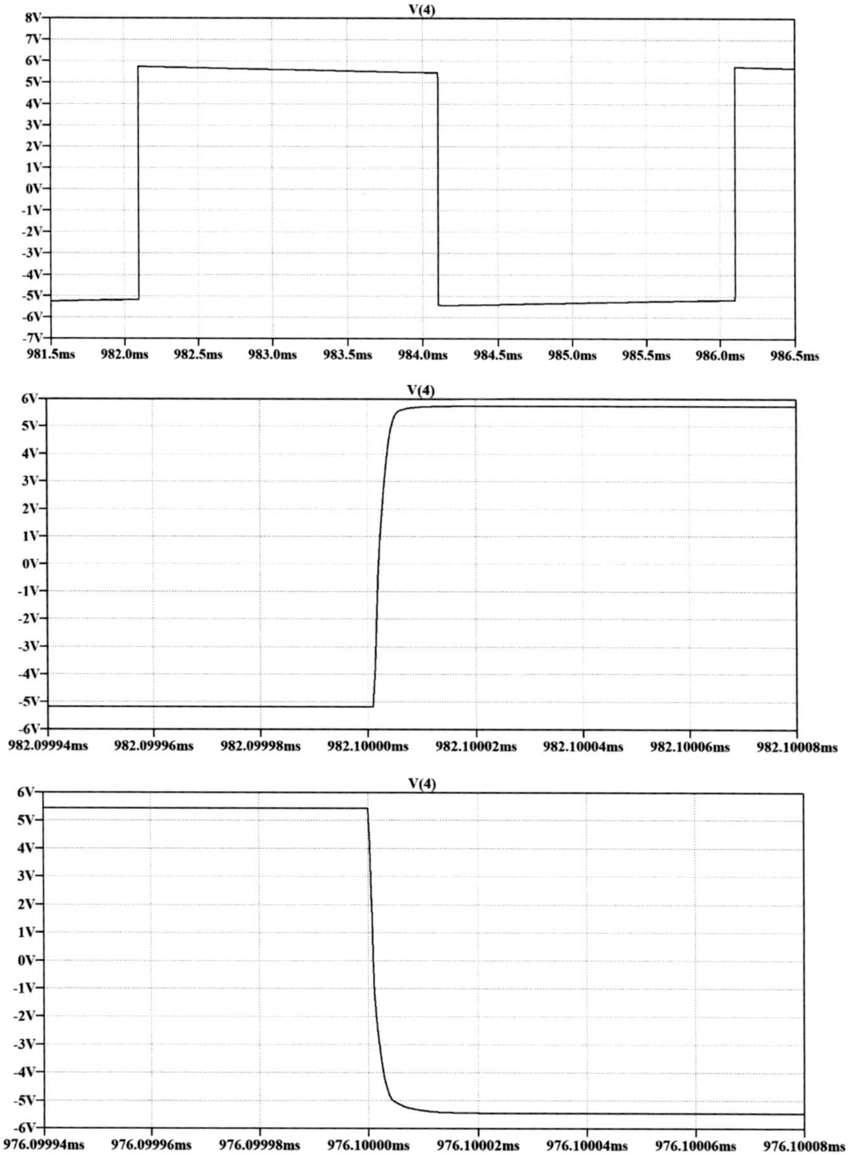

Fig. 4.C.80 Pulse responses. Top: overall response after equilibrium; middle: response to the front edge of the input pulse; and bottom: response to the back edge of the input pulse

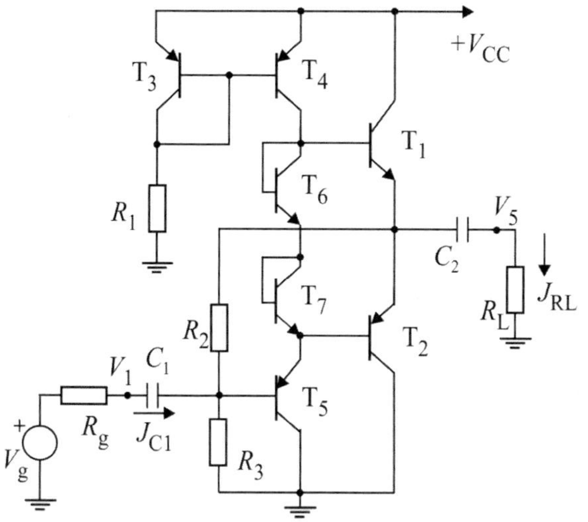

Fig. 4.C.81 Class B amplifier example

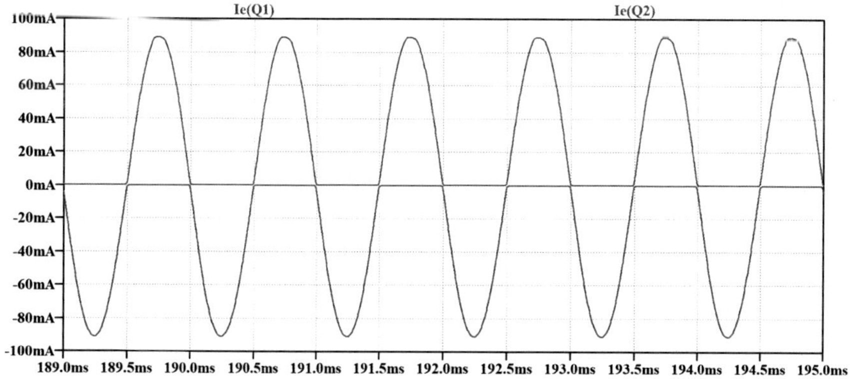

Fig. 4.C.82 Time domain responses of I_{E1} and I_{E2}

$$\begin{aligned}
A_{\mathrm{V}} &= V_5/V_1 = 4.522/16.70\mathrm{e}-03 = 270.78;\\
A_{\mathrm{P}} &= [V_5 * J_{\mathrm{RL}}/2]/[V_1 * J_{\mathrm{C1}}/2]\\
&= [4.522 * 90.45\mathrm{e}-03]/[16.70\mathrm{e}-03 * 0.3979\mathrm{e}-03] = 61,553;\\
P_{\mathrm{L}} &= V_5 * J_{\mathrm{RL}}/2 = 4.522 * 90.45\mathrm{e}-03/2 = 0.2045\,\mathrm{W};\\
P_{\mathrm{CC}} &= V_{\mathrm{CC}} * I_{\mathrm{CC_DC}} = 12 * 0.0326628 = 0.392\,\mathrm{W};\\
\eta &= 100 * P_{\mathrm{L}}/P_{\mathrm{CC}} = 52.2\%.
\end{aligned}$$

$I_{\mathrm{CC\_DC}}$ is the DC component of the power supply V_{CC}.

```
Circuit: Class B amplifier circuit

Direct Newton iteration for .op point succeeded.
N-Period=1
Fourier components of V(5)
DC component:-0.0885055

Harmonic Frequency    Fourier     Normalized    Phase    Normalized
 Number     [Hz]     Component     Component  [degree] Phase [deg]
    1    1.000e+03   4.522e+00   1.000e+00       1.77°       0.00°
    2    2.000e+03   3.573e-03   7.901e-04    -143.17°    -144.94°
    3    3.000e+03   2.401e-02   5.308e-03       3.94°       2.17°
    4    4.000e+03   6.937e-03   1.534e-03     101.61°      99.85°
    5    5.000e+03   7.754e-03   1.715e-03     179.83°     178.06°
Total Harmonic Distortion: 0.583896%(0.661109%)
```

Fig. 4.C.83 Fourier analysis results for the circuit of Fig. 4.C.81

The three diagrams of Fig. 4.C.84 represent the pulse response of the amplifier after equilibrium. Note, due to the large capacitances C_1 and C_2 the transient needs some time to settle (here 2 s). Also, sag is evident so that the top edge of the output pulse is not horizontal.

The top diagram represents a complete pulse and is intended to expose the sag.

The middle diagram depicts the rising edge of the input signal while the bottom one is related to the falling edge. In both cases delay and rising/falling time is evident.

For the falling edge of the output signal, we measured a slew rate of

$\Delta V_5/\Delta t \approx 0.52$ V/ns which is for audio frequency amplification a rather acceptable value.

Finally, Fig. 4.C.85 depicts the frequency response (as a semi-log diagram of the output power). Note the value of the power in this diagram was obtained by frequency domain analysis while earlier we got the same value of P_L through the time domain analysis.

The cut-off frequencies of the amplifier are shown on the figure.

Example 4.C.3 Class B CMOS Push–Pull Amplifier

Given the amplifier circuit of Fig. 4.C.86 it is known $V_{DD} = V_{SS} = 10$ V, $V_{in} = 2$ mV, $V_r = 9$ V, $R_2 = 11.5$ kΩ, $R_3 = 330$ Ω, $C_S = 100$ μF, $R_L = 12$ Ω. Use proper transistor models from an available library. Find:

The waveforms of the output transistors currents and the load current.

The output voltage spectrum, the amplitude of the first harmonic, and the *THD* of the output voltage.

The efficiency of the amplifier

The output voltage waveforms for the following values of the amplitude of the input signal $V_{in} = \{1, 2, 3, 4, 5\}$ [mV].

The frequency response of the circuit up to $f_{max} = 1$ MHz.

SPICE simulations were performed several times, and the first results are depicted in Fig. 4.C.87. There the waveforms of the currents are given. By Id(M1) the drain

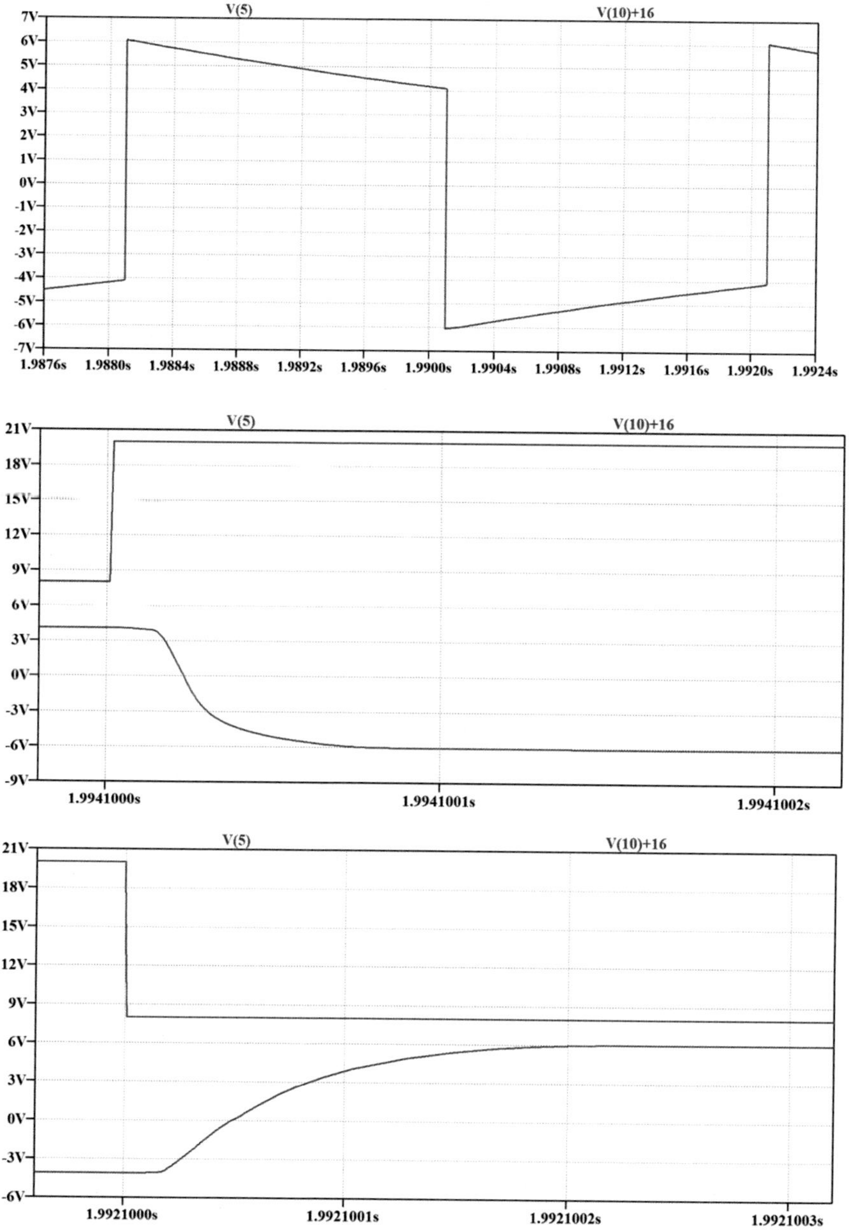

Fig. 4.C.84 Pulse responses. Top: overall response after equilibrium; middle: response to the front edge of the input pulse; and bottom: response to the back edge of the input pulse

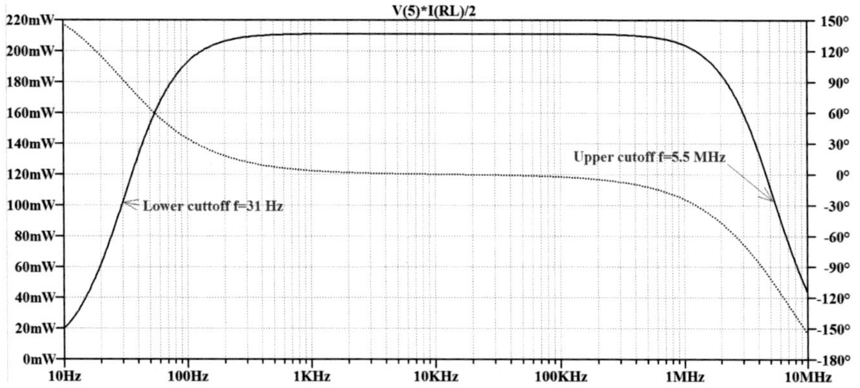

Fig. 4.C.85 Frequency response (output power) of the Class B amplifier

Fig. 4.C.86 CMOS Class B push–pull power amplifier

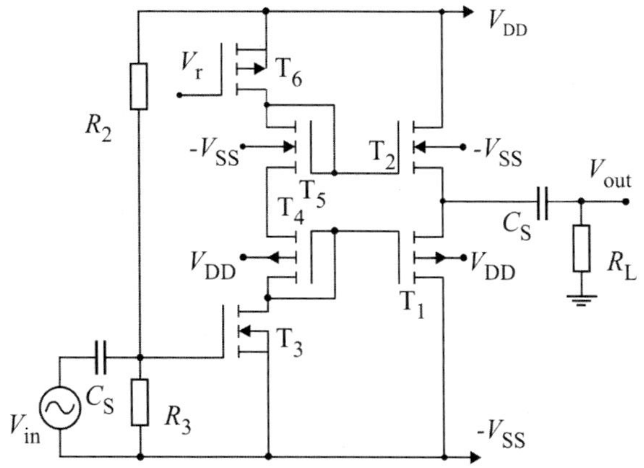

current of the bottom P-channel MOS transistor while by Id(M2) the current of the top N-channel most transistor are denoted. By $I(R_l)$ the current of the load resistor is denoted.

Follows the analysis of the spectrum of the output voltage. It is depicted in Figure 4.C.88. The picture is completed by Fig. 4.C.89 where in the top part numerical values of the first five spectral components are given. Note the *THD* is above 4%.

By using the bottom part of Fig. 4.C.89 we may calculate the efficiency of the amplifier for the signal $V_{in} = 2$ mV.

$$\eta = 100 \cdot P_{load}/P_{batt} = 100 \cdot 0.5 \cdot (\mathrm{V}_{(5)} \cdot \mathrm{V}_{(5)}/R_L)/[V_{SS} \cdot I(\mathrm{vss}) + V_{DD} \cdot I_{(\mathrm{vdd})}]$$
$$= 100 \cdot (0.8649)^2/12)/(2 \cdot 10 \cdot 0.028) = 11.13\%$$

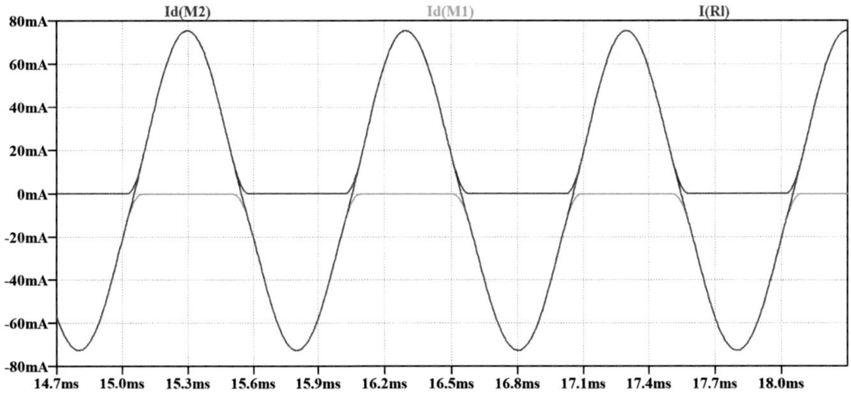

Fig. 4.C.87 Time domain responses of the currents in the output node. By Id(M1) stands for the drain current of the bottom P-channel MOS transistor, Id(M2) for the current of the top N-channel most transistor, and for $I(R_I)$ the current of the load resistor

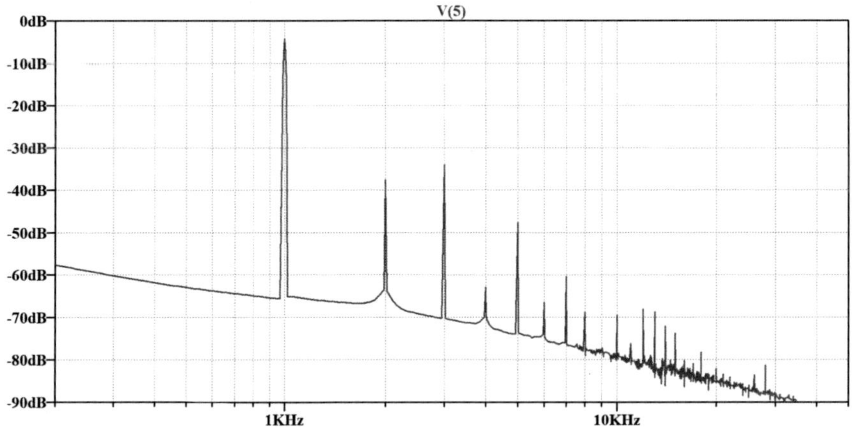

Fig. 4.C.88 SPICE generated spectrum of the output voltage

By analysis of the responses depicted in Fig. 4.C.90 where the output voltage waveforms for five different input amplitudes (of V_g) are shown, one may conclude that in this circuit the maximum amplitude of the input voltage is limited to $V_{gmax} \approx 3$ mV.

The final analysis is related to the frequency response of the amplifier. It is depicted in Fig. 4.C.91 for four temperature values. One can conclude here that the gain at the central frequency changes for about 6 dB for a temperature range from room temperature (here 25 °C) to 100 °C. The opposite may be observed for the pass-band width of the amplifier. It increases for increased temperatures.

Fourier components of V(5)
DC component:6.18914e-005

Harmonic Number	Frequency [Hz]	Fourier Component	Normalized Component	Phase [deg]	Normalized Phase[deg]
1	1.0e+03	8.649e-01	1.000e+00	-18.00°	0.00°
2	2.0e+03	1.953e-02	2.258e-02	-95.99°	-78.00°
3	3.0e+03	2.810e-02	3.249e-02	127.45°	145.44°
4	4.0e+03	1.291e-03	1.493e-03	169.00°	187.00°
5	5.0e+03	5.692e-03	6.581e-03	91.03°	109.03°

Total Harmonic Distortion: 4.013866%(4.017478%)

Fourier components of I(vdd)	Fourier components of I(vss)
DC component:-0.0280838	DC component:0.0280786

Fig. 4.C.89 SPICE generated numerical values of the spectrum of (top) the output voltage and (bottom) the power supply batteries currents

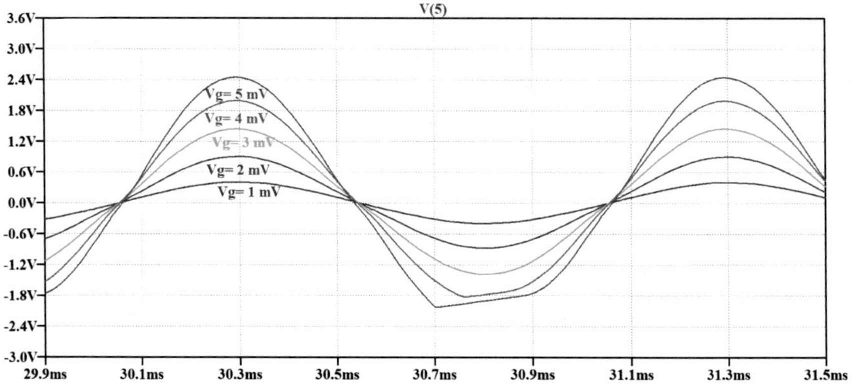

Fig. 4.C.90 Time domain response of the output voltage for different amplitudes for the excitation

Example 4.C.4 Class C Amplifier Using NPN BJT

Given the amplifier circuit of Fig. 4.C.92 find the frequency dependence of the voltage gain. $V_{DD} = 20$ V, $V_{GG} = 0.1$ V, $R_L = 1$ MΩ, $C_S = 100$ μF, $L_S = 0.01$ H, $C = 100$ pF, and $L = 100$ μH. Use proper transistor models from available library.

Find the output waveforms and the spectrum content of V_{out}.

Figure 4.C.93 depicts the frequency dependence of the gain of this amplifier. As we can see it is a narrow-band amplitude characteristic corresponding to the resonant curve of the parallel resonant circuit in the collector.

Note the central frequency was measured form the simulation results to be $f_0 =$ 1.519 MHz which is in very good agreement with the one obtained for the resonant frequency of the parallel resonant circuit.

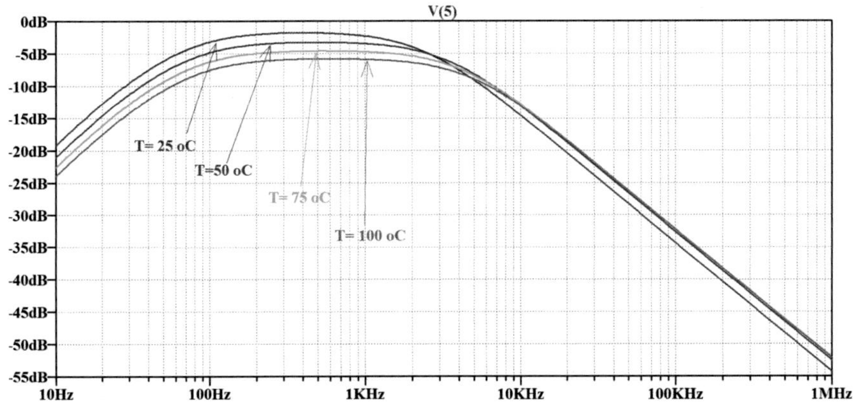

Fig. 4.C.91 Frequency domain response of the output voltage for different temperatures

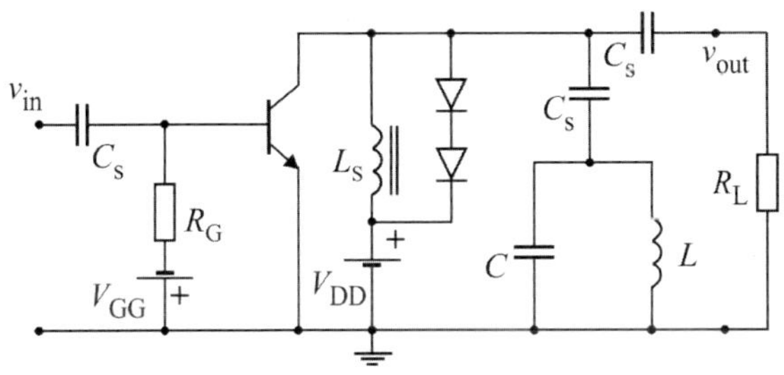

Fig. 4.C.92 A Class C power amplifier circuit

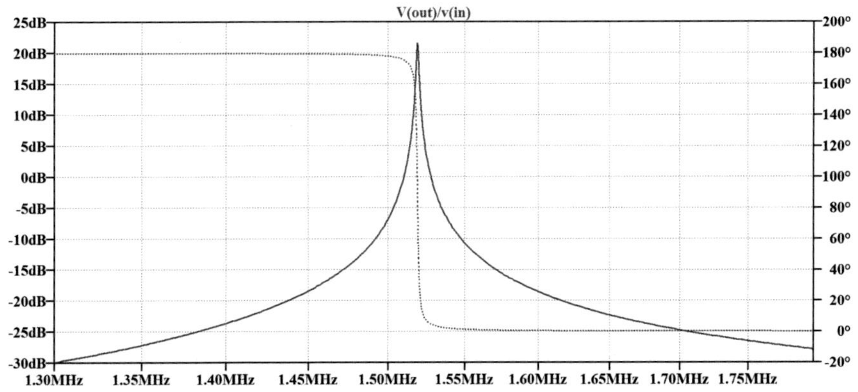

Fig. 4.C.93 Frequency response of the gain

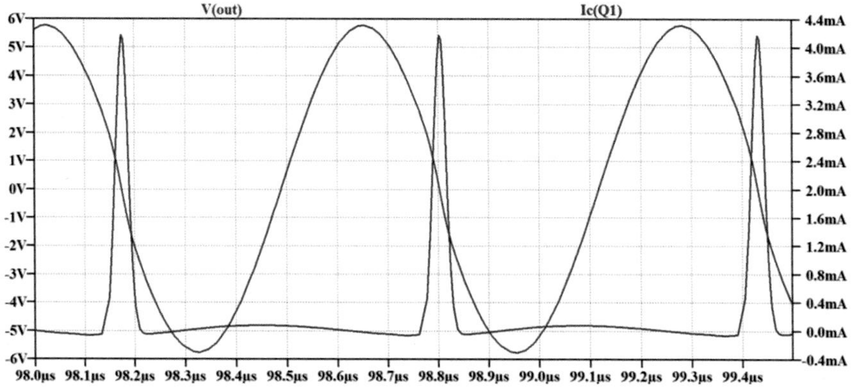

Fig. 4.C.94 Time domain response of the collector current and the output voltage

Fourier components of V(out)
DC component:0.000540174

Harmonic Number	Frequency [Hz]	Fourier Component	Normalized Component	Phase [degree]	Normalized Phase [deg]
1	1.592e+06	5.847e+00	1.000e+00	90.23°	0.00°
2	3.183e+06	2.789e-01	4.770e-02	1.50°	-88.73°
3	4.775e+06	1.417e-01	2.423e-02	-91.36°	-181.59°
4	6.366e+06	9.353e-02	1.600e-02	-177.77°	-268.00°
5	7.958e+06	7.145e-02	1.222e-02	89.43°	-0.80°
6	9.549e+06	4.894e-02	8.371e-03	0.68°	-89.55°
7	1.114e+07	3.722e-02	6.365e-03	-86.36°	-176.59°
8	1.273e+07	2.812e-02	4.810e-03	-179.29°	-269.52°
9	1.432e+07	2.322e-02	3.971e-03	90.08°	-0.15°

Total Harmonic Distortion: 5.845520%(5.861192%)

Fig. 4.C.95 Harmonic content of the output voltage of the

Follows the time domain analysis. First we may see in Fig. 4.C.94 the waveforms of the output voltage [V(out)] and the collector current [Ic(Q1)]. The results are in full agreement with the expected ones.

Finally we have Figs. 4.C.95 and 4.C.96 which are dealing with distortions of the output signal. Figure 4.C.95 depicts the main component and nine harmonic of the output voltage while the same in a graphical form is depicted in Fig. 4.C.96. We may reed that $THD \approx 5.845\%$ while for the second harmonic we have $HD_2 \approx 4.77\%$ meaning that it is dominant.

Example 4.C.5 Class D Amplifier Using Fourth Order Filter
Given the amplifier circuit of Fig. 4.C.97 perform simulation and extract the voltage waveforms of the input and output voltages and their spectrums. Then find the instantaneous value of the power delivered to the load and its average value. Find the battery currents and their spectrums. Use the following element values: $V_{DD} = V_{SS} = 10$ V

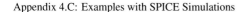

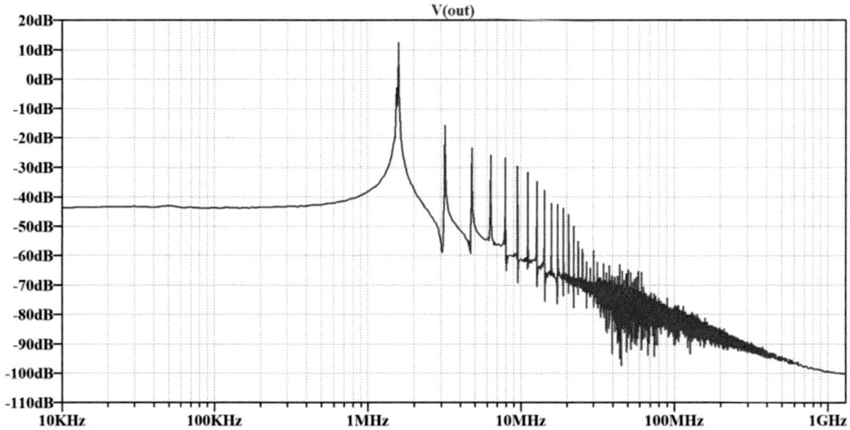

Fig. 4.C.96 Frequency spectrum of the output voltage

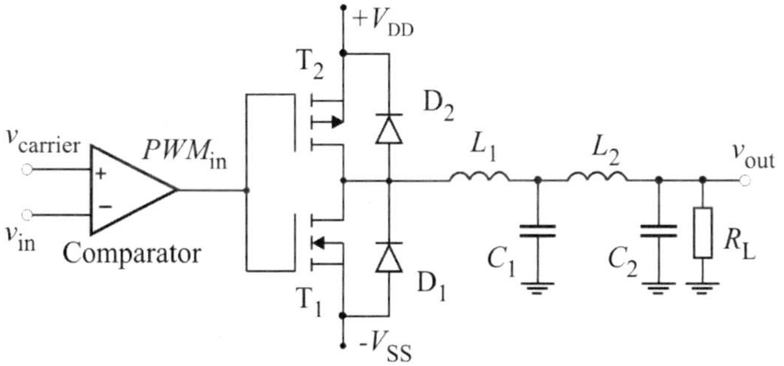

Fig. 4.C.97 Half-bridge switching output stage-based Class D power amplifier using a fourth order low-pass filter

and $R_L = 8\ \Omega$. Use the fourth order LSM filter with 3 dB cut-off frequency of 60 kHz whose element values are $L_1 = 24.17\ \mu$H, $C_1 = 459.93$ nF, $L_2 = 41.93\ \mu$H, and $C_2 = 572.93$ nF. In the PWM modulator use $f_c = 500$ kHz. Use an input signal of the form

$$v_{in} = 3.5 \cdot [\sin(2 \cdot \pi \cdot 500 \cdot t) + \sin(2 \cdot \pi \cdot 5000 \cdot t) + \sin(2 \cdot \pi \cdot 15{,}000 \cdot t)].$$

For the comparator, the transistors, and the (Schottky) diodes use models from any available library.

Compare the results with the ones obtained for the same circuit but with a second order filter using $L_1 = 80\ \mu$H and $C_1 = 90$ nF.

We will start the analysis here by studying the time domain response of the circuit. The voltage waveforms of the input and output circuit are depicted in Fig. 4.C.98.

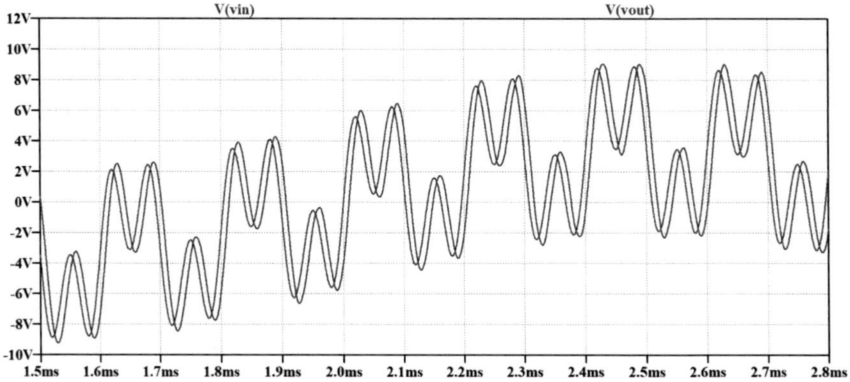

Fig. 4.C.98 Voltage waveforms of the input (red) and output (blue) signal

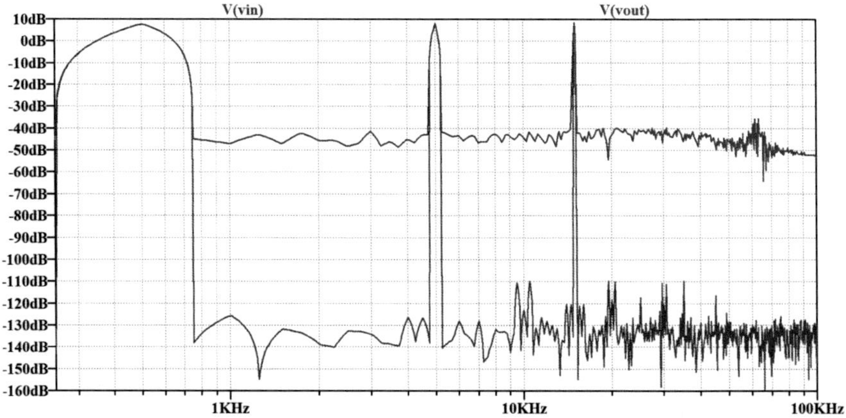

Fig. 4.C.99 FFT analysis results for the input (red) and output (blue) signal

One may see that, apart the small delay (coming from the filter) very good agreement of the input and output waveforms is produced.

From the FFT analysis depicted in Fig. 4.C.98 one may conclude that the output signal has a floor of harmonics at a level of about 50 dB (that is 316.2 times) lower than the signal components (Fig. 4.C.99).

Figure 4.C.100 depicts the values of the main harmonics of the input and output signals. A slight difference in the amplitude values may be noticed. It is of special interest to notice the average (DC) value of the instantaneous power given at the bottom of this figure. It reads $P_{\mathrm{L}} = 1.21261$ W. The waveform of the instantaneous power delivered to the load is depicted in Fig. 4.C.101.

Figure 4.C.102 depicts part of the time domain responses of the battery currents. As expected, short pulses may be observed only the steady part being at very low level of about 1 μA. The spectrum of one of the currents for a specific monochromatic

```
Circuit: pwm_Class_D Amplifier (4th order LSM filter)

Fourier components of V(vout)
DC component:0.00216529

Harmonic Frequency    Fourier    Normalized    Phase   Normalized
 Number     [Hz]     Component    Component  [degree] Phase [deg]
    1    5.000e+02   3.500e+00   1.000e+00      -1.49°        0.00°
   10    5.000e+03   3.551e+00   1.015e+00     -15.17°      -13.67°
   30    1.500e+04   3.734e+00   1.067e+00     -51.26°      -49.77°

Fourier components of V(vin)
DC component:1.46884e-006

Harmonic Frequency    Fourier    Normalized    Phase   Normalized
 Number     [Hz]     Component    Component  [degree] Phase [deg]
    1    5.000e+02   3.500e+00   1.000e+00      -0.00°        0.00°
   10    5.000e+03   3.500e+00   1.000e+00       0.00°        0.00°
   30    1.500e+04   3.500e+00   1.000e+00       0.00°        0.00°

Fourier components of V(bp)
DC component:1.21261
```

Fig. 4.C.100 Results of spectral analysis of the signals. $v(v_{out})$ stands for the output signal, $v(v_{in})$ for the input signal, and $v(\text{bp})$ for the average power at the load

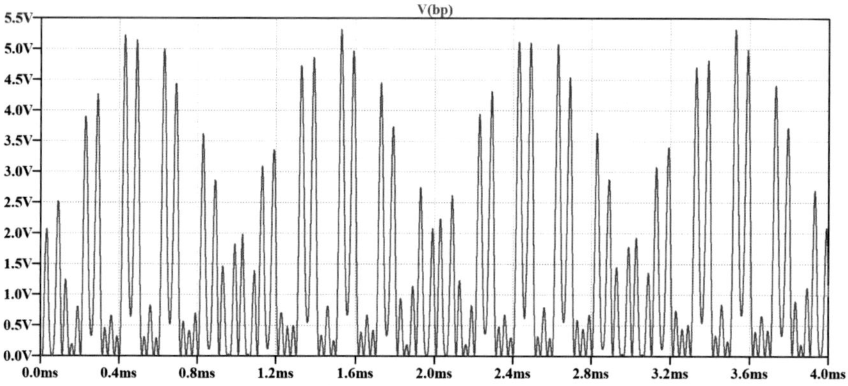

Fig. 4.C.101 Instantaneous value of the load power

excitation of 5 kHz is depicted in Fig. 4.C.103. Abundance of harmonics may be observed up to frequencies of 10 GHz. Hence the worries related to high-frequency radiations (The so-called electromagnetic compatibility—EMC—problem).

Figure 4.C.104 depicts part of the responses of the input and output voltages when second order filter is applied. High (carrier) frequency distortions may be observed impregnated into the output signal. From Fig. 4.C.105 one may conclude that the

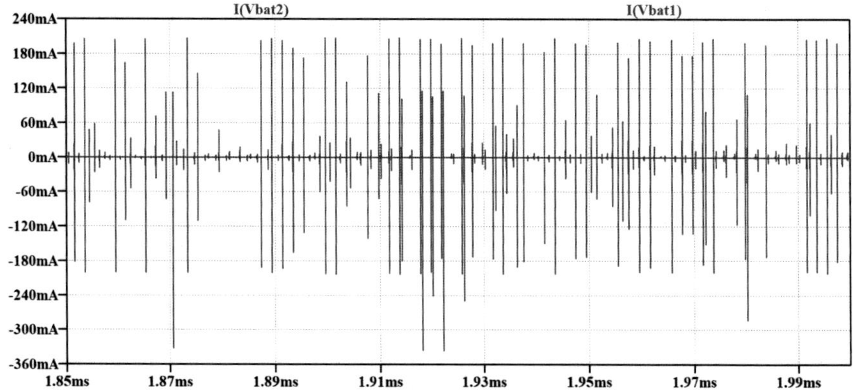

Fig. 4.C.102 Instantaneous values of the supply voltage batteries currents

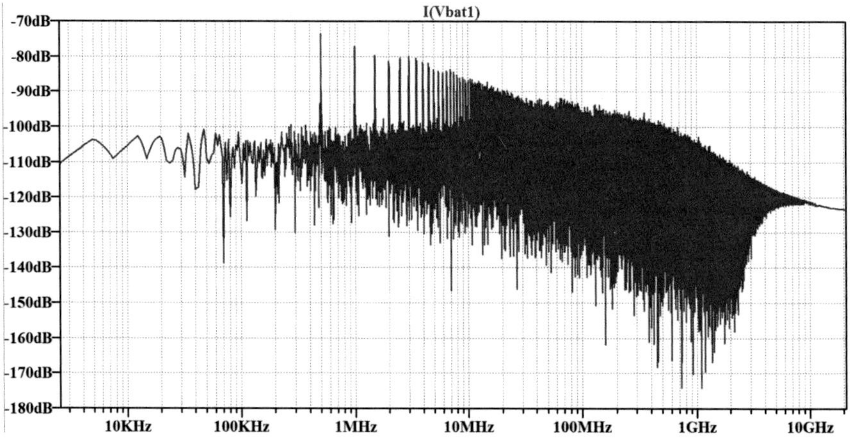

Fig. 4.C.103 Spectrum of one of the battery currents for monochromatic excitation of 5 kHz

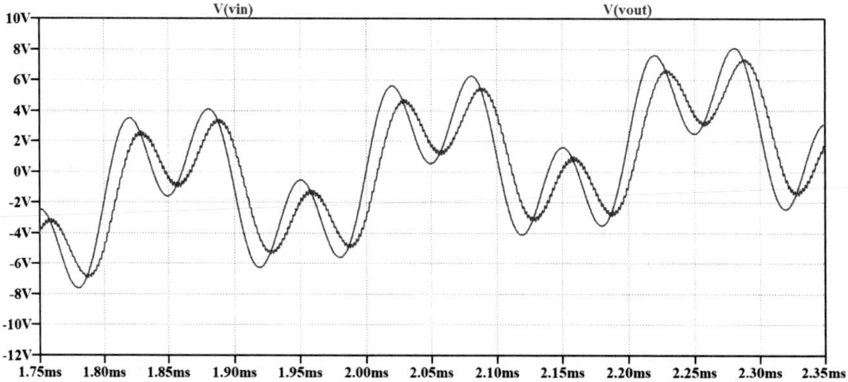

Fig. 4.C.104 Part of the input and output voltage waveforms for circuit with a second order filter

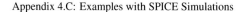

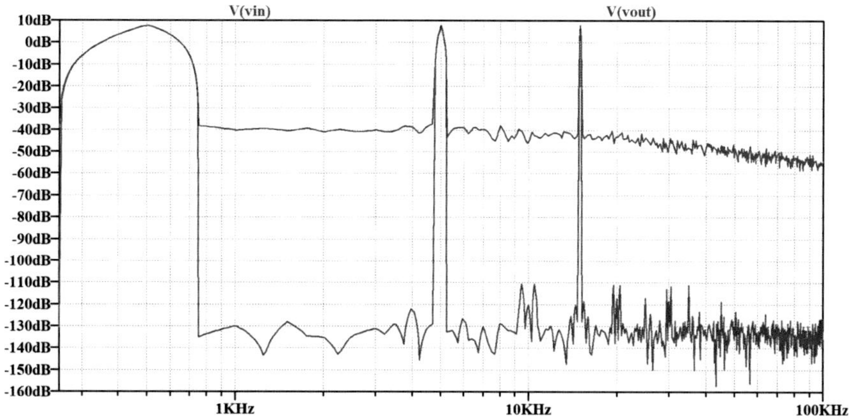

Fig. 4.C.105 The input and output voltage spectrums for circuit with a second order filter

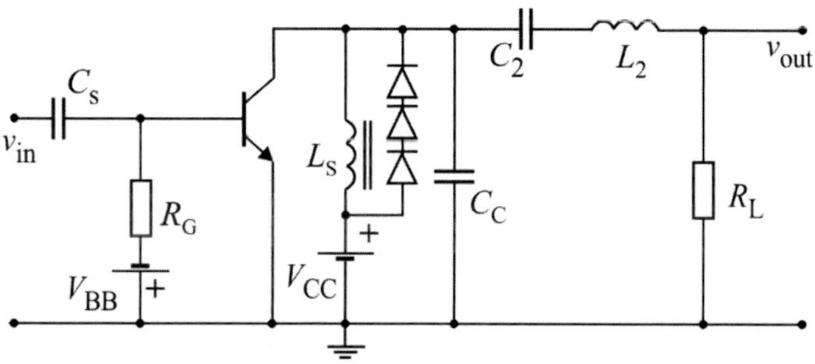

Fig. 4.C.106 The schematic of a Class E amplifier used as example

distortion level is risen for about 6 dB as compared to the circuit implementing fourth order filter.

Example 4.C.6 Class E Amplifier Using BJT

Given the amplifier circuit of Fig. 4.C.106 with $V_{CC} = 5$ V, $V_{BB} = 0.5$ V, $R_L = 12\ \Omega$, $C_C = 100\ \mu$F, $C_2 = 100\ \mu$F, $L_2 = 100\ \mu$H, $C_S = 100$ pF, $L_S = 0.005$ H, and $R_G = 10$ kΩ. Use input pulse train with 50% duty cycle, amplitude of $V_{in} = 1.7$ V, and frequency of $f_0 = 15.81$ MHz.

Find the output waveforms and the spectrum content of the currents and voltage related to the collector node. Study the harmonic content of the output voltage V_{out} and find the efficiency η of the amplifier. Use proper transistor models from available library.

We will start with the voltages at the collector [V(c)] and the output [V(out)]. These are obtained by SPICE simulation and part of the resulting waveform is depicted in Fig. 4.C.107. Note, to allow for as large an amplitude of the collector voltage three diodes in series were connected due to the low breakdown voltage of a single diode.

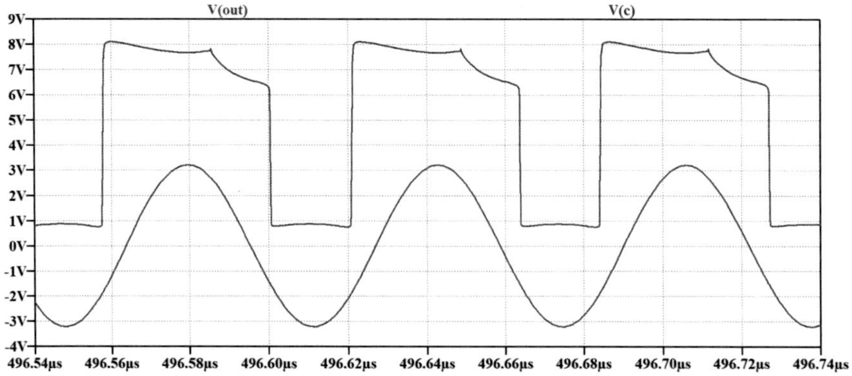

Fig. 4.C.107 Output (red) and collector (blue) voltage waveforms of the class E amplifier

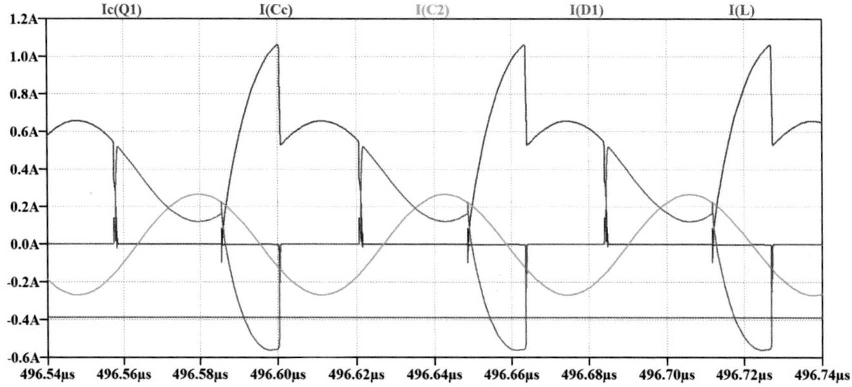

Fig. 4.C.108 Currents of the collector node: collector current [Ic(Q1)]; load current [I(C2)]; shunt capacitor current [I(Cc)]; diode current [I(D1)]; and inductor current [I(L)]

Since the functionality of this circuit is relatively obscure, in Fig. 4.C.108 the waveforms of the currents related to the collector node are depicted. Any of them needs a careful study to get explanation for its run except the current of the load resistor [I(C2)] which is, of course, sinusoidal.

From Fig. 4.C.109 one may read the value of the first harmonic of the input voltage for an amplitude of the excitation pulse of $V_{\text{in}} = 1.7$ V. It reads $V_{\text{out}} \approx 3.214$ V. Its harmonic content is relatively low with $THD \approx 0.523\%$. Here, one is not to forget that the input signal is a square pulse.

We may further read from the same figure that the average (DC) value of the power supply battery current is $I(V_{\text{CC}}) \approx 0.369$ A. That leads to an efficiency of $\eta \approx 46.7\%$.

```
N-Period=1
Fourier components of V(out)
DC component:0.00382837

Harmonic Frequency    Fourier    Normalized    Phase   Normalized
 Number      [Hz]     Component   Component  [degree] Phase [deg]
    1     1.581e+07   3.214e+00  1.000e+00    117.51°        0.00°
    2     3.162e+07   1.397e-02  4.347e-03    -82.40°     -199.91°
    3     4.743e+07   5.890e-03  1.833e-03     51.86°      -65.65°
    4     6.324e+07   6.761e-03  2.104e-03    141.95°       24.44°
    5     7.905e+07   2.562e-03  7.971e-04    -10.68°     -128.19°
Total Harmonic Distortion: 0.522625%(0.545843%)

N-Period=1
Fourier components of I(vcc)
DC component:-0.368641

Harmonic Frequency    Fourier    Normalized    Phase   Normalized
 Number      [Hz]     Component   Component  [degree] Phase [deg]
    1     1.581e+07   3.281e-01  1.000e+00   -160.59°        0.00°
    2     3.162e+07   7.193e-02  2.192e-01    159.43°      320.02°
    3     4.743e+07   1.732e-01  5.278e-01     96.38°      256.97°
    4     6.324e+07   3.320e-02  1.012e-01    -19.35°      141.24°
    5     7.905e+07   2.878e-02  8.773e-02     10.76°      171.36°
Total Harmonic Distortion: 58.697686%(66.863951%)
```

Fig. 4.C.109 Harmonic analysis of the output voltage [V(out)] and the power supply current [I(Vcc)]

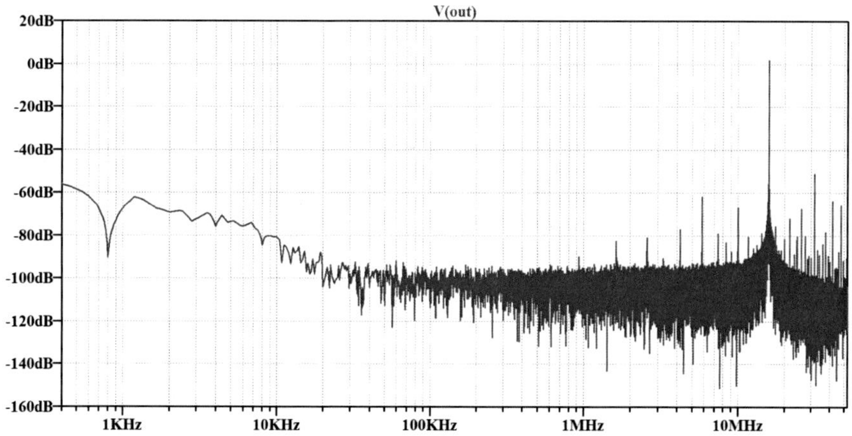

Fig. 4.C.110 FFT analysis of the output voltage

To finish, we may see from Fig.4.C.110 where the graphic representation of the output voltage spectrum is depicted that there are some sub-harmonics which are almost equal to the harmonics of the main signal.

Appendix 4.D
Approximative Calculation of the Harmonic Distortions

As pointed out several times in this chapter, the assessment and minimization of THDs is of great importance from the point of view of their use. In fact, these amplifiers work with large signals, so the nonlinearity of the characteristics of the built-in components is reflected into large amplitudes of the harmonic components of the output quantities. The nonlinearity of the transfer characteristic of the power amplifier in Class B or in Class AB results from the intrinsic nonlinearity of the components and from the asymmetry exposed by the positive and negative half-periods of the input signal. As a consequence, the actual transfer characteristic can be modeled by two different functions so that each of them will be valid for one or the other half-cycle.

The THD, as defined in LNAE_Book 2 and in Chap. 4.33, can be considered as a measure of the energy of the harmonic compared to the energy of the useful signal (basic harmonic). Having in mind the transfer characteristics that we have considered so far, it can be concluded that the second and third harmonics are usually of the greatest importance, so that in this section a method will be developed for approximate calculation of the HD of the second and the third harmonic of a push–pull power amplifier.

Based on the above, we start with a function that approximates the transfer characteristic of the amplifier by the expression

$$y = \begin{cases} f_{\mathrm{p}}(x) = a_1 x + a_2 x^2 + a_3 x^3 & \text{for}\, x \geq 0 \\ f_{\mathrm{n}}(x) = b_1 x + b_2 x^2 + b_3 x^3 & \text{for}\, x \leq 0 \end{cases} \tag{4.D.1}$$

where x is the input signal, y is the response at the output, and **a** are the **b** vectors of the coefficients in the approximate expression.

We will take the usual as the input signal $x(\tau) = X_{\mathrm{m}}\mathbf{sin}(\tau)$, where $\tau = \omega t$. If this signal is substituted in (1) and transform the degrees of sines into sines of multiple angles, the following harmonic amplitudes will occur

V. B. Litovski, *Lecture Notes in Analog Electronics*, Lecture Notes in Electrical Engineering 958, https://doi.org/10.1007/978-981-19-6528-9

$$Y_{1m} = \left| \frac{a_1 + b_1}{2} X_m + \frac{4}{3\pi}(a_2 - b_2) X_m^2 + \frac{3}{4} \frac{a_3 + b_3}{2} X_m^3 \right| \tag{4.D.2a}$$

$$Y_{2m} = \left| \frac{2}{3\pi}(a_1 - b_1) X_m + \frac{1}{2} \frac{a_2 + b_2}{2} X_m^2 + \frac{4}{5\pi}(a_3 - b_3) X_m^3 \right| \tag{4.D.2b}$$

$$Y_{3m} = \left| \frac{4}{15\pi}(a_2 - b_2) X_m^2 + \frac{1}{4} \frac{a_3 + b_3}{2} X_m^3 \right|. \tag{4.D.2c}$$

We notice that here, as before, the amplitudes of the harmonics depend on the coefficients in the approximation function (Taylor's expansion) and on the amplitude of the signal.

To determine the coefficients in the approximation function, we introduce the following

$$y'_{p0} = (dy/dx)_{|x=0^+} \tag{4.D.3a}$$

$$y'_{n0} = (dy/dx)_{|x=0^-} \tag{4.D.3b}$$

$$y_{pM} = y(X_M) - y(0) \tag{4.D.3c}$$

$$y_{nM} = y(-X_M) - y(0) \tag{4.D.3d}$$

$$y'_{pM} = \frac{dy}{dx}_{|x=X_M} \tag{4.D.3e}$$

$$y'_{nM} = \frac{dy}{dx}_{|x=-X_M} \tag{4.D.3f}$$

where

$$a_1 = y'_{p0} \tag{4.D.4a}$$

$$a_2 = 3\frac{y_{pM}}{X_M^2} - \frac{y'_{pM} + 2y'_{p0}}{X_M} \tag{4.D.4b}$$

$$a_3 = -2\frac{y_{pM}}{X_M^3} - \frac{y'_{pM} + 2y'_{p0}}{X_M^2} \tag{4.D.4c}$$

$$b_1 = y'_{n0} \tag{4.D.5a}$$

$$b_2 = 3\frac{y_{nM}}{X_M^2} - \frac{y'_{nM} + 2y'_{n0}}{X_M} \tag{4.D.5b}$$

$$b_3 = -2\frac{y_{nM}}{X_M^3} - \frac{y'_{nM} + 2y'_{n0}}{X_M^2}. \tag{4.D.5c}$$

In these embodiments, the quantity X_M was introduced, which represents the largest value of the amplitude of the input signal that we expect. Accordingly, the normalized value of the amplitude would be

$$x_n = X_m / X_M. \tag{4.D.6}$$

By substitution of (4.D.5) and (4.D.6) into (4.D.2) one gets

$$Y_{1m} = [p_{10}S_0 + p_{11}S_1 + p_{12}S_2]X_m \tag{4.D.7a}$$

$$Y_{2m} = [p_{20}D_0 + p_{21}D_1 + p_{22}D_2]X_m \tag{4.D.7b}$$

$$Y_{3m} = [p_{30}S_0 + p_{31}S_1 + p_{32}S_2]x_n X_m, \tag{4.D.7c}$$

where we used

$$S_0 = \left(y'_{p0} + y'_{n0}\right)/2 \tag{4.D.8a}$$

$$S_1 = \left(y_{pM} - y_{nM}\right)/(2X_M) \tag{4.D.8b}$$

$$S_3 = \left(y'_{pM} + y'_{nM}\right)/2 \tag{4.D.8c}$$

$$D_0 = y'_{p0} - y'_{n0} \tag{4.D.9a}$$

$$D_1 = \left(y_{pM} + y_{nM}\right)/2 \tag{4.D.9b}$$

$$D_3 = y'_{pM} - y'_{nM} \tag{4.D.9c}$$

and

$$p_{10} = 1 - 16x_n/(3\pi) + 3x_n^2/4 \tag{4.D.10a}$$

$$p_{20} = 2/(3\pi) - x_n/2 + 4x_n^2/(5\pi) \tag{4.D.10b}$$

$$p_{30} = -16x_n/(15\pi) + x_n^2/4 \tag{4.D.10c}$$

$$p_{11} = 8x_n/\pi - 3x_n^2/2 \tag{4.D.10d}$$

$$p_{21} = 3x_n/4 - 8x_n^2/(5\pi) \tag{4.D.10e}$$

$$p_{31} = 8x_n/(5\pi) - x_n^2/2 \tag{4.D.10f}$$

$$p_{12} = -8x_n/(3\pi) + 3x_n^2/4 \tag{4.D.10g}$$

$$p_{22} = -x_n/4 + 4x_n^2/(5\pi) \tag{4.D.10h}$$

$$p_{32} = -8x_n/(15\pi) + x_n^2/4. \tag{4.D.10i}$$

Now, for the harmonic distortions we get

$$\mathrm{HD}_2 = \frac{p_{20}d_0 + p_{21}d_1 + p_{22}d_2}{p_{10} + p_{11}s_1 + p_{12}s_2} \tag{4.D.11a}$$

$$\mathrm{HD}_3 = \frac{p_{30} + p_{31}s_1 + p_{32}s_2}{p_{10} + p_{11}s_1 + p_{12}s_2} \tag{4.D.11b}$$

where the following was introduced

$$d_i = D_i/S_0 \tag{4.D.12a}$$

$$s_i = S_i/S_0. \tag{4.D.12b}$$

These terms represent the complete analysis needed to calculate the harmonic distortions. In cases when x_n converges to unity, i.e., when the actual signal is approximately equal to the maximum signal we have chosen, it turns out that the following approximate expressions can be used

$$HD_2 \approx p_{21}d_1/s_1 \tag{4.D.13b}$$

$$HD_3 = (p_{30} + p_{32}s_2)/s_1. \tag{4.D.13b}$$

For the sake of illustration, we assume that the transfer characteristic of the output stage of the amplifier with MOS transistors from Fig. 4.3.30 can be approximate by the expression

$$i_{\text{out}} = \begin{cases} I_B(1-x)^2 & x < -1 \\ I_B(1-x)^2 - I_B(1+x)^2 & -1 \le x \le 1 \\ -I_B(1+x)^2 & x > 1 \end{cases} \tag{4.D.14}$$

Table 4.D.1 Quantities needed for calculation of the THD

Quantity	Value	Amplitude	Value
$f'_{p0} = f'_{n0}$	$-4I_B$	S_0	$-4I_B$
$f_{pM} = -f_{nM}$	$-I_B(1 + X_M)^2$	S_1	$-I_B(1 + X_M)^2$
$f'_{pM} = f'_{nM}$	$-2I_B(1 + X_M)$	S_2	$-2I_B(1 + X_M)$

where $x = v_{in}/\sqrt{I_B/A}$, and i_{out} is the short-circuit current (when the output terminal is connected to ground). The required numerical values for the following calculations are obtained on the basis of Table 4.D.1.

The value of HD_2 is equal to zero while for HD_3, if $x_n = 1$, one gets

$$\mathrm{HD}_3 = \frac{-\frac{16}{15\pi} + \frac{1}{4} + \left(\frac{8}{5\pi} - \frac{1}{2}\right)\frac{(1+X_M)}{4X_M} + \left(-\frac{8}{15\pi} + \frac{1}{4}\right)\frac{1}{2}}{1 - \frac{16}{3\pi} + \frac{3}{4} + \left(\frac{8}{\pi} - \frac{3}{2}\right)\frac{(1+X_M)}{4X_M} + \left(-\frac{8}{3\pi} + \frac{3}{4}\right)\frac{1}{2}} \quad (4.D.15)$$

If we assume that X_M is a large number we get $HD_3 \approx 0.2$.

References

1. Abe S, Yang S, Shoyama M, Ninomiya T, Matsumoto A, and Fukui A (2012) Operating mechanism of SiC-SIT DC circuit breaker in 400V-DC power supply system for data center—in case of grounding fault. In: Twenty-seventh annual IEEE applied power electronics conference and exposition (APEC), Orlando, FL, pp 2189–2194
2. Abedinpour S, Shenai K (2011) Insulated gate bipolar transistor. In: Rashid MH (ed) Power electronics handbook, devices, circuits and implementation. Elsevier, Oxford
3. Agarwal A et al (2015) 600 V, 1–40 A, Shottky diodes in SiC and their applications. http://www.cree.com/~/media/Files/Cree/Power/Articles%20and%20Papers/PWRTechnicalPaper1, last visited 18 Dec 2015
4. Ambacher O et al (2000) Two dimensional electron gases induced by spontaneous and piezoelectric polarization in undoped and doped AlGaN/GaN heterostructures. J Appl Phys 87(1):3224–3233
5. Araújo SV (2013) On the perspectives of wide-band gap power devices in electronic based power conversion for renewable systems. Kassel University Press
6. Armaou A, Christophides PD (2001) Chrystal Temperature control in the Czochlarski crystal growth process. AIChE J 47(1):79–106
7. Bahat-Treidel E (2012) GaN-based HEMTs for high voltage operation: design, technology and characterization. Göttingen, Cuvillier
8. Baik KH (2004) Design, fabrication, and characterization of GaN high power rectifiers. Ph. D., thesis, Univ. of Florida, Gainesville
9. Brown M, Weis M (2015) 4.5kV 4H-Sic diodes with high current capabilities. EPE-PEMC, Dubrovnik and Cavtat, 1–4
10. Cai Y, Zhou Z, Lau KM, Chen KJ (2006) Control of threshold voltage of AlGaN/ GaN HEMTs by fluoride-based plasma treatment: from depletion mode to enhancement mode. IEEE Trans Electron Dev 53(9):2207–2015
11. Carroll E, Oedegard B, Stiasny T, Rossinelli M (2001) Application specific IGCTs. ICPE'01—international conference on power electronics, Seoul, South Korea
12. Carroll E, Klaka S, Linder S (2015) Integrated gate-commutated thyristors: a new approach to high power electronics. In: International electric machines and drives conference (IEMDC), Milwaukee
13. Carusone TC, Johns DA, Martin KW (2012) Analog integrated circuit design. Wiley, New York
14. Cordell B (2011) Designing audio power amplifiers. McGraw Hill, New York
15. Czochralski J (1918) Ein neues Verfahren zur Messung der Kristallisations-geschwindigkeit der Metalle (A new method for the measurement of the crystallization rate of metals). Z Phys Chem 92:219–221

V. B. Litovski, *Lecture Notes in Analog Electronics*, Lecture Notes in Electrical Engineering 958, https://doi.org/10.1007/978-981-19-6528-9

16. Dapkus II DA (2015) Using MOS-gated power transistor in AC switch applications. International rectifier, technical paper DT 94-5
17. Dimitrijević M, Milojković J, Bojanić S, Litovski V (2011) ICT and power: synergy and hostility. In: Proceedings of the 10th IEEE international conference on telecommunications in modern satellite, cable and broadcasting services, TELSIKS 2011, Niš, Serbia, pp 186–194
18. Eden R (2012) SiC and GaN electronics: where, when, and how big? Compound semiconductors, http://www.compoundsemiconductor.net/article/89752-sic-and-gan-electronics-where-when-and-how-big.html
19. Eden R (2014) SiC & GaN power market valued at $2.5 billion in 2023. https://technology.ihs.com/489338/sic-gan-power-semiconductors-2014
20. Extance A (2013) SiC and GaN power devices jostle to grow their role. Power Dev,' 9
21. Fontserè A (2013) Bulk temperature impact on the AlGaN/GaN HEMT forward current on Si, sapphire and free-standing GaN. ECS Solid State Lett 2(1):4–7
22. Franco F (2002) Design with operational amplifiers and analog integrated circuits. McGraw-Hill, New York
23. Gaalaas E (2006) Class D audio amplifiers: what, why, and how. Analog Dialogue, 40
24. Galeckas A, Grivickas P, Grivickas V, Bikbajevas V, Linnros J (2002) Temperature dependence of the absorption coefficient in 4H- and 6H-Silicon carbide at 355 nm laser pumping wavelength. Phys Stat Sol (a) 191(2):613–620
25. Germain M et al (2015) Advanced development of III-nitride material for high power components. IMEC ESA contract 20073/06/NL/PA—Abstract
26. Grebennikov A (2008) High-efficiency power amplifiers: turning the pages of forgotten history. High Freq Electron 7(9):18–26
27. Grebennikov A (2010) Power amplifier design fundamentals: more notes from the pages of history. High Freq Electron 9(5):50–57
28. Grey P, Hurst P, Lewis S, Meyer R (2001) Analysis and design of analog integrated circuits. Wiley, Hoboken
29. Griffin TE (2006) Super gate turn-off thyristor. report AR-TR-3884, Army Research Laboratory, Adelphi
30. Hasan M (2017) Fundamental limit of analog multiplication in linear discriminant classifier. Master's thesis, University of Tennessee
31. Hashimoto S et al (2010) Epitaxial layers of AlGaN channel HEMTs on AlN substrates. SEI Tech Rev 71:83–87
32. Juncai M et al (2012) Characteristics of AlGaN/GaN/AlGaN double heterojunction HEMTs with an improved breakdown voltage. J Semiconduct Publ Chin Inst Electron 33(1): 014002–1/014002-5.
33. Kapoor R, Shukla A, Demetriades G (2012) State of art of power electronics in circuit breaker technology. In: Proceedings of Rec. IEEE ECCE conference, 615–622
34. Kolessar R, Nee H-R (2001) An experimentally validated electro-thermal compact model for 4H-SIC power diodes. In: Proceedings of ISIE 2001. IEEE international symposium on industrial electronics, Pusan, Korea, 2, 1345–1348
35. Lendenmann H, Mukhitdinov A, Dahlquist F, Bleichner H, Irwin M, Söderholm R, and Skytt P (2001) 4.5 kV 4H-SiS diodes with ideal forward characteristic. Proceedings of the 2001 international symposium on power semiconductor devices and ICs, ISPSD 2001, Osaka, Japan, 31–34
36. Lenka TR, Panda AK (2011) Characteristics study of 2DEG transport properties of AlGaN/GaN and AlGaAs/GaAs-based HEMT. Fizika i Tehnika Poluprovodn 45(5):660–665
37. Li K (2014) Wide bandgap (SiC/GaN) power devices characterization and modeling: application to HF power converters. Ph.D. thesis at: Université Lille 1—Sciences et Technologies, École Doctorale Sciences pour l'Ingéenieur
38. Lin H (2015) Market and technology trends in WBG materials for power electronics applications. In: CS MANTECH conference Scottsdale, Arizona, 33–36
39. Litovski V, Mrčarica Ž, Ilić T (1997) Simulation of non-linear magnetic circuits modelled using artificial neural network. J Simul Pract Theory 5(6):553–570

40. Litovski VB (2006) Basic electronics. Akademska Misao, Belgrade, Serbia, (in Serbian)
41. Liu W (1999) Fundamentals of III–V devices-HBTs, MESFETs, and HFETs/HEMTs. Wiley, Hoboken, 330–336
42. Liu S et al (2012) Threshold voltage dependence on channel width in nano-channel array AlGaN/GaN HEMTs. Phys Status Solidi C 9(3–4):879–882
43. Macfarlane DJ (2014) Design and fabrication of AlGaN/GaN HEMTs with high breakdown voltages. Ph. D. thesis, School of Engineering, University of Glasgow
44. Mastro M (2011) II-V compound semiconductors: integration with silicon-based microelectronics. In: Li T, Mastro M, Dadgar A (eds) (2011) Fundamentals and the future of semiconductor device technology. CRC Press
45. Milošević D (2009) High-efficiency linear RF power amplification, a class-E based EER study case. Ph. D. thesis, Eindhoven University of Technology
46. Milosevic D, Van Der Tang J, Van Roermund A (2003) On the feasibility of application of class E RF power amplifiers in UMTS. In: Proceedings of the 2003 international symposium on circuits and systems, ISCAS'03, Bangkok, Thailand, vol 1, 149–152
47. Milosevic D, Van Der Tang J, Van Roermund A (2002) Investigation on technological aspects of class E RF power amplifiers for UMTS applications. ProRISC workshop on circuits, systems and signal processing, 351–356
48. Mitova R, Ghosh R, Mhaskar U, Klikic D, Wang M-X, Dentella A (2014) Investigations of 600-V GaN HEMT and GaN diode for power converter applications. IEEE Trans Power Electron 29(5):2441–2452
49. Nagel LW (1975) SPICE2: A Computer program to simulate semiconductor circuits. EECS Department University of California, Berkeley, Technical Report No. UCB/ERL M520
50. Ndjountche T (2019) CMOS analog integrated circuits, high-speed and power-efficient design. CRC Press
51. Neudeck PG, Okojie RS, Chen L-Y (2002) High-temperature electronics—a role for wide bandgap semiconductors? Proc IEEE 90(6):1065–1075
52. Afolayan OA (2021) Design and construction of audio power amplifier. Int J Adv Multidisciplinary Acad Res Dev 1(1):1–6
53. Ejiofor OS, Silver AC (2015) Design and construction of A 300 Watt audio amplifier. Int J Eng Manag Res 5(6):9–14
54. Rakús M, Stopjaková V, Arbet D (2017) Design techniques for low-voltage analog integrated circuits. J Electr Eng 68(4):245–255
55. Rashid MH (ed) (2011) Power electronics handbook, devices, circuits and implementation. Elsevier, Oxford
56. Redouté J-M, Steyaert M (2010) EMC of analog integrated circuits. Springer, Berlin
57. Ren F, Zolper JC (2003) Wide energy bandgap electronic devices. World Scientific, Singapore
58. Reusch D(2013) Enhancement mode GaN on silicon increased performance ad new applications. IEEE workshop on wide bandgap power devices and applications, Columbus, Ohio
59. Rutenbar GA, Gielen GGE, Antao BA (2002) Computer-aided design of analog integrated circuits and systems. Wiley, Hoboken
60. . Ruppel B (2000) Final project report stereo audio amplifier. The George Washington University School of Engineering And Applied Science Department Of Electrical And Computer Engineering
61. Self D (2002) Audio power amplifier design handbook. Newness
62. Shenai K, Dudley M, Davis RF (2013) Current status and emerging trends in wide bandgap (WBG) semiconductor power switching devices. ECS J Solid State Sci Technol 2(8):N3055–N3063
63. Simoes MG, (2011) Power bipolar transistors. In: Rashid MH (ed), Power electronics handbook, devices, circuits and implementation. Elsevier, Oxford
64. Smorchkova IP et al 2000 Polarization-induced charge and electron mobility in AlGaN/GaN heterostructures grown by plasma-assisted molecular-beam epitaxy. J Appl Phys 87(1):334–344

65. Smorchkova IP et al (2001) AlN/GaN and (AlGa)N/AlN/GaN two-dimensional electron gas structures grown by plasma-assisted molecular-beam epitaxy. J Appl Phys 90(10):5197–5201
66. Soclof S (1985). Application of analog integrated circuits. Prentice Hall, Hoboken
67. Tan WS, Houston PA, Parbrook PJ, Wood DA, Hill G, Whitehouse CR (2002) "Gate leakage effects and breakdown voltage in metalorganic vapour phase epitaxy AlGaN/GaN heterostructure field-effect transistors. Appl Phys Lett 80(17):3207–3209
68. Teisseyre H, Perlin P, Suski T, Grzegory I, Porowski S, Jun J (1994) Temperature dependence of the energy gap in GaN bulk single crystals and epitaxial layer. J App Phys 76(4):2429–2434
69. Valencia-Ponce MA, Tlelo-Cuautle E, de la Fraga LG (2021) On the sizing of CMOS operational amplifiers by applying many-objective optimization algorithms. Electronics 2021(10):3148
70. Wambacq P, Sansen W (1998) Distortion analysis of analog integrated circuits. Cluwer
71. Wambacq P, Gielen GGE, Kinget PR, Sansen W (1999) High-frequency distortion analysis of analog integrated circuit. IEEE Trans Circ Syst—II Anal Digit Sig Process 46(3):335–345
72. Whaling C (2015) Market analysis of wideband gap devices in car power electronics. IEEE Transp Electrification, eNewsletter, July/August 2015
73. White Ml et al (2002) Characterization of 1.8 MeV Proton irradiated AlGaN/GaN field-effect transistor structures by nanoscale depth-resolved luminescence spectroscopy. IEEE Trans Nucl Sci 49:2695–2701
74. Wu Y-F et al (2005) Field-plated GaN HEMTs and amplifiers. In: 2005 IEEE compound semiconductor integrated circuit symposium, Palm Springs, CSIC 2005 Digest, pp 170–172
75. Xiao-Guang H, De-Gang Z, De-Sheng J (2015) Formation of two-dimensional electron gas at AlGaN/GaN heterostructure and the derivation of its sheet density expression. Chin Phys B 24(6)
76. Xu P et al (2012) Analyses of 2-DEG characteristics in GaN HEMT with AlN/GaN super-lattice as barrier layer grown by MOCVD. Nanoscale Res Lett 7:141
77. Ye PD et al (2005) GaN metal-oxide-semiconductor high-electron-mobility-transistor with atomic layer deposited Al_2O_3 as gate dielectric. Appl Phys Lett 86(6):1–3
78. Yole Development Report (2015) Status of power electronics industry. (abstract)
79. Zhang A (2001) Gallium nitride-based electronic devices. Ph. D. thesis, University of Florida

Index

V. B. Litovski, *Lecture Notes in Analog Electronics*, Lecture Notes in Electrical Engineering 958, https://doi.org/10.1007/978-981-19-6528-9